THE POLITICS OF
THE SOLAR AGE

HAZEL HENDERSON is an internationally-published futurist, lecturer and consultant to organizations in over thirty countries. She is also an activist and founder of many public service groups and has authored over two hundred articles. She is a Director of Worldwatch Institute of Washington, D.C.; an advisor to The Cousteau Society, and the Calvert Social Investment Fund of Washington, D.C.; a member of the World Futures Studies Federation (Rome), and the World Social Prospects Study Association (Geneva); a Fellow of the Lindisfarne Association (Berne) and the Elmwood Institute (Berkeley, California). She serves on the Editorial Boards of *Technological Forecasting and Social Change* (New York), *The Journal of Humanistic Psychology* (San Fransisco), *Futures Research Quarterly, New Options Newsletter* (both of Washington, D.C.), *Resurgence* and *The Journal of Inter-disciplinary Economics* (both of Britain). She holds an honorary doctorate from Worcester Polytechnic Institute, Massachusetts for her work in alternative economics and technology, and was a Visiting Lecturer at the University of California, Santa Barbara (1979) and held the Horace Albright Chair in the Department of Forestry, University of California, Berkeley in 1982. She is a Senior Research Fellow at the Research Center for Economic, Technological and Social Development of the State Council of the People's Republic of China (Shanghai) and also a member of the Speaker's Advisory Committee on the Future, of the Florida State Legislature. From 1974 to 1980, she served as a member of the U.S. Congress Office of Technology Assessment Advisory Council and on advisory committees of the National Science Foundation and the National Academy of Engineering. Henderson, who was born in Britain, lives in St. Augustine, Florida and is associated with the Center for Governmental Responsibility and the Division of Continuing Education at the University of Florida in Gainesville.

Also by Hazel Henderson
Creating Alternative Futures: The End of Economics

THE POLITICS OF
THE SOLAR AGE

Alternatives to Economics

HAZEL HENDERSON

Knowledge Systems, Inc.
Indianapolis, Indiana
1988

The Knowledge Systems edition of *The Politics of the Solar Age* contains a new Introduction by the Author and a Foreword by Fritjof Capra.

Portions of Chapter 1 appeared in the *Newsletter of the Association for Humanistic Psychology,* © 1979, and *The Journal of Current Social Issues,* Fall, © 1978. Reprinted by permission of the publisher.

Portions of Chapter 2 are reprinted from editorials by Hazel Henderson in *The Christian Science Monitor,* "Proposition 13 and the End of the Age of Keynes," August 9, 1978, and "Thanks OPEC We Needed That," December 29, 1978, © 1978 by *The Christian Science Monitor.* Used with permission.

Portions of Chapters 3 and 4 are excerpted from an editorial by Hazel Henderson in *The Christian Science Monitor,* "Jumping to the Safety-Net Economy," October 10, 1979, © 1979 by *The Christian Science Monitor.* Used with permission. Portions of Chapter 4 appeared in a somewhat different form in book reviews contained in *Business and Society Review,* © 1973, 1976, 1979. Reprinted by permission.

Material in Chapter 5 is excerpted with permission from *Human Resources Management* Volume 17, No. 4, Winter 1978, © 1978 by the University of Michigan, and from a paper given by Hazel Henderson before the North American Society of Corporate Planners, October 12, 1977, Ottawa, Ontario, Canada.

Portions of Chapter 7 are a revised version of a speech given by Hazel Henderson at the University of Toronto on the occasion of the posthumous Wilder Penfield Award to E. F. Schumacher by the Canadian Vanier Institute for the Family, and reprinted in *Environment,* May, © 1978. Used by permission of the publisher.

Chapter 8 is adapted from a background paper prepared for Dr. Fritjof Capra for his book *The Turning Point,* Simon and Schuster, © 1982. Used by permission.

Chapter 11 appeared in a somewhat different form in *Best's Review,* © 1978 by A. M. Best Company, Inc., and in *Risk Management,* May, © 1978. Reprinted by permission of the publishers.

Chapter 12 appeared in a somewhat different form in *Technological Forecasting and Social Change,* © 1978 by Hazel Henderson. Reprinted by permission of the publisher.

Library of Congress Cataloging-in-Publication Data
Henderson, Hazel, 1933–
    The politics of the solar age.
    Includes index.
    1. Economic History--1971– . 2. Economic policy.
  3. Renewable energy sources--Economic aspects. I. Title.
HC59.H383   1988   330.9'048   88-13571
ISBN 0-941705-06-4

For Ali and all the world's young people;
for the Earth-keepers,
the servants of Gaia
and the planetary citizens of
the dawning Solar Age

Contents

PART THREE
COMING HOME: FROM REDOUBLING OLD EFFORTS TO
RECONCEPTUALIZING OUR "PROBLEMS"

List of Illustrations

Foreword

As we are approaching the end of the 1980s, it is becoming increasingly apparent that the major problems of our time cannot be understood in isolation. The threat of nuclear war, the devastation of our natural environment, the persistence of poverty along with progress even in the richest countries — all these are systemic problems, which means that they are closely interconnected and interdependent. They cannot be solved with the fragmented approaches typical of our academic institutions and government agencies. Their thinking is tied to the concepts and values of an outdated paradigm, the mechanistic world view of seventeenth-century science and the patriarchal value system, which are inadequate for dealing with the problems of our overpopulated, globally interconnected world.

Nowhere is this more apparent today than in the field of economics, where the conceptual framework underlying the discipline has become so narrow that it has driven economists into an impasse. Most economic concepts and models are no longer adequate to understand economic phenomena in a fundamentally interdependent world, and current economic policies can no longer solve our economic problems.

Hazel Henderson, futurist, environmentalist and economic iconoclast, has been driving home this point for over a decade with an intensity, brilliance, and originality that are still unmatched today. She has challenged the world's foremost economists, politicians, and corporate leaders with her well-founded critique of their fundamental concepts and values. Because of a special talent for presenting her radical ideas in a disarming, nonthreatening manner her voice is heard and respected in government and corporate circles; she has held an impressive number of advisory positions and has cofounded and directed numerous organizations, in which her new ways of thinking are elaborated and applied.

This book is a new and updated edition of Henderson's essays, published originally in 1981. As she illustrates with many trenchant insights in the special Introduction to this edition, her basic points are still as valid as they were then, and her "call for a complete overhaul of economics" is as urgent as ever.

In her early work, Henderson was very inspired by her friend E.F. Schumacher, author of the pioneering book *Small is Beautiful*, and prophet of the ecology movement. She helped arrange his first lectures in the U.S. and he wrote the Foreword to her first book. Like Schumacher, Henderson criticizes the fragmentation in current economic thinking, the absence of values, the obsession of economists with unqualified economic growth, and their failure to take into account our dependence on the natural world. Like Schumacher, she extends her critique to modern technology and advocates a profound reorientation of our economic and technological systems, based on the use of renewable resources and attention to the human scale.

But Henderson goes considerably beyond Schumacher both in her critique and in her outline of alternatives. Her essays offer a rich mixture of theory and activism. Each point of her critique is substantiated by numerous illustrations and statistical data; each suggestion for alternative futures is accompanied by countless concrete examples and references to books, articles, manifestos, projects, and activities of grass-roots organizations. Her focus is not limited to economics and technology but deliberately includes politics, as evident from the book's title. In fact, she asserts: "Economics is not a science; it is merely politics in disguise."

Henderson's style of writing is unique. Her sentences are long and packed with information, her paragraphs collages of striking insights and powerful metaphors. In her efforts to create new maps of economic, social, and ecological interdependence, she constantly seeks to break out of the linear mode of thinking. She does so with great verbal virtuosity, showing a distinct flair for catchy phrases and deliberately outrageous statements. Academic economics, for Henderson, is "a form of brain damage," Wall Street is chasing "funny money," and Washington is engaged in "the politics of the Last Hurrah," while her own efforts are directed toward "defrocking the economic priesthood," announcing "the end of flat-earth economics," and promoting a "politics of reconceptualization."

The key point of Henderson's critique is the striking inability of most economists to adopt an ecological perspective. The economy, she explains, is merely one aspect of a whole ecological and social fabric. Economists tend to divide this fabric into fragments, ignoring social and ecological interdependence. All goods and services are reduced to their monetary values and the social and environmental costs generated by all economic activity are ignored. They are "external variables" that do not fit into the economists' theoretical models. Corporate economists, Henderson points out, not only treat the air, water, and various reservoirs of the ecosystem as free commodities, but also the delicate web of social relations, which is severely affected by continuing economic expansion. Private profits are being made increasingly at public cost in the deterioration of the natural environment and the general quality of life.

To provide economics with a sound ecological basis, Henderson insists, economists will need to revise their basic concepts in a drastic way. She illustrates with many examples how these concepts are narrowly defined and have been used without their social and ecological context. The gross national product, for example, which is supposed to measure a nation's wealth, is determined by adding up indiscriminately all economic activities associated with monetary values, while all nonmonetary aspects of the economy are ignored. Social costs, like those of accidents, litigation, and health care, are added as positive contributions to the GNP, rather than being subtracted. Henderson speculates that those social costs may be the only fraction of the GNP that is still growing.

In the same vein she insists that the concept of wealth must shed some of its present connotations of capital and material accumulation and give way to a redefinition of wealth as human enrichment, and that profit must be redefined to mean only the creation of *real* wealth, rather than private or public gain won at the expense of social or environmental exploitation. Henderson also shows with numerous examples how the concepts of efficiency and productivity have been similarly distorted. "Efficient for whom?" she asks with her characteristic breadth of vision. When corporate economists talk about efficiency, do they refer to the level of the individual, the corporation, the society, or the ecosystem? Henderson concludes from her analysis of these basic economic concepts that a new ecological framework is urgently needed, in which the concepts and variables of economic theories are related to those used to describe the embedding ecosystems.

In outlining her new ecological framework Henderson does not limit herself to its conceptual aspects. She emphasizes throughout this book that the reexamination of economic concepts and models needs to deal, at the deepest level, with the underlying value system. Many of the current social and economic problems, she submits, will then be seen to have their roots in the painful adjustments of individuals and institutions to the changing values of our time.

A fundamental economic problem that has resulted from an imbalance in our values, according to Henderson, is our obsession with unlimited growth. Continuing economic growth is accepted as a dogma by virtually all economists and politicians, who assume that it is the only way to ensure that material wealth will trickle down to the poor. Henderson shows, however, by citing abundant evidence, that this "trickle-down" model of growth is totally unrealistic. High rates of growth not only do little to ease urgent social and human problems but in many countries have been accompanied by increasing unemployment and a general deterioration of social conditions. Henderson also points out that the global obsession with growth has resulted in a remarkable similarity between capitalist and Communist economies, both being dedicated to industrial growth and technologies as well as increasing centralism and bureaucratic control.

Henderson realizes, of course, that growth is essential to life, in an economy as well as in any other living system, but she urges that economic growth has to be qualified. In a finite environment, she explains, there has to be a dynamic balance between growth and decline. While some things need to grow, others have to diminish so that their constituent elements can be released and recycled. With a beautiful organic analogy she applies this basic ecological insight also to the growth of institutions. Just as the decay of last year's leaves provides humus for new growth the following spring, she argues, some institutions must decline and decay so that their components of capital, land, and human talents can be used to create new organizations.

Throughout this book Henderson makes it clear that economic and institutional growth are inextricably linked to technological growth. She points out that the masculine consciousness that dominates our culture has found its fulfillment in a certain "macho" technology, a technology bent on manipulation and control rather than cooperation, self-assertive rather than integrative, suitable for central management rather than regional and local application by individuals and small groups. As a result, Henderson observes, most technologies today have become profoundly anti-ecological, unhealthy, and inhuman. They need to be replaced by new forms of technology, she affirms, technologies that incorporate ecological principles and correspond to a new set of values. She shows with abundant examples how many of these alternative technologies—small scale and decentralized, responsive to local conditions and designed to increase self-sufficiency—are already being developed. They are often called "soft" technologies because their impact on the environment is greatly reduced by the use of renewable resources and constant recycling of materials.

Solar energy production in its multiple forms—wind-generated electricity, biogas, passive solar architecture, solar collectors, photovoltaic cells—is Henderson's soft technology par excellence. She contends that a central aspect of the current cultural transformation is the shift from the Petroleum Age and the industrial era to a new Solar Age. Henderson extends the term "Solar Age" beyond its technological meaning and uses it metaphorically for the new culture she sees emerging. This culture of the Solar Age, she explains, includes the ecology movement, the women's movement, and the peace movement; the many citizen movements formed around social and environmental issues; the emerging counter-economies based on decentralized cooperative, and ecologically harmonious lifestyles; "and all those for whom the old corporate economy is not working."

Eventually, she predicts, these various groups will from new coalitions and develop new forms of politics. Since the original publication of this book, Henderson has continued to advocate the alternative economies, technologies, values, and lifestyles that she sees as the foundation of the new politics of the Solar Age. Her work and her life exemplify the unique blend of theory and activism that has become the hallmark of the emerging new paradigm. As Hazel Henderson would put it herself, she walks her talk.

FRITJOF CAPRA
BERKELEY, CALIFORNIA
AUGUST, 1988

Total Productive System of an Industrial Society
(Three-Layer Cake with Icing)

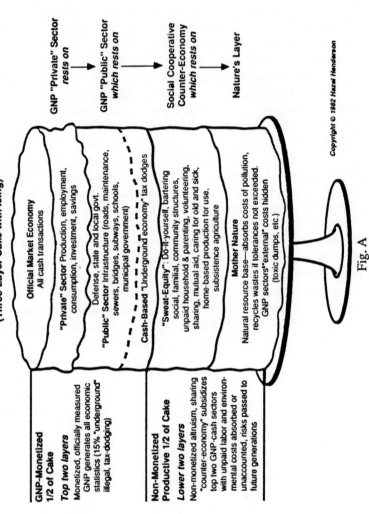

GNP "Private" Sector
rests on

→ GNP "Public" Sector
which rests on

→ Social Cooperative
Counter-Economy
which rests on

→ Nature's Layer

Official Market Economy
All cash transactions

"Private" Sector Production, employment, consumption, investment, savings

Defense, state and local govt.
"Public" Sector Infrastructure (roads, maintenance, sewers, bridges, subways, schools, municipal government)

Cash-Based "Underground economy" tax dodges

"Sweat-Equity": Do-it-yourself, bartering social, familial, community structures, unpaid household & parenting, volunteering, sharing, mutual aid, caring for old and sick, home-based production for use, subsistence agriculture

Mother Nature
Natural resource base—absorbs costs of pollution, recycles wastes if tolerances not exceeded. GNP sectors'"external" costs hidden (toxic dumps, etc.)

GNP-Monetized
1/2 of Cake
Top two layers
Monetized, officially measured GNP generates all economic statistics (15% "underground" illegal, tax-dodging)

Non-Monetized
Productive 1/2 of Cake
Lower two layers
Non-monetized altruism, sharing "counter-economy" subsidizes top two GNP-cash sectors with unpaid labor and environmental costs absorbed or unaccounted, risks passed to future generations

Copyright © 1982 Hazel Henderson

Fig. A

Introduction

The Politics of the Solar Age which I described and advocated in this book —written almost a decade ago—has been on hold. In face of the unprecedented challenges of the 1980s, the American political process has produced instead a Politics of Denial which soon drifted into the Politics of the Last Hurrah. Today, a new era is dawning in the world: The Age of Interdependence, and the new printing of this book is in the hope that our political process is now ready to expand its horizons and take up the challenge of the new agenda set forth in these pages.

Most of this book is about dealing with change. Bewildering uncertainties are now affecting most people, whether Americans, Russians, Chinese or others, as human societies restructure and realign themselves in the many global transformation processes now occurring. There are at least seven great globalization processes changing all our lives:

1. production and technology;
2. employment, work and migration;
3. finance, debt and information (which has become interchangeable with money);
4. the arms race and militarization;
5. global pollution and resource-depletion;
6. culture and consumption; and
7. the multiple realignments and restructurings driven by the prior six globalization processes.

These processes are circular and interactive and all are accelerating due to their interactivity, and they are irreversible. I have made them the subject of my seminars and speeches for the past several years and have summarized their effects in "The Three Zones of Transition: A Guide to Riding The Tiger of Change," in *Futures Research Quarterly,* Spring, 1986.

I wrote this book as an early warning about the imminence and the causes of these great changes and to point out the happier side of all this uncertainty: that things can also change for the *better!* For example, I could not have predicted that within a decade, a conservative U.S. President would return from Moscow announcing that the post-World-War-Two, Cold War era was drawing to a close—or the advent of Mikhail Gorbachev's "perestroika," although I did identify the *reasons* that a U.S.–Soviet detente was likely: in particular, that the two superpowers' military rivalry would lead only to the mutually-assured destruction of each others' *economies*. Japan is now the clear winner of the Cold War, and, as I, Duane Elgin and other futurists have predicted, the new world game, Mutually-Assured Development, has already begun to shape the rest of this century. (See my article in *Plowshare Press*, Autumn, 1987, Center for Economic Conversion, Mountainview, CA.) Japan, with a quarter of all the world's capital, is the key player in this new game—with China, India, the Four Asian Tigers (Korea, Taiwan, Singapore and Hong Kong), Brazil and the other newly-industrializing countries shifting some of the "action" from Europe and North America to the Southern Hemisphere.

This book outlines many of the underlying causes of the changes we are experiencing in the late–1980s and is still predictive of what we can expect in the 1990s. The specifics are unknowable, but I believe that major trends I described will continue: toward a world of ever-closer interdependence and greater cooperation, with shifts toward production systems based on renewable resources and managed for long-term sustainability. I believe that these trends will continue to drive value-shifts, cultural confusion, changing family structures, expanding grassroots movements and a basic reconceptualization of our knowledge and academic disciplines.

The *relative* decline of the U.S.'s formerly dominant position as the world's policeman and locomotive of economic growth is no tragedy. Paul Kennedy, author of *The Rise and Fall of The Great Powers* (1988) makes this case, as do others, including James Chace, David Calleo, Mancur Olson, Walter Russell Mead, George Kennan, of the "solvency school" as well as a wide genre of global/environmental researchers including myself, Jay Forrester, Lester R. Brown, Amory and Hunter Lovins, Jan Tinbergen, Donella and Dennis Meadows. Rather, as the U.S. takes its place in a new multipolar world, it is clear that such broader sharing of the burdens of common global security is a prerequisite for building a peaceful, ecologically sustainable, equitable world. Today there are claims that, in fact, the U.S. *won* the Cold War, citing Mr. Gorbachev's perestroika in the U.S.S.R. and Mr. Deng Xiaoping's new China as evidence for the superiority of capitalism and the failures of communism. This book belies such simple interpretations, points to the exhaustion of *both* these ideologies and explores their historic competition over what turned out to be merely different approaches to the same goal: *industrialism*. It reviews the brief 300-year history of industrialism, itself a powerful ideology, a system which transformed the face of our planet and unleashed forces which are now transforming all human societies and the biosphere itself.

This transformation beyond industrialism has been named the "Post-Industrial Era" (Daniel Bell), the "Third Wave" (Alvin Toffler), the "Information Age" (Yoneji Masuda, John Naisbitt), the "Communications Era" (Robert Theobald), the "Turning Point" (Fritjof Capra) and my own term, the "Solar Age." The Solar Age signifies much more than a shift to solar and renewable resource-based societies operated with more sophisticated ecological sciences and biologically-compatible technologies. It entails a paradigm shift from fragmented "objective" reductionist knowledge and the mechanistic, industrial worldview to a comprehensive awareness of the interdependence of all life on earth—what is now well-known as the *Gaia* hypothesis: that our planet is a living organism and we humans are participants (not just observers) in its evolutionary unfolding. Thus, the Solar Age is also a new Age of Light with our human technologies learning ever more from Gaia's own genius in capturing and utilizing the daily flow of photons from the sun; from Gaia's mighty cycling of all elements, water, atmosphere, soils, plants and animals; and the myriad ways of cooperating with each other and joining the overall symbiosis of these planetary processes in a new age of enlightenment. This new Age of Interdependence is one of *mutual* development — far beyond the narrow concepts of *economic* growth or development, which are proving disastrous in Africa and elsewhere and leading to hunger and desertification.

The new agenda, implicit in the new talk about "level playing fields," is nothing less than creating some new rules to manage the "global ballpark" itself and maintaining the global "commons": our oceans, atmosphere, ecosystems, space and now the newest commons we have created, the global economy. While competitive rules (win-lose) work well where marketplace conditions exist (i.e., as Adam Smith described: where buyers and sellers meet each other with equal power and equal information and no side-effects are experienced by innocent bystanders), in a *commons*, cooperative rules are needed (win-win) because with common ownership, nobody wins unless *everybody* wins and individuals are better off if they consider the needs of others and are worse off if they act selfishly. For example, if some people insist on standing up to see better in a crowded football stadium, everyone else is forced to stand up, and while no one sees any better, everyone is more uncomfortable.

Competition *and* cooperation are both appropriate strategies under certain circumstances and nature employs both equally and in balance. Competition between species, groups, organizations and individuals, as well as ideas, keeps unhealthy over-growth at bay. Cooperative strategies between all these same players are equally important—creating the "glue" which keeps everyone orchestrated and functioning within the agreed upon rules of interaction. The fact that both the Soviets and the Chinese are reinstating market rules where feasible, has little to do with the ideologies of capitalism, since humans have used markets for millennia. Rather than whether an economy is socialistic, market-oriented or mixed (as most are), it is more relevant to know to what extent it is organized *cybernetically* to take advantage of feedback, not just in the form of prices (which often are rigged, or do not reflect full social costs), but also feedback from voters (i.e., democracy) and from nature (such as acid rain or climate change). The more a society is structured to use a variety of these multi-dimensional feedbacks—to *learn* from them, modify structures, behavior patterns, as well as values—the better they can also adapt to new conditions and survive. Clearly, as we move beyond the economic view, whether "left" or "right," it is time to give a decent burial to the two European philosophers of industrialism and economics, Karl Marx and Adam Smith, and reintegrate other useful disciplines into a new multi-cultural view of human and ecologically-sustainable development. A new politics of values and a reordering of priorities is needed to address this larger transition.

This book still represents the general framework of my thinking and few of its recommendations have been adopted in the U.S., although many have been in other countries. My strategic goal in this book was to explore in further depth the reasons why economics had colonized public policy making to the exclusion of so many other, more appropriately-equipped disciplines, including general system theory, cybernetics, ecology, game theory, anthropology, engineering, biology and others. I wanted to offer a guide to help citizens demystify economic policy analyses and daily pronouncements on Gross National Product growth, inflation, trade balances, deficits, interest rates, etc. which obscure vital debates about new priorities and creating positive futures for the human family. Figure A provides an overall framework for thinking about productive societies, their GNP-measured, money-denominated sectors and their equally large non-money sectors of unpaid production and vital cooperative work.

Today, more than ever, citizens are aware of the need to overhaul the entire discipline of economics, from "left" to "right" and top to bottom, and replace its narrow, incorrect formulas with broader policy tools (see Chapter 13). Since 1984, and the founding of The Other Economic Summit (TOES) in London, an ever-larger group of concerned citizens' organizations and social scientists have convened "alternative economic summits." Such "teach-ins" for the assembled financial press have tried to show how the seven Summit leaders (of the U.S., Canada, Japan, Britain, France, Germany and Italy) and their economic decisions affect our less-fortunate neighbors in the Southern Hemisphere and the global environment, as well as to offer alternative policies to build a more equitable, peaceful and sustainable world. I am happy that this book continues to serve as a guide to such groups and to students in many countries enrolled in economics courses who struggle to make sense of this obsolete discipline. I am also glad that the book continues to help them critique their professors' assumptions and those in their textbooks, and that it even changes the direction of their studies! Much of it is a decoding of economic thought and a new slant on economic ideas, past and present, and is therefore unchanged by recent events and can continue to serve these purposes.

However, my fervent hope that the United States of America, my beloved adopted country, would expand its political debate to shed light on all these momentous planetary changes, has not been fulfilled. The many reasons—low voter participation, lack of proportional representation, the still-alienated "Vietnam generation," high adult illiteracy, the inability of candidates to have free access to the public airwaves and the consequent corruption of political campaigns by the need to raise vast sums for TV advertising—are the subjects of numerous books. In addition, the traditional isolation of a historically rich endowment of the continents' resources created a wealthy, self-contained economy and a mass-consumerist set of individualistic values, which led to complacency. For all these reasons, the growing groups of "planetary citizens"—the activists for human rights, corporate and government accountability, environmental and consumer protection, peace and global equity—have continued to build their grassroots movements, preferring "politics by other means" rather than affiliation with either Republicans or Democrats.

Thus, the malfunctioning political process produced only the Politics of Denial and, like many other cultures facing rapid change, it slew the messenger, President Jimmy Carter, who had tried to describe the new world challenging the U.S. in his administration's *Global 2000 Report* (see the Epilogue). The Report discussed new threats to common planetary security—the environmental destruction in the name of economic growth—while urging global cooperation in conserving energy and resources and shifting to more sustainable forms of development. Meanwhile, Ronald Reagan, with his supply-side economics—promising huge tax cuts, a balanced Federal budget and deregulation of business—had just been elected when this book went to press. The politics of reconceptualization, which I hoped would characterize the 1980s, did not reach the mainstream but remained as a growing social movement while the political process involved ever fewer eligible voters. President Reagan's much-touted 1980 "landslide" captured the White House with a mere 26% of the vote—which in some more participatory democracies might almost have been viewed as a coup d'etat! Without a presence in electoral politics, such as Germany's Green Party enjoys, the social movements grew. But the two parties continued offering similar platforms, taking campaign contributions from similar interest groups and acting like "two football teams owned by the same owner" as one political satirist saw it.

Despite these barriers to voter participation, many millions of U.S. citizens vote for small party and independent candidates at all levels of government, particularly in local races. Moreover, millions of them belong to citizen movements: for peace, social justice, human rights, corporate accountability, alternative technology/conservation and environmental protection, as well as organic gardening, holistic health and other less political concerns. The pervasive nature of these citizen movements, both on the progressive and the reactionary spectrum of concerns, attests to the lack in the U.S. of more satisfying political processes. For example, the socially–responsible investment movement, which organized much of the pressure for U.S. economic divestiture in South Africa, has spread from the earlier and now very successful mutual funds—Calvert Social Investment Fund (on whose Advisory Board I serve), Dreyfus Third Century Fund, Working Assets, Pax World Fund, New Alternatives and others, together with college, union and church stockholders—into a powerful social voice, with over $200 billion worth of assets, or some 10% of all stocks traded on the New York stock market. Significantly, these socially-screened portfolios have outperformed the market—Calvert dipped only 8% in the October 1987 crash versus the 24% drop on the Dow Jones Average.

Some of the more transient conditions which buttressed the Reagan Administration included a predictable backlash from concerned and fundamentalist groups disoriented by the rapid pace of change and the disruption of family and traditional values. Many of these groups are not inherently reactionary, but await political leadership that can more accurately diagnose and prescribe for the conditions they face: as neglected older people, lonely and uprooted singles, threatened white males competing in a shrinking job market and rural traditionalists.

As the 1980s unfolded, it became more obvious that to pursue all the old goals—spur economic growth and consumption, cut taxes, increase U.S. military influence around the world—would bog the country down in an increasing backlog of social and environmental costs that would end up undermining our real wealth. My prediction was that inflation would become the main symptom of this unsustainable "politics of the last hurrah." Instead, the symptom was treated with another round of "old time religion"monetarism which squeezed the U.S. into a near-depression in 1980-81, with the worst bankruptcy and unemployment rates since the 1930s. The deficit–financing mounted and along with significant worldwide energy conservation and a lessening of OPEC's power, inflation in the U.S. was reduced. These factors contributed much more to the U.S. economy than to other countries. Since petroleum prices are denominated in dollars, the high-interest rate, strong dollar policies of the Reagan Administration kept petroleum prices lower than in most other countries. The price of oil since 1981 has declined 23% in U.S. dollars, but only 11% in Japan, while in the Federal Republic of Germany, the real cost of oil increased 7%, in India it rose 11% and in France it rose 38% (*World Oil: Coping with the Dangers of Success*, Worldwatch Inst., Washington D.C., July 1985). Not only did the strong dollar substantially lower the U.S. inflation rate and stimulate the economy, it also lowered the oil import bill. The U.S. trade imbalance would have been far worse if not for a sharp fall in the oil import bill, from $61 billion in 1981 to an estimated $32 billion in 1985. While the falling oil prices boosted the "Reagan recovery," they accelerated the decline of Third World economies and made the plight of their citizens more desperate, while increasing the likelihood of debt defaults.

In Chapter 5, I focused systematically on the energy-intensity of industrialism and the new vulnerabilities all industrial economies face as hostages to the overall decline in quality and availability of fossil fuels and the nuclear trap, as well as the need to focus on renewable energy alternatives. I also describe the now-familiar "net energy" formulas developed by Amory Lovins, Howard T. Odum, and others and show why they measure energy-efficiency more accurately in thermodynamic, rather than economic terms. The global energy situation is no less serious today than it was in 1980, in spite of world conservation efforts and the temporary oil glut. This is illustrated by the fact that whenever the oil-strategic Middle East situation changes even slightly, oil prices on the spot market spurt upward and the world's stock markets sell off. The fact is that according to the U.S. Department of Energy's Energy Research and Advisory Board, the U.S. in 1987, at current prices and technology, had proven oil reserves of only 30 billion barrels—sufficient to meet U.S. consumption for five years. In addition, the U.S. "oil patch" has been decimated and imports are up to 40% from 33% at the time of the 1973-74 Arab oil embargo. While the Reaganites repealed the 55-mile speed limit and higher mileage standard for automobiles, *Business Week* editorialized (July 18, 1988), "Despite falling oil prices, the U.S. is in the middle of an energy crisis far more drastic than the price shock of the 1970s." Noting that the U.S. energy bill in 1986 still came to a hefty 10% of GNP while Japan's was only 4%, the magazine called for a stiff tax rather than waiting for a price mechanism to work—adding that "today's energy price must reflect future dangers" and that a more natural shift to an energy-efficient economy must be encouraged. In this book, I have reviewed many of the policy options to achieve this still vital goal, and have noted that the energy-efficiency of Japan and European countries is a key, often unreported factor, in their superior economic performance and the U.S. negative trade balance.

Throughout this book, I had expected that creeping inflation would continue to be the primary barometer (as it was in the 1970s) of the overall vulnerability of industrial societies and their inability to come to grips with the transformation to leaner energy and resource economies. In the case of the U.S. I was wrong as the Reaganites forced down the highly visible and politically embarrassing "inflation barometer" in one of the greatest political "magic shows" in history. With brilliant TV skills and "government by press release," Mr. Reagan syphoned the inflationary bubble into other areas and racked up more budget deficits than all those incurred before him in U.S. history, while exhorting a balanced budget, doubling military expenditures, passing a huge tax cut and calling it all the "Reagan Recovery." The administration taking power in 1989 will have such a daunting task of picking up the pieces and paying for the colossal consumption sprees of the past that many able politicians opted to stay on the sidelines in case they would find themselves presiding over a global depression.

Not surprisingly, this power vacuum created by obsolete economic ideology is being filled by state and local governments, private enterprisers and independent activists who are pragmatically pursuing post-economic strategies around *real* technological choices in *real* communities involving *real* people and *real* resources. States and cities are busy setting up their own foreign policies, trade policies and relationships, including "sister-cities" programs with the Soviet Union and other countries. Travel agencies, small businesses and citizens are arranging tours and conducting citizen-diplomacy with the U.S.S.R. Media impresarios are producing everything from rock concerts, on the successful "LIVE AID" model which provided relief to African famine victims, to two-way video conferences and "global town meetings" linking citizens all over the world. Such post-economic, post-cold war pragmatism is also flourishing among independent politicians at the state, local and national levels, including the League of Elected Officials (LEO) which explores direct diplomacy and innovative foreign policies. Over 370 city councils and 70 county councils endorsed the nuclear freeze and 120 towns refused to cooperate with the Federal Emergency Management Agency's crisis relocation planning for nuclear evacuations. Pragmatism is flourishing all over the U.S. from this mushrooming of local government initiatives to the boom in small alternative businesses—small businesses owned by women have grown from 700,000 in 1976 to 3 million in 1985—to barter clubs, skills exchanges and international counter-trade, with one quarter of all global trade in 1985 being conducted in barter.

In Chapter 9, I discussed the growing policy in many countries of seeking "common ground" between working people, unions and environmentalists, and how such coalitions can expand to include larger issues of human rights and justice—since an environmentally–benign economy is usually one of power and resource–decentralization and is more people-intensive than capital and energy-intensive. As equity concerns went underground in the Reagan years and as greed enjoyed new levels of reward on Wall Street, this common ground on a whole range of environment/equity issues became clear: peace and environmental protection are unachievable without justice and equitable resource-sharing—within and between countries. Ever-broader coalitions emerged: to deal with the global debt-crisis and the destruction of the equatorial rain-forests, to "re-educate" the World Bank about the devastation to indigenous peoples and the environment that its economic "development" projects incurred. This Gordian knot of economic policy impacts forced an in-depth examination of economists' assumptions about capital-labor ratios and why industrial societies had drifted so far toward ever-greater capital and energy intensity, with all its centralizing effects, automation and insensitivity to human rights and ecological imperatives.

I discussed in Part Three the reframing of these "capital-labor ratio issues" beyond the old view of "labor-productivity" and the skewed statistics it generated, which led to the belief that the relative decline in the performance of the U.S. economy was the fault of labor unions and greedy workers. Even labor union economists were trained in the same labor-productivity view! I warned that the pressures would mount to "restore" productivity via more layoffs, plant-closings, mandated job-sharing, further decline in real incomes. Also, the familiar "fudging" of the unemployment figures would begin to make it appear that things were better due to the vigorous growth in creation of *new* jobs. However, as I expected, this upbeat job-creation rate largely consisted of the huge growth of part-time jobs, where what was counted previously as *one* well-paying job would continually be replaced with *two* or *three* part-time jobs at much lower wages. I was proved right, but I underestimated the demoralization of unions, which led to the widespread "give-backs" of the 1980s, nor did I see the extent to which this shift to minimum-wage part-time employment would lead to the increasing poverty, homelessness and hunger, or the accompanying despair, drug abuse and crime.

Even *fears* of renewed inflation among investors are re-awakened at the slightest hint of increasing employment levels—sending the stock market down and bringing hair-trigger tightening by the Federal Reserve Board. I had noted how the full-employment "flashpoint" which triggers such inflation fears, had in the post-war period drifted up from 2-3% to 6%, as structural unemployment and stagflation grew during the 1970s. Since 1980, due to the lowering of overall job quality and wages and the huge increase in part-time jobs, the "flashpoint" has moved back into the 5% range. On the plus side, from the point of view of small and entrepreneurial businesses, hiring more people at these lower, part-time rates, became

easier and small businesses created the lion's share of new jobs in the
1980s. All these factors made it possible for the Reverend Jesse L. Jackson
to create a "common ground" among workers, farmers, environmentalists,
minority groups, women, peace activists and middle-class Democrats —
a "Rainbow Coalition" which also reaches out to a global rainbow coalition, as the issues of global equity, peace and environmental protection are
more clearly seen as inseparable.

Thus, in the U.S. of the 1980s, inflation was traded for 1) unemployment, under-employment, declining real incomes, poverty and homelessness; 2) federal deficits; 3) trade deficits making the U.S. the world's
largest debtor nation; 4) neglected maintenance of public infrastructure,
education, social services; and 5) further environmental depletion. For the
first time in 30 years, the U.S. experienced in 1988 a deficit in investment
earnings, i.e., more overseas investors were repatriating profits from the
U.S. than U.S. investors were bringing home. Meanwhile, in other countries, the chief symptoms of industrialism's unsustainability were the
world's crushing debt burden and the "inflation barometer."

The passivity of labor unions and most other citizens in accepting the
bizarre course of events of the Reagan years can only partially be understood as problems of denial and lack of participation by voters. Naturally,
it was difficult for what was the post-war world's richest and most envied
nation to face a new situation of relative decline, as other countries in
Europe and Japan grew, many with our Marshall Plan assistance—even
though we could have thus claimed their success, *our* success. However,
this widespread passivity of the U.S. public is fundamentally explainable
as a *paradigm* problem, where even leaders and traditionally-trained
policy analysts are looking at the world through the obsolete spectacles of
macro-economic management and no longer appropriate sets of statistics.
This hampers redefinition of "problems" which might be opportunities in
disguise. For example, in Chapter 10, I dissected the "declining productivity" flap as a need for new, redefined productivity *measures* and *indicators*,
beyond the old labor-productivity formula (output per man-hour, sic). I
argued for a broader context and disaggregating to show
capital–productivity, energy–productivity, and management–productivity
as well as deriving from ecological scientists their view of
ecosystem–productivity. Just as GNP figures are adjusted for inflation by
the price deflator, an ecological deflator is also needed to correct for over-
stated productivity gains. Most productivity measures in economics are

still over-stated in per person terms. Even though there was some study of these broader factors, they were not integrated and unaccounted increments to such productivity measurements were often mysteriously labeled "knowhow." As an example of how these issues are being gradually reframed, largely by interdisciplinary policy groups, the U.S. Office of Technology's 1988 Report, *Technology and the American Transition: Choices for the Future*, points the way by shifting focus to "amenities." (I felt gratified that my six-year term on its Advisory Council was not spent in vain.) Similarly, Britain's *The Economist* has covered the need for new "total-factor" productivity approaches and in a December 10, 1986 article "Countries In Trouble," a series of social indicators were used to review fifty developing countries and *economic* data was used for only one third of these measures. One of the conclusions was that the real wealth of nations is educated citizens, a point I have made repeatedly. Few researchers are yet attempting to integrate *ecosystem*-productivity, not because it is not readily measurable, but economists can't understand bio–productivity without some further training. Nor do they clarify the difference between wealth and money (which they confuse), or between wealth and "illth,"or goods and "bads," i.e., between genuine, *useful* innovations and *healthy* products, or mere product novelties such as new brands of cigarettes or ozone-destroying packaging. For these reasons, there is little clarity to investment decision-making or research and development funding.

I also pointed out in Chapters 10 and 11 the ways in which tax policies favor increasing capital and energy intensity (and still do) while employment is still over-taxed relative to other factors, thus leading to *mindless* automation. I believed that, barring these skewed subsidies to automation and penalties toward employment, labor had probably become the more efficient factor of production, and that a people-intensive economy was inherently more efficient on environmental grounds—for example, family farms (now suffering from faulty economic policies). A recent cover story in *Business Week* (June 6, 1988) entitled "The Productivity Paradox," at last bears me out. The article points out that after investing billions ($17 billion in 1987 and $19 billion in 1988), the payoffs to automation are as elusive as ever and the U.S. still trails Japan in productivity by a wide margin. For example, General Motors, after spending "more on automation than the Gross National Products of many *countries*," had little to show for it, and is now down-sizing. These new conclusions come from a "new math of productivity" which throws out traditional accounting formulas, and substitutes the broader framework which is outlined in this book. Today, what I and others termed "total-factor productivity" is gradually replacing the old "labor-productivity" measurements. As many management textbooks now agree: the key to productivity is the *overall* effectiveness of the organization, from capital investment strategies and management quality to employee skills and commitment and, most important, paying attention to the human factors and customer satisfaction. *Business Week* concluded that narrow economic formulas focused attention too closely on the production and labor costs, which still blinds decision-makers to these broader factors and often leads to wrong conclusions and to funding the wrong investments.

Today, in the U.S., automation has whittled direct labor down to 8-12% of total production costs—and some companies no longer even account for labor as a separate cost category, but merely fold it into overhead. The new math of productivity now also includes *time* as a manufacturer's most precious commodity, while product life-cycles are shrinking too. I also pointed in Chapters 10 and 11 to the overall dilemma as companies and whole economies become "innovation junkies" (as *well* as "energy junkies"), since the other side of the coin of innovation is *obsolescence* and the need to constantly *replace* perfectly usable equipment and product lines before they are even paid for. Today, in many high tech markets, a whole generation of technologies come and go in less than three years. No company or economy can afford that pace of innovation/ obsolescence for long, and the speed-up explains why so many state and regional development strategies are devastated by plant openings and closings and the constant shifting of production locations which are *also* subsidized in the tax code by accelerated depreciation, tax credits, etc.

The automation–unemployment trade-off is still very much a live issue, with computer companies citing all of the new jobs created in the much ballyhooed "services and information economy." My argument was that without sharper measures of what was really "productive" and "efficient," much automation would be counter-productive, and actually *pyramid* costs (with back-up systems, poor system configuration, incompatible systems, inadequate human-machine interface, etc.) and that the huge retraining costs, displacement, rapid turnover, relocations, etc. would fall on the taxpayers, rather than the corporate sector. Similarly, I warned that the increase in "services sector" employment would create much additional "paper-shuffling" transaction costs, as well as many deskilled, lower paying jobs and more structural unemployment among those left behind. Some of the evidence is now in on all this, including the limits of automation on its own terms. As the *Business Week* article stated, "Replacing people with machines and chipping away at waste and inefficiency no longer go very far," while "slash-and-burn" plant-closings are now clearly responsible for the "hollowing of the American economy" as we become a nation of hamburger-flippers, taking in each other's laundry. Thermodynamicists, engineers and general systems theorists understand far better than economists that such changes are not reversible, that a society must have a solid productive base and that a "services economy" can not rest on thin air for very long. Even the "Big Eight" U.S. accounting firms noted that structural change and technology are not built into current accounting systems "so we have no numbers on them." In fact, some companies are now achieving productivity gains by *de-automating*, such as Cal-Comp Inc. which makes graphic plotting machines. It streamlined and simplified its operation by throwing out its assembly line and redesigning its graphic plotters with 50% fewer parts. Its assembly costs were cut by 30% enabling it to cut the price of the machine by 45%, while *retaining* its 40% profit margin. I have kept a whole file of examples of excessive capital–intensity and over–automation including a 1988 in-house audit of 1000 World Bank Development projects which suffered from conventional economics and unfounded technological optimism.

In Chapters 10, 11 and 12, I described other macro-effects of these technological processes, their effects on insurance, inflation and national accounts, not to mention the increasingly severe long-term impacts on the environment, which is *still* left out of even the most sophisticated productivity indicators. For the longer-term, complex impacts, I urged the shift to what is now called "chaos theory" (*Chaos: Making a New Science,* by James Gleick, 1988) with its more dynamic, biological change models which are crucial to the needed paradigm shift beyond economics. I have always urged my economist friends wishing to expand their horizons, "Take a biologist to lunch!" Recently, several professional meetings have taken place which have brought economists together with other professionals in the biological, ecological sciences, as well as non-equilibrium thermodynamicists, information, decision and game theorists, and many of the new "chaos theory" pathbreakers. Unfortunately, many economists see the potential of chaos theory not so much to improve overall technological choices and public policy, but as simply a sharper tool to beat the world's ever-more turbulent stock and financial markets and to "win" in the new global economy.

It is now essential to learn the larger lesson: that the world's economy is another "commons" and the rules must be changed from those of zero-sum, winners and losers, to the win-win rules of cooperation—a new "Bretton Woods." But this time include *all* the players—both the rich creditor nations and the debtors—so that a new level playing field can enable everyone to "win." Instead of nations engaging in head-on competition over a narrow range of consumer items—cars, TV sets, computers, toys, etc.—the new cooperative strategy would be to look for the *real* "niches" (as even Adam Smith recommended in his often misunderstood *comparative,* not competitive, advantage theory). For example, a natural "niche" export for the Dutch is the technology of dyking out the seas. They have had thousands of years of experience and have no competition in selling this technology to such low-lying areas as the City of New Orleans, Bangladesh, and perhaps many of the world's coastal cities if we do not stop adding "greenhouse gases" to the atmosphere.

Today, countries need to find their *creative* advantage, whether it's Britain's highly successful export of "eccentricity" (i.e., The Beatles, Boy George and punk fashions) which was ignored and unsubsidized, while the DeLorean automobile flop cost British taxpayers some 200 million pounds. Similarly, my home state of Florida, rather than continuing to export raw phosphate rock in competition with Morocco and suffer severe environmental damage, could gear up to enter a potentially vast and needed export area: the desert-greening business. Florida has some two thousand varieties of halophyte grasses along its over 2,000 mile coastline which can be developed by agricultural bio–technologists at its major universities. These plants, when scientifically hybridized for various climatic regions, can be planted in areas where soils are ruined due to salination over-use and over-irrigation. The plants produce a crop much like wheat, except more nutritious, which can be marketed for flour and bread, while the roots and stems suck the salt and other excessive minerals from the soil so that it can be restored. As I pointed out in *Creating Alternative Futures* (1978), we need to remember that plants mine millions of tons more minerals each year than all of the mining activities of humans. There are many similar forms of bio–development needed all over the world and Nature is the basic resource, as the CHEMRAWN (Chemical Research Applied to World Needs) conference in Toronto in 1978 made clear. The bio-scientists at that conference stated that there were natural plant analogues and substitutes for every current input into industrial production processes, and all that was needed to move to renewable, sustainable forms of production was a change in our mechanistic, industrial mindset, and the redirection of research away from less useful areas, particularly weapons production.

It is now clear to many politicians that the competitive, expansionist, GNP-growth-oriented, resource-intensive industrial system is also destabilizing to every locality and region on the planet and is catastrophically war-prone. As political will coalesces and a new *global* trade and financial system is negotiated—as it must be if disaster is to be avoided—politicians face a difficult task of balancing their responsibilities to their own citizens with the need to ride the global roller coaster of world trade. Political leaders today realize that their lot lies with their electorate, rather than with the footloose investors and financiers who have bankrolled many of their campaigns in the past. It is politicians who will be left holding the bag if the world trade roller coaster leaves their regions dislocated and their voters unemployed, or if they have to break the bad news that all the future may offer in the global "fast lane" is the prospect of competing for jobs with Taiwanese and Asian workers at ever-lower wage rates, while local resources and environments continue to be depleted. Few politicians will want to run for election on such platforms while at the same time fewer of their former rosy, high-tech, export scenarios are believable.

What is emerging amongst many politicians faced with this policy crunch is a set of more realistic, balanced scenarios which combine local and regional self-reliance and indigenous development with more thoughtful, finely-tuned strategies for playing in the fast lane of world trade. These scenarios do not require withdrawing from the global scene and dis-inventing technologies or trade, but instead, they require thinking harder about what mix of global and local strategies are best to maintain the viability of regional businesses, resources and entrepreneurs, while enhancing the skills of their people and fostering the traditions and distinctive qualities of their own cultures. A new level of visioning of alternative futures is needed to catalyze the millions of people all over the world whose movements for peace, human rights and ecological sanity attest to their readiness for new leadership. Such visions will be beyond the left and right categories of the old flat-earth politics based on economic models. One example is currently underway in my home state of Florida with many state and local level efforts to envision desirable future scenarios. Processes will be facilitated by futures research think tanks and consultants, borrowing from the "search conference" techniques pioneered by Eric Trist and Fred Emery at Britain's famed Tavistock Institute and the "anticipatory democracy" projects advocated by futurist Alvin Toffler. Numerous areas in the U.S. and Canada have conducted such projects over the past ten years, including the still-relevant Canadian Conserver Society Report, which, if implemented, would make Canada a model for the world in preparing for the twenty-first century. Likewise, the only sure way to deal with today's internecine bickering between nations about obsolete bilateral trade statistics (now swamped by financial flows and the other energy and productivity effects I have mentioned) is to pool their precious national sovereignty and rewrite the rules governing the global game.

The best way to level the global playing field is to put a floor under it: by working together on the longer-term goal of equivalent global minimum wage standards, and worker safety and environmental standards. Much work on this agenda has been achieved already through United Nations agencies and treaties. Only in this way can we curb the current rewards to global "pirates," who still have a competitive advantage over more responsible companies and nations. As I have documented, it is the citizens' organizations from all over this planet which have pushed these kinds of long-term agendas onto the national agendas of politicians—from the Chipko Movement in the Indian sub-continent which restores denuded forestlands and the greenbelt Movement founded by Prof. Wangari Matthai in Africa, to Amnesty International, and consumer and workers groups. Meanwhile, both *defensive* strategies are needed to cushion new globe-girdling effects and *proactive* strategies to out-think the game played by most multi-nationals and investors, who can call up computer programs daily to find out which nation-state is foolish enough to be offering a 35% return to investors. Citizens' groups and nation-bound politicians are hard put to play against the multi-national actors in such high-risk games, since the former must deal with long-term processes while the latter can move at the speed of electronic funds transfers.

Only when politicians start offering voters realistic and more attractive scenarios which balance local and regional development with life-enhancing high-technologies, from cooperative space-exploration, satellite-based, people-to-people video conferencing, local and global telecommunications, to the most advanced resource-conserving, sustainable technologies and people and skills-intensive enterprises, can they expect to regain the attention of voters. The lessons of industrialism are in, and even diehard development economists now admit that their advice was disastrous, as more countries learn from Africa's miseries, that it is agriculture and enhancement of local resources that are the prerequisites for healthy development.

During the 1980s, particularly in the U.S. and Britain, the old economic formulas and rear-view mirror policies led to ever-greater financial manipulation and the "global casino" (the tightly-woven 24-hour asset-management of the electronic marketplace) reduced real wealth and even money to meaningless blips on hundreds of thousands of computer-trading screens. Massive waves of "hot money" sloshed around the planet every day, swamping domestic economic management and making a mockery of trade policies, employment strategies, monetary and fiscal measures. Wall Street's rush to greed produced insider-information scandals, program trading and the warning scare of October, 1987. The leveraged-buyout (LBO) craze led to further piles of debt and a *narrowing* of share-ownership, rather than the worker-owned, ESOP (employee stock ownership plan) based economy for which my old friends, Louis and Patricia Kelso, inventors of the ESOP and the LBO, still labor. In fact, many concerned Wall Streeters, including Felix Rohatyn who bailed out New York City in the 1970s, agreed that the original purpose of capital markets—to provide capital efficiently to entrepreneurs and industry—was being turned on its head. As *Business Week* described in a June, 1988 cover story on "power investing," the new strategy of Wall Street's brokerages, investments banks and institutional investors (whose profits were hit in the October 1987 crash) was that of assembling ever-larger LBO war chests to buy up more industrial assets and whole companies in order to bolster up their *own* sagging balance sheets. As I wrote in *The Futurist*, March-April, 1988, we are all now learning the difference between money and real wealth, and such shenanigans only spur the already proliferating flight to the informal economy of bartering, counter-trade and local, limited-purpose "currencies" such as the LETS (local exchange trading systems) designed by Michael Linton of British Columbia, Canada (see Chapter 7).

My call for a complete overhaul of economics is as urgent as ever, both at the level of macro-economic management, where order-of-magnitude errors are skewing policies, as well as on the micro, local levels already mentioned. Many of the post-economic policy tools I described in Chapter 13 are slowly being accepted by governments: futures studies, technology assessments, environmental-impact statements and social indicators. My colleague, Lester R. Brown, founder and President of the Worldwatch Institute and his research team are still pioneering this field with their *State of the World Reports* issued annually. The 1988 Report estimates that global expenditures of some $150 billion per year are necessary between now and the year 2000 as a "down payment" on sustainability. By re-prioritizing global military security budgets to include sustainable development, the military portions could be reduced gradually each year from $900 billion in 1990 to approximately $751 billion in 2000. The new South Commission, which includes many high-level representatives from developing countries, is interested in creating its own social indicators of the much broader development strategies it favors. Such indicators can be vital in redefining the more unique culturally-specific, *non*-Western development models now being pursued by China, for example. I visited China in 1987 and returned there in 1988 as a Senior Research Fellow of the Research Institute for Economic, Technical and Social Development of the State Council, in Shanghai, and wrote a brief report, "China: Key Player in A New World Game" for *Futures Research Quarterly,* Fall, 1987. The first task of development theory today is to decode the "cultural DNA," i.e., the value-system of society, in order to understand what its goals are and *what* it seeks to optimize. During the past eight years, I have visited dozens of countries and consulted with thousands of citizens' organizations, government officials and business people, and I have learned as much or more from them as they may have from me.

A new era is at last dawning in which all nations *can* pool their sovereignty and cooperate in building a mutually-secure, win-win world. In the U.S. the "baby-boomers" and the "Vietnam generation" are moving into mainstream, influential positions. New coalitions are emerging, such as Jesse Jackson's Rainbow Coalition, here and in other countries. However, the needed reconceptualization cannot come too soon. For example, Walter Mondale's Democratic presidential race was derailed by traditional economic models caught in the "tax-increase trap" which voters rightly reject. The issue is still to *re-order priorities within the national budget* in the U.S. and in other countries. The priority-reordering issue became clearer to Americans with the Iran-Contra affair and the defense-procurement scandal of 1988 involving annual Pentagon waste and fraud, estimated by some witnesses in the $50 billion range. Lastly, few political leaders or candidates anywhere have found alternatives to the hollow promise of per-capital-averaged, GNP-measured "economic growth," which until redefined, will make matters worse. The 1987 Report of the World Commission on Environment and Development, chaired by Prime Minister Gro Harlem Brundtland of Norway, entitled *Our Common Future* called for an integration of economics and ecology, in much the same terms as my *Harvard Business Review* article, "Ecologists versus Economists" in 1973, and called for a more realistic form of "sustainable development." Canada's Environment Minister, Tom McMillan, spearheaded a National Task Force on Environment and Economy in 1986, whose Report addresses many of the issues squarely and is available from the Canadian Council of Resource and Environment Ministers (Downsview, Ontario, 1987).

It remains for "outsiders" to redefine development, such as those researchers whose work is faithfully reported in the monthly political newsletter *New Options,* edited by Mark Satin, himself a leading researcher in the sustainability debate. The May 30, 1988 issue of *New Options* challenged all the U.S. political candidates to address the reordering of priorities and the huge federal budget deficits that had become an all-purpose excuse for the status quo, by pointing out that *the way to balance the budget* was precisely *to build a sustainable society.* Amen! It also remains to be seen how long it will take for governments, the World Bank, the International Monetary Fund and other "economic development" agencies to reshape their paradigms and programs and address the urgent task of retraining their economists and development officers. However, the task has begun and I hope that this book will continue to prove a useful reference as the debate goes forward.

I want to thank again all of the scholars, scientists, activists, researchers, alternative investors and entrepreneurs, many of them my valued friends and allies, whose work is mentioned in this book. It would be as impossible to name them all as to cite only a few. With them, I still celebrate the evolutionary potential of all the planetary citizens in our human family, and the evolution of Gaia, our lovely, mysterious blue planet — mother of us all.

HAZEL HENDERSON
ST. AUGUSTINE, FLORIDA
JULY, 1988

PART ONE

The Coming Era of
Posteconomic Policy Making

CHAPTER 1

The Politics of Reconceptualization

In the summer of 1978, I spent my customary few days with Buckminster Fuller and his World Game. For the first time, a Pentagon official who ran the Department of Defense's war-gaming and modeling program had asked to attend the World Game. With the unconscious arrogance of bureaucrats and other heroic conceptualizers, he explained to the World Game players how he and the joint chiefs model the world: a world of human decision makers (leaders in the Kremlin, the White House, and the capital cities of other nation-states), with their relative firepower, equipment arsenals, supplies, transportation, communications, and industrial bases. His model was driven by these human decision makers and, further, assumed them to be "rational actors" on the global stage.

We World Gamers began to probe his assumptions. We asked if he had considered a "force-field"-driven model, in which much larger, external forces are now making puppets out of even our most powerful world leaders. We suggested, for example, climatic cycles operating over millennial time scales, the global geological distribution of petroleum, the buildup of carbon dioxide in the planet's atmosphere, topsoil loss rates and advancing desertification, the increasing rates of extinction of animal and plant species—many of which have been set in motion by our own fertility and past activities. Might not these have now become the *dominant* forces at work, or shouldn't they at

Parts of this chapter appeared in the *Newsletter of the Association for Humanistic Psychology,* San Francisco, January 1979, and in *The Journal of Current Social Issues,* New York, Fall 1978. Reprinted with permission.

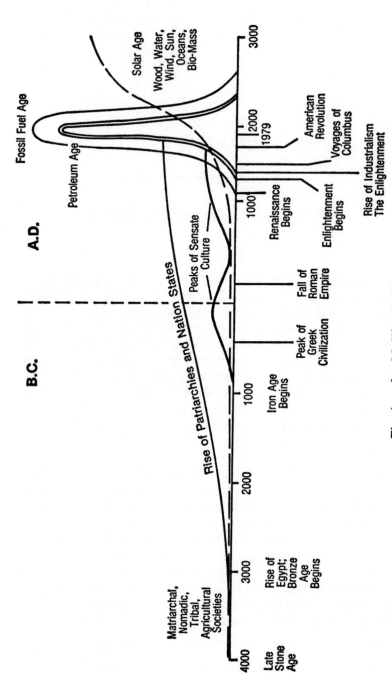

Fig. 1   after M. King Hubbert

least be included in his model? No, said the DOD expert, they were not in his model.

Suddenly, I saw in a new light the task for all of us involved in citizen movements for social change—human rights, corporate accountability, economic justice, consumer and environmental protection, holistic health, appropriate technologies—and those promoting simple living, personal growth, and greater awareness of the interdependence of the human family on this blue planet. We now must help create greater understanding of the fact that *today's "leaders" and "decision makers" are no longer in charge of events,* even though they still imagine themselves the "rational actors" of their decision models, firmly in command from their "war rooms," as they once believed in simpler, slower times.[1] They are like ancient kings who commanded the ocean tides to come in, or the early priests and priestesses whose incantations "caused" the sun to rise. They, like all of us, are also puppets of all these larger forces. Thus the "spontaneous devolution" of their institutions has begun.

The task for all of us committed to these social-change movements is to see that *we are one coalition* in the larger politics of reconceptualization. Together we must demystify today's counterfeit priesthood of "puppet" leaders, and map and align our own energies with these larger-field forces and the energies that, in reality, drive our planet: the daily solar flux, which in turn drives our planetary weather system; the cycles of oxygen, of nitrogen, and of hydrogen, and the plant photosynthesis that is our *primary economic system.* Only if all the "little people," the laity, the growing numbers of planetary citizens, can align themselves as a correspondingly powerful "force field" of growing human awareness of these realities can we join together and surround the narrow logic and rationalizing of today's puppet leaders, trapped in their institutional and conceptual bunkers and enmeshed in the intellectual baroque of the old order. We can see ourselves and our diverse social-change activities as part of a living orchestration, generating larger patterns, out of which grow new paradigms of knowledge, policy, and personal behavior. This for me is the politics of reconceptualization.

For many of us, activities in various movements for social change have helped us understand our own and each other's inner space and to tap the deeply coded knowledge of the creation. This inner/outer search provides a base for healing the body politic. Some of us, in the environmental movement for example, began with the objective manifestations of human pathology or, as in my case, with diagnosing the pathology of economics. Now we are coming together in a growing coalition with the potential for "wholing" ourselves *and* recycling our culture.

It was only about a decade ago that the public became aware that the environment was being strained by industrial society. Today we find these stresses are now being felt within the economy, and the situation is much more pressing. The business press, for example, publishes more articles on resources and their depletion. For the first time in much of the world press, there is discussion of whether more economic growth is sustainable. This suggests that things are really coming into a crunch that we are going to have to face.

We see today the increasing inability of all mature industrial societies to manage themselves. These strange new diseases of structural inflation and structural unemployment seem all too evident. Economists trying to deal with these diseases are continually rationalizing away the real reasons for them. Those reasons are better understood outside the discipline of economics. They are rooted in the way we use resources and raw materials, and they are rooted in the particular capital- and energy-intensive type of technology that we have developed.

Industrial leaders confuse us when they talk about being able to consume our way back to prosperity while at the same time trying to convince us that there is an energy crisis and that raw materials are becoming scarce. The problem is that most of these industrial societies are thought of as a monstrous abstraction: as an "economy." If we think all the dimensions of a human society can be reduced to an abstraction called an "economy," is it any wonder we are losing control of the society?

The U.S. inflation rate is currently back at double-digit levels, and unemployment remains tragically high, and the situation will persist until a wholly new view of events emerges. As I shall discuss further, we are entering an era of "posteconomic" policy making. Economists now argue publicly about all the macroeconomic options: their inability to control the money supply, and whether it is time to try wage and price controls again—all in their vain efforts "to maintain steady economic growth."[2]

It is never a matter of "growth" versus "no growth"; it is a matter of *what* is growing, *what* is declining, and *what* must be maintained. Of course, what is declining is the particular type of industrial society based on excessive resource consumption and on nonrenewable resources, and what is growing is the renewable-resource economy, even though statisticians are not monitoring its growth, because they have not yet conceived of it. So we have debates about the decline of industrialism, the birth of "postindustrialism" and even "reindustrialization."

The term "postindustrial" has no content at all. It connotes that we are looking into a rearview mirror and pretending that it is a crystal ball. What Daniel Bell meant by the term "postindustrial" was what he called a "tertiary, services and knowledge-based" economy. Such an economy would flower from continuing the current trends of increased productivity in the manufacturing and agricultural sectors. It is very much an extrapolative view growing out of the existing society.[3]

This kind of extrapolative view of the future also characterizes the work of Herman Kahn, of the Hudson Institute: onward and upward, more and better. We might call it a hyperindustrial future. Some economists and many futurists now include the idea of discontinuity or breaking with this trajectory. The British author James Robertson uses the images of breakdown and breakthrough as the two alternative paths that we might travel. He talks about three types of alternative future in *The Sane Alternative* (1979).[4]

The first is the continual extrapolation that he calls the "HE-future," a hyperexponential industrial future. The second is the "SHE-future," a sane, humanistic, ecological future, which implies a shift in direction. The third is the "TC-future," which is the totalitarian conservationist future, similar to that described in Robert Heilbroner's book *The Human Prospect*. Both Robertson and Heilbroner fear we will hit the crisis of resource scarcity without sufficient preparation and that the only way to arrest the situation will be to institute totalitarian control.

Then there is E. F. Schumacher's view of the future, which he described in *Small Is Beautiful,* in which industrial societies begin to decentralize after the extent of the institutionalization of all our needs reaches its logical conclusion. Similar views come from Amory Lovins, expressed in *Soft-Energy Paths,* Kirkpatrick Sale in *Human Scale* and others, and from Karl Polanyi in his marvelous book published in 1944, *The Great Transformation*, wherein he pointed out all the logical inconsistencies of the industrial revolution and how it would finally have to be transformed.

The past three hundred years in the West—the age of scientific enlightenment—have been based on the logical positivism and instrumental rationality inherited from the French philosopher René Descartes. This Cartesian logic, which leads us to believe that we can understand wholes by examining their parts, has led to the Tower of Babel of reductionism now fractionating our knowledge and policy making. It has given rise to today's welter of special-purpose agencies, institutions, and corporations, all trying to maximize narrow goals. They lack any meaningful coordination or a clearly enunciated set of values, goals, or principles other than the single-minded pursuit of "efficiency"—itself an ill-defined concept that has become reified as the key slogan of our utilitarian era. Rarely do we ask the larger questions "Efficiency for whom?" "Efficiency over what time period?"

In *Creating Alternative Futures*, I tried to present my scenario of a transformation focusing on the end of economics. I believe that the discipline of economics is not viable. It obscures the way people in industrial cultures talk to each other about *what is valuable* under drastically changed conditions. In this book, I have found the need to go deeper in exploring the roots of our current crises. In Part Two, I explore the development of our values and how they represent the culmination of a very successful three-hundred-year period of technological innovation in a specific direction. I have also felt the need to investigate more thoroughly how economists took over the driver's seat in public policy—a very old story, I discovered.

The economic difficulties our industrial society faces are symptomatic of the transition from economies that maximize production and are based on nonrenewable resources, to economies that minimize waste, recycle everything, maximize renewable resources, and are managed for sustained-yield productivity. Farmers have always understood what sustained-yield productivity means—now we have to teach it to economists.

One remedy involves correcting a long-standing conceptual error propagated by economists long before Keynes: the *equating* of our society's total socioeconomic productivity with that portion of it based on competitive, market-based cash transactions and the flows of money they generate—measured as the gross national product (GNP). Economists plot only this "formal," "official" economy of market-based production of goods, commodities, and services and the jobs they provide in the private sector, along with the taxes, jobs, services, subsidies, and transfer payments that make up the public sector. But we are so used to this "money veil" (as economists admit) and its statistical illusions that we forget that alongside this "official" economy there is and always has been a shadowy, "unofficial" or "informal" economy. It is based on our traditional heritage of cooperation, reciprocity, barter, and use-valued (rather than market-valued) productive activities. It includes home remodeling and fix-ups, mechanical repairs, home-workshop and craft production, furniture refinishing, food growing and canning, and all the vital community-based voluntary and unpaid household production (including parenting children, caring for the old and sick, ameliorating the stresses of the marketplace competitors, and cleaning up the messes left by careless production and consumption).

Such socially indispensable work, though unpaid, has always provided the essential cooperative social framework that has allowed the highly rewarded competition of the marketplace to be "successful." According to Scott Burns, in *The Household Economy*, this "informal" economy was estimated in 1969 as equivalent to some $300 billion annually (more than all the wages and salaries paid out by all the corporations in the United States) if it were "monetized" and included in the GNP. As the GNP-measured "formal" economy declines, this "informal" emerging countereconomy will continue its rapid growth, providing a safety net for many and a bridge to a more balanced socioeconomy for the future.

In the future there may be a whole new rationale of production and consumption. The American home has always been seen as the basic *consumption* unit, and we are now beginning to see emerging the American home as a *production* unit, the way it used to be before the industrial revolution. That goes from the solar collectors that people are putting on their roofs so that they can unhook themselves from the power company, to home canning, to crafts, and to the rise in home repair. This is basically an understanding of "use value," and this is what the countereconomy is about. It's about use value, rather than market value—the value of products for one's own use, rather than for sale. Today we have the rise of these small activities, which simply could not make it during an era of cheap energy. They couldn't compete. Now they are coming back into their own. Burns notes that the American government statistics value the household only when it breaks down. They know the cost of welfare. They know the cost of aid to dependent children and social services. They *could* input 'the value of viable productive households; this is something that should be measured too, Burns says. He also notes that while income-tax laws allow corporations to deduct and depreciate items of capital equipment, householders are forced to treat their own productive assets—whether sewing machines, ovens, freezers, yogurt makers, or home tools—as consumer items.

The particular evolution of industrial society has brought us to the point where we are going to have to transform ourselves. This means we will have to reconceptualize our situation and reshape our values. Everybody says that it is impossibly idealistic to imagine that human beings can change their values. However, it is not up to us alone. The planet is gently nudging us along in the direction that we have to go. The planet can be seen as a Skinnerian box with all the positive and negative reinforcers telling us which way we have to go. The programmed learning environment of the planet is signaling what its operating principles are: principles of cooperation, honesty, humility, and sharing. It is not going to be a matter of our having to do it all by ourselves. Furthermore, value changes are the stuff of all human history!

These new values are emerging not only because people are beginning to question the competitive, high-technology, and urban rat race. That's only part of it. A far more important reason is that the goal of ever-rising material wealth in mass-consumption industrial societies is simply no longer very realistic, despite what advertising tells the public. Inflation, of course, is the pervasive symptom that is slowly bringing us down off the joyride. There is a growing rejection of the sort of Big Brother, computerized technocracies George Orwell depicted in *Nineteen Eighty-four*.

There is also a growing rejection of competition as the basic way of fueling our kind of economy. As far back as 1937, in *The Neurotic Personality of Our Time,* the psychologist Karen Horney described what she called the peculiar American neurosis brought about by excessive competition. She said that there were three characteristics of this American neurosis: first, that of aggressiveness so stimulated that it began to conflict with the tenets of Christian brotherhood; second, the desire for material goods so vigorously stimulated that it could never be satisfied, leading to widespread dissatisfaction; and third, with expectations for untrammeled freedom soaring so high, people could not square them with the societal limitations that eventually surround us all.

Our task, it seems to me, is nothing short of recycling ourselves and recycling our culture. Unfortunately, our social imagination has been preempted by all the existing technological furniture that surrounds us. These manifestations of our industrial value system insulate us from the primary reality of the biosphere. This technological environment is so intrusive and ever-present that it presents instant answers to questions that we have not even asked ourselves. It suggests to us the technological fix, the quick way out. In order to restore and sustain our imaginative vision, which is now our crucial capability, we are going to have to re-vision our situation in time and space.

For example, I find that it is a very useful exercise to imagine that *we* are extraterrestrials sent here to visit this planet for a while. If we can imagine ourselves as extraterrestrials, we gain a whole new view of our situation. Another image that I like to use is to imagine ourselves, the human species, as a termite colony. For generations, we have lived in a particular beam in the basement of a particular house. We have multiplied and, finally, with this current generation, we have reached the extent of that beam. We are now emerging on its surface. All the time that we were living in that beam, we developed termite geography, termite mathematics, termite economics, and termite physics, which fitted that reality. Suddenly we look around and see that not only have we been living for all our generations within this beam, but that the house has collapsed around us and the roof has blown off. We encounter enormous vistas of time and space that require us to revise our economics, our mathematics, our physics, and all our other disciplines. Only such new metaphors and visions can help us organize the welter of information in which we are all drowning.

Similarly, we need once again to explore our myths, because myths have always been the most efficient coding of human experience. One might say that myths are social DNA. There is much wisdom that we have to remember from these traditions. So, for our next evolutionary step we are going to have to bring our total selves—body, mind, and spirit—to new levels of awareness and aliveness. Dr. Jean Houston, the author of *Mind Games,* calls this process the "quickening" of human beings. Computer people might call it bringing ourselves on line and bringing ourselves up to real time.

We have to have faith that each of us has the capability of being much more than we have ever been called upon to be before. We have the requisite complexity needed to embrace the new realities of our interdependent situation, but in order to deal with these awesome global interdependencies that we have created, and to deal with the first law of ecology—"everything is connected to everything else"—we must first reintegrate ourselves. Whether it is mind or body, "we" or "they," subjective or objective, science or religion, male or female, the new world view has to be nonlinear, dynamic, contextual, and systemic. It has to deal with the mutual causality of all relationships.

The first thing that has to go is linear economics, which is based on competition, rather than cooperation. For years, economists have used the concept of "externalities" to explain those social costs of production that they did not want to include in their balance sheets and accounting. I always like to call the concept of "externalities" a Freudian slip, because it shows so clearly the economists' own logic and mind-set.

Now we have come to the realization that these "externalities"—the social costs of a polluted environment, disrupted communities, disrupted family life, and eroded primary relationships—may be the only part of our GNP that is growing. We are so confused that we add these social costs into the GNP as if they were real, useful products. We have no idea whether we are going forward or backward, or how much of the GNP is social costs and how much of it is useful production that we intended.

We need a complete restructuring of economics and of all the statistical illusions by which we are trying to manage this abstraction called "the economy." We must include all kinds of data from many other disciplines, including psychology, biology, and physics. Economists must learn this or simply be swept away.[5]

The social-cost side of the ledger is almost a mirror image of GNP. In other words, we can evaluate these industrial societies by the social costs that they create, whether the costs of cleaning up the

Love Canal or the Three Mile Island nuclear plant. Some of these social costs are now beginning to be quantified, although very ineptly. For example, you could make a model of an industrial society that would show all the various industrial sectors that are producing more "bads" than "goods."

An obvious example of this is the drug industry, where we are getting more "diseconomy" and "disservice" than real production. It is clear that we must begin to separate "goods" from "bads" and wealth from "illth." We must reconceive the entire system. Resource-intensive industries are going to decline in the future. Many of the companies in these areas may have to resort to "de-marketing," as I suggested in *The Futurist* in 1974. That is, as high resource costs begin to price some of these products out of the market, some of these companies may actually have to unhook people from them and get rid of them, whether it's TV dinners or whatever becomes priced out of the market because of the basic resource costs. De-marketing campaigns began a few years ago with the electric utilities that had to start to persuade people not to use electricity. Perforce, the oil companies are also taking up de-marketing and pushing conservation.

Politicians vied with each other in recent elections in offering voters phony tax cuts and escapism, rather than helping us face the inevitable austerity period ahead as all industrial economies make the painful transition to less resource-intensive forms of production and consumption. Today we need to understand this transition and how the Soaring Sixties bogged down in the Stagflation Seventies. The Economizing Eighties will be a period of belt-tightening and hard choices during which we can redeploy our enormous assets and lay the groundwork for the sustained-yield productivity and renewable resource-based economies of the dawning solar age of the 1990s. The eighties will be a period of reconceptualization and innovation, redirected investments, recycling, redesign for conservation, rehabilitation and reuse of buildings for new life, revival of small towns and small businesses, and resurgence of neighborhood-based and local enterprises, co-ops, and community development, which release human energy and potential in new local and regional efficiencies of scale.

Already, these growing shoots of the decentralized, informal countereconomy are booming: 50 million Americans belong to co-ops; in 1977, 32 million grew $14 billion worth of their own vegetables, and

5 million belonged to self-help health-care groups; do-it-yourself renovation accounted for some $18 billion worth of building-supply sales, while 10 percent of the increase in total employment in 1978 was not provided by the "formal" economy but was due to the increase in self-employed people. Stanford Research Institute's 1976 report on "voluntary simplicity" noted that 5 million Americans have already dropped out of the industrial rat race, reducing their cash needs in favor of simple life-styles and inner enrichment, rather than keeping up with the Joneses.[6]

As the defectors from the formal economy increase, or opt for part-time work and less cash income, they will relieve some excess demand pressures and open up more jobs for those who want *into* the industrial economy's newer sectors. The growth of this "underground" economy (which few of the statistics in Washington, D.C., will tell you about) is now measured more carefully, since an article entitled "The Subterranean Economy," by Peter Gurman, chairman of the Department of Economics at Baruch College, appeared in the *Financial Analyst Journal* in November 1977. The article estimated that $200 billion a year was "off the books," outside the GNP, and later studies in the United States and Europe confirm the trend. Much is just plain tax dodging, such as not reporting dividends or cash transactions, but some portion consists of those who have dropped into bartering, self-help, mutual-support life-styles. This new countereconomy is much more decentralized, uses less transportation, and is community-oriented.

Population statistics, too, are now showing the return to smaller towns or to rural areas. In 1975, for the first time since the turn of the century, the outward migration from metropolitan areas was greater than that into the cities. People were not going to the suburbs —they were going into the woods or back to small towns and rural areas.

Similarly we see the growth of home gardening. Statistics from the Worldwatch Institute, of Washington, D.C., show that 43 percent of all U.S. families raised some of their own fruits and vegetables in 1977. There is also the growth of alternative media. *Mother Earth News* now has 3 million subscribers, *Rolling Stone* has a 1.5 million circulation, and *Prevention Magazine* 2 million. Another part of this picture is the growth of the human potential and the magazines that cater to it—from *Psychology Today* to *New Age*

Journal and *East-West Journal.* The growth of preventive health care and the movements for jogging, fitness, organic foods, and health centers can be measured in the growth of such magazines as *Prevention.* Another example is the tremendous success of Frances Moore Lappé's *Diet for a Small Planet,* which sold several million copies and was the first vegetarian cookbook that made a global connection between world hunger and the need to go to more-vegetarian lifestyles in rich countries.

There is also the growth of alternative technologies: those technologies that are based on renewable resources such as solar power, wind power, and geothermal resources. A recent study by the Mitre Corporation expected the solar-energy industry to be a $10 billion industry in 1985. In fact, the extent to which traditional economics has ignored the real world system is illustrated by the fact that fully 80 percent of all the world's *savings* is local and informal and not monetized—thus also uncounted in economists' statistics on savings and capital formation (according to Orio Giarini's *Dialogue on Wealth & Welfare,* Pergamon Press, New York, 1980).

We need to think globally and act locally, as elaborated in Chapter 13. We need to clarify our vision so as to remove all the narrow boundaries our dichotomizing, either/or logic has erected, and see that they do not exist in nature—but only in our minds.

This kind of reintegration of perception needs to include our view of our communities, nations, and ethnic differences. We have to go beyond racism, sexism, and nationalism, and beyond the sort of stereotyping that the Cartesian world view encourages: stereotyping by which we oppress each other with our definitions. Imaginative revisioning must mean empathy, and empathy draws our attention to the great concern with equity. Equity is crucial to the environment. Environmental protection is impossible without greater social equity. We know that if the pie of material goodies cannot go on growing, then we must share it more equitably.

We need the growth of alternative marketing channels like Oxfam's Bridge, in which the idea is to link the consumers who care about the planet and their concern for the plight of Third World countries. Oxfam's Bridge links those kinds of consumers together with craft producers in villages in Bangladesh, India, Africa, and South America—linking them together directly, using their own types of direct-mail catalogs. This kind of marketing bypasses corporate structures. These new communications networks are typified by the many "advertisements" in this book from *The Alternative Celebrations Catalogue* (formerly the *Alternative Christmas Catalogue*).[7] This direct-mail catalog sells psychic gifts to people. Their pitch is "Don't buy your friends at Christmas all these ticky-tacky plastic novelties and materialistic junk. Buy your friends a membership in Friends of the Earth, Action for Children's TV, SANE, and other peace groups and public-interest organizations."

We need advertising agencies modeled after the Public Media Center in San Francisco, which sells only "social issues," not products. They sell campaigns to stop nuclear power or to boycott the products of companies that are behaving in ways that people consider irresponsible. This type of counteradvertising, which uses the Fairness Doctrine to get its messages through, is proliferating.

The ecology that must be restored and rebalanced is our human ecology of values. This healing process has already begun in mature industrial societies. In movements like the holistic health movement, we see people taking responsibility for their own health and not expecting a vast, high-technology medical-industrial complex to keep a few of us healthy at the expense of creating a great deal more sickness and pathology for most other people. Many of the citizen movements for racial and sexual equality, for the rights of handicapped citizens, for social justice, and for environmental protection are all part of the healing process that is now going on in industrial society. The proliferation of citizen action in these areas in all aging industrial societies is leading to many new political parties.[8]

Another example of local action occurred in the town of Herkimer, New York. There a large, multinational company decided to close a perfectly good furniture factory because it simply did not contribute enough to their profits. The townsfolk got together and decided that the town would buy the plant. On that scale, the plant was profitable enough, and they preserved the jobs of all of the workers. It is now a town-owned and town-operated enterprise.[9] I believe we are going to find many new and small-scale ventures in the countereconomy.

There is a rise in the direct farm-to-consumer food marketing and a rise in the number of health-food stores. A study recently done by the Agricultural Council of America showed 89 percent of those responding said they wanted more direct farm-to-consumer-type marketing and 73 percent said they would exchange frills for lower prices.

The overblown expectations that people in all industrial societies had are running into this sort of reality. It's almost as if many, many people had the sneaky idea that the party would be over sometime and that it's not going to be that difficult for people to adjust. In fact, I found Europeans who were very excited by the idea of carless Sundays. They were saying, "Oh, wouldn't that be wonderful? The air would be clear, and people could go out into the street without being afraid of being killed, and people could get on their bicycles and not be mowed down by automobiles." We see people taking responsibility once they realize the limits of the institutionalizing of all human needs. Industrial societies have given us the false promise that all our needs can be institutionalized. The result breaks down the fabric of local community cohesion. Of course, it is very necessary to institutionalize a great many needs in an interdependent society, but there is a limit to how much you can institutionalize before you starve the more informal ways of relating that people have always used and found satisfying.

The balanced, renewable-resource-based socioeconomy of the future must be designed and capitalized now, so that it can provide satisfying work and rewarding life-styles for all our people. We now need economists who can see our economy whole: both the older, GNP economy, which is running out of steam, and the emerging countereconomy, which will broaden the way to a viable alternative future.

Other signs that this is going on in all industrial societies are indicated in such books as *The Silent Revolution,* by Ronald Englehardt (1978), who writes about the movement in all industrial societies to what he calls "postmaterialist values." Examples are efforts to control the spread of nuclear power and to achieve environmental protection and social justice, leading to more and more sophisticated and skilled political intervention and participation in government. These surging new political energies have emerged from the exhaustion of industrial culture, the single-minded materialism that George Leonard, president of the Association for Humanistic Psychology, described as an unprecedented experiment in human history: the nearly complete secularization of life. But we do not live by bread alone, and the striving for new meanings has burst out of the narrow politics of Left and Right.[10] Party labels have lost their meaning.[11] New coalitions arise in unlikely places over formerly irreconcilable issues not only in the United States and the other maturing industrial societies but in the centrally planned industrial countries of the Eastern European bloc and in the developing countries as well. We are at a great political watershed. It will be a decade or more before the dust settles and we are able to perceive even dimly the pattern of a new consensus within and between the countries of this interdependent planet. The politics of reconceptualization has begun.

NOTES — CHAPTER 1

[1] Recent events in Iran, Afghanistan, El Salvador, Colombia, and elsewhere, as well as the increase in terrorism, nuclear proliferation, and the resurgence of Islam, all bear witness to the increasing inability of national governments to control their domestic affairs, let alone "manage" international situations or predict global resource depletion and ecosystem perturbations. The traditional responses, such as escalating arms expenditures, are becoming tragically absurd, as former UN Ambassador Charles W. Yost commented on the Carter administration's Defense Department budget increase of $15.3 billion: "We shall find after some experience with the current escalation that both sides [the U.S.A. and the U.S.S.R.] have raced each other to new levels of useless power and debilitating waste, without either having increased significantly its invulnerability or its security" (*The Christian Science Monitor*, February 8, 1980). While the United States was saber-rattling over protecting its oil "lifeline" in the Middle East and coping with the domestic backlash of students protesting the prospect of being sent to fight for the inglorious goal of allowing Americans to continue to guzzle an unfair share of the world's petroleum, the Russians ran into similar hornets' nests with their miscalculations in Afghanistan and the outraged reactions of censure by Islamic and Third World nations, not to mention their own energy-supply problems (summed up in *Fortune*, February 25, 1980, pp. 82–88) and fresh outbreaks of nationalism in the Soviet Baltic states of Estonia, Latvia, and Lithuania (*The Christian Science Monitor*, January 29, 1980).

[2] Still blinded by the economists' simple models of supply and demand, the U.S. presidential debate over how to control inflation at the unprecedented levels of up to 18 percent remained murky, vacillating between drastic cuts in government spending, pie-in-the-sky tax cuts, even higher interest rates, and mandatory wage-price controls. Even though the role of energy in inflation was increasingly recognized, policies of mandatory conservation, of gasoline rationing, were avoided by most candidates, with the exception of Edward Kennedy. Jimmy Carter stalled in spring 1980 with statements that his anti-inflation policy was "under review," while Republicans favored massive public and private investment in energy supply projects, decontrol of energy prices, tax credits, faster depreciation, more tax cuts (on the theory that they would "stimulate" private spending and saving), and generally began sounding like old-style Keynesian

Democrats, since all these policies would be highly inflationary. The Federal Reserve Board's "old-time religion" of higher interest rates to reduce inflation, increasingly backed by candidates of both parties, provided the proof that macroeconomic policy levers have decoupled—no longer driving policy. As I predicted earlier, inflation will closely *follow* interest-rate increases, rather than be reduced, as discussed further in Chapter 2.

2 Daniel Bell seems to be rethinking his position regarding the smooth transition to the postindustrial stage in his earlier books *The End of Ideology* and *The Coming of Post-Industrial Society*. Bell now pays much closer attention to energy as the base of the rapid rise of industrialism and agrees with the long-asserted position of environmentalists that energy conservation is the basic and cheapest mode of dealing with our current economic transition (see *Energy and Growth in America: Energy and the Way We Live*, a syndicated newspaper series by the University of California, 1980—funded by the National Science Foundation).

4 James Robertson, *The Sane Alternative*, U.S. edition with a Foreword by Hazel Henderson, is available from River Basin Press, P.O. Box 30573, St. Paul, Minn. 55175, $4.95.

5 For those wishing to see these statistical absurdities for what they are, an indispensable book is the little classic by Darrell Huff *How to Lie with Statistics*, first published in 1954 and now in its thirty-first printing (W. W. Norton). Amusingly illustrated, it is the best $1.95 value for citizens concerned with exposing bureaucratic and corporate doublethink.

6 Since the 1976 Voluntary Simplicity Study, the Stanford Research Institute has initiated the Values and Lifestyles Program (VALS) and anticipates that by 1990, one fourth of the U.S. population will have shifted their values in this new direction of nonconformity with mass-consumption tastes.

7 *Alternative Celebrations Catalogue*, ed. Bob Kochtitsky, $5.00 from Alternatives, Inc., Box 429, Ellenwood, Ga. 30049.

8 New political parties such as the Citizens Party have formed in the United States, coalescing around issues of equitable and ecologically sound resource use, consumer activism, local control of the economy, and restraint of big companies. Similar efforts that are less focused on immediate electoral victories include the California-based Campaign for Economic Democracy, spearheaded by Jane Fonda and Tom Hayden, which focuses on a transition to safe, renewable energy and worker control of businesses, and the increasingly powerful Solar Lobby, in Washington, D.C. The more visionary, global movement coalescing around the prodigious communication efforts of Mark Satin, author of *New Age Politics* (Delta, 1979), has now incorporated as the New World Alliance. Citizens seem to be learning that although short-term, narrow political victories are possible via sharply focused, single-issue campaigns, they become disastrously divisive, as we have seen with the antiabortion extremists. Many are turning to the more painstaking but rewarding tasks of building broader coalitions such as those linking workers and environmentalists over in-plant pollution health hazards and the even broader coalitions for corporate accountability *and* social justice *and* wiser resource use *and* local economic control typified by the Progressive Alliance, Americans Concerned About Corporate Power, which promotes the Big Business Day teach-in. Earth Day '80 expanded the scope of the earlier environmentalism to embrace broader concerns with challenging economics and narrow, reductionist science and championing social justice, citizen-based action, and community-based innovation of more appropriate

technologies. All these latter groups are based in Washington, D.C., and I have the honor of serving them as a volunteer adviser.

[9] Many such fascinating case histories of new forms of economic enterprise are documented in *Democracy at Work* by Daniel Zwerdling, available for $5 plus $.50 postage from the Association for Self-Management, 1414 Spring Rd. N.W., Washington, D.C. 20010.

[10] An example of the "new age" politician, concerned with multidimensional political issues that will characterize the new coalitions in industrial societies for the rest of this century, is California's Assemblyman John Vasconcellos, author of *A Liberating Vision: Politics for Growing Humans*, Impact Publishers, P.O. Box 1094, San Luis Obispo, CA 93406, $6.95.

[11] Another view of the contemporary citizen movement is expressed in *Anticipatory Democracy: People in the Politics of the Future*, ed. Clement Bezold (Vintage, 1978).

CHAPTER 2

The End of Flat-Earth Economics

Today inflation, unemployment, and the "stagflation syndrome" affect all mature industrial societies, from Britain (the oldest) and the U.S.A. (the most elaborated) to Canada, Japan, Sweden, Germany, France, and the other Western European countries, to Australia and New Zealand (all in different stages of the industrialization process). Many of the socialist, centrally planned industrial societies of Eastern Europe and the U.S.S.R. itself are bogging down in similar syndromes of resource depletion, overcentralization, bureaucratization, and the growing unsustainability of militarism, expansionism, and competition in the existing world trade system. In the socialist countries, the symptoms are not classified in the "stagflation" interpretation, since this is an expression of the problem in the terminology of market economics, which they reject. We are witnessing the crises of industrialism itself, which is simply further along in the Western-style, mixed-market societies and accounts for their more pronounced set of symptoms, while the socialist countries still lag behind in their drive toward the same set of unsustainable goals. E. F. Schumacher recounted to me his many conversations with leaders of developing countries and those leaning toward socialism: "These leaders will tell you that the Western, capitalist countries are like express trains heading toward a precipice of self-destruction—and then they add: but we shall overtake them!" Indeed, as

Portions of this chapter are excerpted from my editorials in *The Christian Science Monitor*, August 9, 1978, and December 29, 1978. Used with permission.

we shall explore further, Marxist theory holds that socialism is simply a later stage of capitalist development and would grow out of the processes of capitalist production and accumulation. In fact, the new vulnerabilities of industrialism itself now have produced an even greater irony: that the fierce and supposedly fundamental debate between Marxism and capitalism of the nineteenth and early-twentieth centuries has turned out to be a surface argument. Both systems are systems of industrialism, dedicated to maximizing material production and narrowly conceived technological "progress"; both shortchange the considerations of ecological tolerances and the fundamental needs of human beings that go beyond material sufficiency to needs for philosophical meaning and concerns of the spirit. Today some of the best critiques of Marxism and its bureaucratic expression in the Stalinist model of state socialism or even state capitalism (where the supposedly socially concerned state begins operating just like a multinational corporation) are now coming from Marxists and socialists! For example, this new wave of criticism in France is typified by Bernard-Henri Lévy's *Barbarism with a Human Face* (Harper & Row, 1979) and the columns in *Le Monde* that deal with decentralizing technology and reducing the centralizing tendencies beloved of both large corporations and socialist commissars, by Marxist theoretician Roger Garaudy.

Thus, since the exhaustion of industrialism's logic is now more apparent in the oldest and most elaborated examples of Western Europe, North America, and Japan, we shall focus on the developing symptoms of "stagflation" in these countries. In addition, this may help us clarify our own political debates and reformulate problems more fruitfully, while at the same time allowing us to anticipate the course of events as the same syndrome progresses in the less-elaborated socialist countries and "less-developed" countries still trying to overtake us or imitate our mistakes. Many of the most astute thinkers in the "less-developed" countries are now able to see that differing development models offered by the Western, market-oriented societies and the Eastern European, centrally planned societies produced a new set of problems and symptoms that are very similar: catastrophic urbanization, unsustainably resource-intensive production methods, costly centralized technologies requiring huge bureaucracies, unattainable levels of specialization, technological dependence, lost food self-sufficiency, and disruption of their own cul-

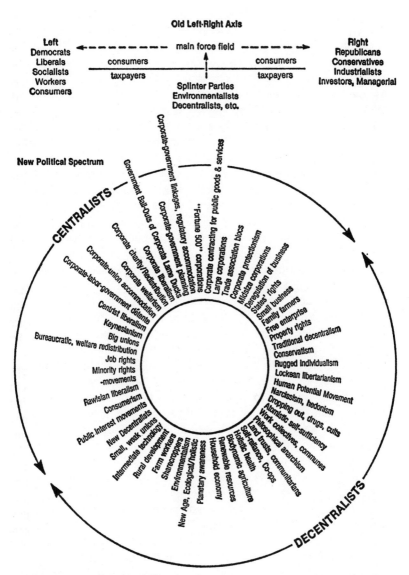

Old Left-Right Axis

Left
Democrats
Liberals
Socialists
Workers
Consumers

← — — — — — — main force field — — — — — — →

consumers   consumers

taxpayers   taxpayers

Splinter Parties
Environmentalists
Decentralists, etc.

Right
Republicans
Conservatives
Industrialists
Investors, Managerial

New Political Spectrum

CENTRALISTS

DECENTRALISTS

Corporate-government linkages, regulatory accommodation
Government Bail-Outs of Corporate Lame Ducks
Corporate-government planning
Corporate liberalism
Corporate charity/Redistribution
Corporate welfarism
Corporate-union accommodation
Corporate-labor-government détente
Centrist liberalism
Keynesianism
Big unions
Bureaucratic, welfare redistribution
Job rights
Minority rights
-movements
Rawlsian liberalism
Consumerism
Public Interest movements
New Decentralists
Small, weak unions
Intermediate technology
Rural development
Farm workers
Sharecroppers
Environmentalism
New Age, Ecological/holistic
Planetary awareness
Household economy
Renewable resources
Biodynamic agriculture
Holistic health
Self-reliance, Co-ops
Philosophical anarchism
Work collectives, communes
Land trusts, communitarians
Atomistic self-sufficiency
Dropping out, drugs, cults
Narcissism, hedonism
Human Potential Movement
Lockean libertarianism
Rugged individualism
Conservatism
Traditional decentralism
Property rights
Free enterprise
Family farmers
Small business
States' rights
Deregulation of business
Midsize corporations
Corporate protectionism
Trade association blocs
Large corporations
Corporate contracting for public goods & services
"Fortune 500" corporations

Fig. 2 Changing Political Configurations in Industrial Democracies

tures. Many now are looking for their own, third way, realizing that industrialism, technocracy, and the acculturation process underlying these styles of development are all expressions of white, Eurocentric cultures, and are now clearly failing in any case.[1] Leaders of "less-developed" countries who have been fascinated with such hybrid development models as the Yugoslavian worker self-management approach to smaller-scale, decentralized production and the China of Mao Tse-tung have tried to draw from all the available global experience of "development"—itself an ambiguous term. African countries such as Tanzania (under the leadership of Julius Nyerere) have made painful attempts to forge unconventional development paths, and many other countries oscillate between the excesses of both the capitalist, multinational corporate models and those of the repressive, bureaucratic, doctrinaire models of communism. However, the difficulties of forging new paths to human and societal development on a globe dominated by the antagonisms of the world's two chief expressions of industrialism—communism and capitalism, embodied in the global struggle of the U.S.A. and the U.S.S.R.—leave little room for maneuverability or experimentation. Indeed, the world is so polarized and destabilized by this cold war between the giants, that it claims countries as new victims each day: Iran, where Western-style capitalism was thrown out with the bathwater in a bewildered revolt against "modernization" and a revival of seventh-century Islam; and Kampuchea, where the devastation of U.S. military action destroyed the economy and led to an apparent rejection of industrialization so mindless and confused by conflicting forces that it has reduced the country to starvation and destroyed much of its once refined culture. El Salvador and Afghanistan have fallen victim to the same deadly struggle. The two poles of the cold war will continue to generate such catastrophic instabilities as pull smaller, weak, dependent states this way and that, with increasing inability to govern internally or deal with the impossible conditions of competition imposed by the present world monetary and trade system. Even China, with well over a fifth of all the world's population, has now adopted the destabilizing path of industrialism, with her "Four Modernizations" program of agricultural mechanization, technological development, industrial investment, and increased defense and world trade linkages and is firmly on the path of material "progress" at whatever social and ecological costs. However, it is possible that the world's 900 million Moslems may provide a wedge in the U.S.A.-U.S.S.R. confrontation, since they excoriate both "Godless Marxists" and "capitalist infidels."[2]

Another key ideology underlying both market-oriented capitalistic development and socialist models of industrialism is the belief in the scientific, quantitative, reductionist world view inherited from Descartes and Aristotle, the preoccupation with materialistic values as opposed to more metaphysical values, which will be discussed in Chapter 6. These preoccupations are profoundly intertwined with the crisis of economics in its development of quantitative tools for measuring human social progress and welfare, as for example the gross-national-product indicators, largely accepted by both capitalist and socialist countries. The crux of the crisis of economics of *all* schools of thought—Keynesianism, monetarism, Marxism, "laissez-faire," neo-classical and "supply-side" economics—is that they all share the hypnotism of money, looking only at those sectors of production and consumption in their countries that are monetized and involve cash transactions. This colossal error of equating the monetized half of most industrial economies with the whole system of production, consumption, and maintenance is common to all branches of economics and accounts for its one-dimensional, linear, partial view, as opposed to the wider realities of seeing economies *whole*. Indeed, it is essential for us to remind the one-eyed economists that most of the world's production, consumption, and maintenance still occurs outside the monetized economies, as it always has. Most of the world's people are sustained by growing their own food, tending their own animals in rural areas, and living in small, cooperatively run villages and settlements or as nomads following herds, harvesting wild crops, fishing, and hunting in economies based on barter, reciprocity, and redistribution of surpluses according to customs such as feasts and potlatches. One of the aspects of the crises of industrial development is that it begins to suck all such informal, use-value production and consumption into the monetized economies, drawing populations into the cities, denuding rural agricultural areas, dissolving the cultural glue of village life and reciprocal community systems of food-sharing, care of the young and elderly, and folk medicine, and destroying inherited cultural wisdom learned in coping with diverse ecological conditions. Thus, industrialism and the economic logic underpinning it tacitly view the industrialization process as also one of monetization of all production and consumption and the accumulation of investment "capital" or "surplus" (viewed as money whether denominated as dollars, rubles, pounds, or yen). As industrialization and monetization spread and colonize more and more of the informal, use-value production, consumption, accumulation, and exchange systems that are nonmonetized, limits to this process are reached. Symptoms of these limits show up as anomalies in trying to "monetize" environmental resources and place cash value on air, water, open space, and even human life and the loving, caring relationships that allow humans in all societies to provide services to each other free of charge. The industrial model in its mass-consumption, global-advertising stage of titillating rural populations with visions of city lights, cars, flashy clothes, cigarettes, booze, Coke, and rock music now creates mass migrations throughout the planet in search of money, jobs, and status symbols. At the same time, traditional values of sharing, respect for nature, and reciprocal unremunerated services appear old-fashioned, boring, and backward, described by Marx, who despised peasant culture, as "the idiocy of rural life." But after tempting every rural community on the planet that can be reached by mass media and transistor radios with the consumer model of the industrialized "good life," industrialism is caught in a cruel hoax: it cannot deliver. It can deliver for only *some* of the population at the expense of *others*—and as we are now seeing, it can deliver only some of the time (while there are cheap, abundant environmental resources to be used up). But it cannot work in the long run.

Thus today we see a planet in the turmoil of destabilized values and cultures with explosive and unrealistic expectations of the promised land of industrialization and "modernization" as it is projected in movies, television, and advertising. Not surprisingly, in mature industrial countries, the monetized view has won out to the point where housework, child raising and nurturing, and caring for the sick and old in one's own family are all becoming monetized and institutionalized activities, while the environment itself demands "repayment" in cash as we are forced to clean up chemical dumps and purify water (sometimes even in order that it can be used in manufacturing processes, let alone for drinking) (see Plate 26). At the same time, the well-intentioned missionaries of economic development have persuaded rural villagers that they need hospitals and European styles of education and training so that they can get "jobs," and that the services they used to perform for each other communally, such as nursing the sick, should be remunerated in cash. Furthermore, they are told by foreign economists and those economists of their own countries who have been trained at Harvard, Stanford, the London School of Economics, or at the University of Moscow that their welfare can be measured in average, per-capita money terms, and that since their countries' GNPs are low, that they are "poor" and must export more to earn foreign exchange. The price of entering this world-trade rat race is either borrowing foreign capital (i.e., money) at usurious interest rates, permitting multinational companies to build factories to exploit their cheap labor and resources, or to accept heavy-handed, Big Brotherly "assistance," with its own kind of strings attached, from nations of the Eastern European and Soviet bloc. An extraordinary example of the pervasiveness of the monetized, economic world view was the meeting in Nairobi in 1979 under the auspices of the United Nations Environment Program to discuss how to improve the methods of economics of cost/benefit and risk/benefit analyses for better application to "management" of environmental resources and pollution control. Soviet economists addressed the issue of how to reduce the living productivity of the biosphere to quantitative calculus in money units in terms similar to those of economists from the market-oriented economies. When asked for advice by the U. S. State Department concerning the formulas that economics might develop, my response was that the economic method was entirely inappropriate, since economic models do not take account of bio-productivity, the requirement for diversity in ecosystems, the widely differing approaches to production and consumption in each culture and value system as *resources* (as important as coal or oil), as well as the fact that economics is still not firmly grounded in the basic laws of thermodynamics, as explained more fully in Chapters 8 and 10.

Thus we see the current dilemmas of maturing industrial societies in a wholly new light. It is said that those who live by the sword shall die by the sword; and similarly those societies who live by the monetized view may decline because of their monetized view of allocating their resources and managing their governments. But today we see that the tyranny of the monetized sectors over the nonmonetized sectors and the ecosystem is collapsing of its own weight. To give Marxian critiques of capitalism their due, they did raise the issues of how the definitions of economists and the monetized system were used to read some groups out of their fair share of the rewards for their con-tributions to overall "productivity," e.g. the workers who had to sell their labor in the rigged, monetized market, as described in Chap-ter 8. But the problem today not only concerns who owns the means of production; it also concerns the unsustainability and destruction of human values inherent in the means of production themselves. Thus today we can see that workers are no longer the only victims—indeed, many of them have bought into the existing industrial system and benefited greatly from it. But the relative improvement of the situa-tion of organized workers in industrial countries has been bought at the price of denying full participation in the cash rewards of the monetized sectors to *other* groups: women, minorities, floating popu-lations of migrant and "guest" laborers, and all those arbitrarily defined out of the game by customs and caste systems, such as In-dia's "untouchables." Feminist critiques of even socialism (which has at least promulgated a more humane set of goals for sharing "wealth" socially) are devastating. Women in the maturing industrial societies now challenge the focus of socialism on the plight of the worker. True, they say, the workers under laissez-faire capitalism were oppressed—but today many are well organized and often ex-clude women and other minorities from their unions, and when they go home at night they often bully their wives and children and resist the progress of those even more oppressed.[3] Similarly, the progress made by workers (and other groups who derive incomes from wages, rather than property ownership) in the mature industrial countries has also rested on the ability of these societies to exploit the formerly abundant resources of nature and those of "less-developed" coun-tries.

Today, therefore, the stage is set to play out all these older conflicts among groups *within* nations and those *between* nations on the stage of the now inextricably interlinked global "economy," i.e., viewed exclusively by national governments as the world trade and

The *Total Society* (the whole of the sphere) is shown with many other dimensions mapped by other disciplines and methods using "cuts" along many different planes and axes, e.g., political science, sociology, biology, anthropology, physics, thermodynamics, chemistry, ecology, information theory, general systems, etc. New, more comprehensive methods such as technology assessment, environmental impact studies, and future studies attempt to integrate many approaches.

In this schematic representation, the economic view is depicted by the rear cross-section labeled on the left. The economic view is characterized as: linear concepts of the monetized sector, measured by GNP, market "supply" and "demand" equilibrated by price. The Cartesian view studies discrete aspects of "the economy" in micro-detail, and then adds and averages data to form an aggregated macro-model. Social, environmental, and other non-monetized interactions are modeled as "exogenous factors" or "externalities." See fore cross-section labeled on the right. There is generalization from the micro- to the macro-level, rather than starting with a systemic model. This is a pseudo-empirical approach since gathering micro-data already implies the assumption of a macro-model.

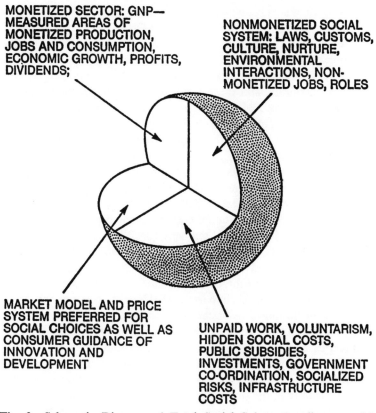

MONETIZED SECTOR: GNP—MEASURED AREAS OF MONETIZED PRODUCTION, JOBS AND CONSUMPTION, ECONOMIC GROWTH, PROFITS, DIVIDENDS;

NONMONETIZED SOCIAL SYSTEM: LAWS, CUSTOMS, CULTURE, NURTURE, ENVIRONMENTAL INTERACTIONS, NON-MONETIZED JOBS, ROLES

MARKET MODEL AND PRICE SYSTEM PREFERRED FOR SOCIAL CHOICES AS WELL AS CONSUMER GUIDANCE OF INNOVATION AND DEVELOPMENT

UNPAID WORK, VOLUNTARISM, HIDDEN SOCIAL COSTS, PUBLIC SUBSIDIES, INVESTMENTS, GOVERNMENT CO-ORDINATION, SOCIALIZED RISKS, INFRASTRUCTURE COSTS

Fig. 3 Schematic Diagram of Total Social Sphere (nonlinear, multi-dimensional, dynamic) hypothetically sectioned to show limited area mapped by economics

monetary system. A key to understanding how the drama will unfold is to observe how the definitions of which activities are to be monetized and which demonetized will change. In a very real sense, as mentioned, the "success" of nations' monetized, GNP-measured economies has always rested on their ability to draw definitional boundaries, so that the monetized sector could always "externalize" those social and environmental costs it did not want to account for by imposing them on groups whose jobs could be defined as "nonproductive" (for example, maintenance work such as cleaning, garbage collection, nursing, and other human services) as well as on women in the family whose vital daily maintenance activities and nurturing of the "productive breadwinners" and their offspring are still unpaid and not accounted for in the GNP. Significantly, the crucial nature of maintenance work in a system—which thermodynamicists understand better than economists—is now commanding attention in all maturing industrial societies. As the system gets more complex, elaborate, and interlinked, more and more people have to be diverted to these maintenance activities. They include the growing number of white-collar workers in government and corporate bureaucracies trying to coordinate all the diverse and conflicting production activities (which I describe as the syndrome of the "entropy stage" of industrialism) as well as the burgeoning "human services" which replace formerly nonmonetized work, such as day-care centers, social counseling and welfare agencies, family therapists and psychiatrists, drug rehabilitation centers, crime prevention, and institutionalized care for the aged and sick (see Plate 26). Today, surrogate "homemakers" provide vital daily maintenance of decent living conditions and food preparation—activities now monetized, as women are forced into the cash economy as breadwinners or simply reject the role of unpaid supporters of the cash economy and accept the dollar definition of self-esteem themselves. Thus, as all these formerly unaccounted maintenance costs are flooding back over the boundaries into the monetized economy together with the environmental reparations mentioned earlier, the monetary economy, not surprisingly, begins to suffer from overall "declining productivity," and particularly in the services sector, as discussed further in Chapter 9. An example of the booming "social maintenance" industries is the growth of the day-care business, now a multimillion-dollar industry (*The Christian Science Monitor,* October 17, 1979).

Thus the crises of governance in all mature industrial economies today are inseparable from the fact that too much of their national policy making and the modeling of their problems are reduced to the abstraction of "managing their economies," using the monetized data and tools provided by economists and econometricians and the concepts of macroeconomic management.[4] This obscures all the other dimensions of the reality of dynamic societies, structured in now unsustainable modes of interaction with their resources and energy and based on technologies and institutions that now must be redesigned to fit totally new situations. These new realities imply a shift of focus from the monetized, GNP-measured economies and the economic policy approaches to a new era of "posteconomic policy making." From now on, as the economic and price-system levers become ever more divorced from reality, industrial societies will need to refocus their attention on policy levers that are nonmonetary, nonfiscal, and nonprice-oriented, as we shall explore further in this book. Meanwhile, one of the greatest problems that will be encountered is that "upstream" in the academic world: the enormous overproduction of economists themselves. Economics as a discipline has "colonized" all kinds of public-policy debates in areas for which its methodologies are totally unsuited and for which they yield no useful policy directives, as spelled out subsequently (see Plate 18).

There is perhaps no better way of summing up what economists call "inflation" and its systemic character in today's "stagflation" syndrome than to define inflation, from beyond economics, as simply all the factors and variables that economists have left out of their models. Thus even the term "inflation" has become a mystification, covering the economists' lack of understanding of the new situations facing societies where industrialism has matured. Today's inflation, structural unemployment, and spreading tax revolt signal the end of the age of Keynes in the United States and all other mature industrial societies—in Western Europe, Canada, Japan, and Australia. Indeed, the bogging down of these societies in the "stagflation" syndrome augurs the late phase of industrialism itself and its transition to some new, undefined "postindustrial" phase. The basic nature of the painful transition involves a necessary shift from economies based on maximizing labor productivity (and thereby continually increasing the capital and energy intensity of their industrial production), which heretofore has been based on *non*renewable energy, to economies that must now conserve capital, energy, and materials and more fully employ their human resources. This shift will require the development of a newly designed, more efficient production system based on renewable resources and managed for sustained-yield productivity.

This is a tall order for today's economists acculturated during a brief period of fossil-fuel-based abundance, which provided the slack that allowed massive Keynesian pump priming and demand-stimulation without instant inflation. In Chapter 9, and in *Creating Alternative Futures,* this "trickle-down" model of economic growth is described further, a model in which capital investments in more efficient, automated industrial plants were encouraged to raise per-capita productivity even though many workers were disemployed. The total economic pie was expected to keep growing fast enough to create new jobs for labor-force entrants, while the "services sector" was supposed to grow and provide more white-collar jobs for the upwardly mobile. All this worked quite well as long as abundant cheap resources from all over the planet were available to fuel this type of economic growth. Today, in a turbulent, global economy with rising population and expectations combining with dwindling resources (especially petroleum), this Keynesian game is up, and the excessive energy and resource dependency of the late-stage industrial societies has become their most crucial vulnerability. High rates of inflation and rapidly eroding real incomes now indicate that the GNP growth measure is becoming little more than a statistical illusion. Economists are now engrossed in watching their own policy interventions show up in short-term oscillations of what used to be called "business cycles." They are now entangled and hypnotized by bouncing interest rates and unpredictable levels of employment, saving, spending, investment, etc.—all now *effects* of economic policy, rather than objective phenomena. This short-circuiting of economic management and government inability to model the complexities or distinguish causes from effects is now visible to the average citizen. For example, in 1979, small savers brought a class action, filed by Public Advocates of San Francisco, on behalf of all citizens who have bought U. S. Savings Bonds. The suit charged the U. S. Treasury with deceptive practices, advertising Savings Bonds by claiming that they "put your financial worries to rest," whereas they would actually be worth only fifty cents on the dollar at maturity if current inflation persists.

Similar disarray exists in the debates about "declining productivity" and innovation and the calls for "deregulation," supply-side economics and ceilings on the federal budget, all examined later in this book. Leaders still quarrel over which industrial countries should try to "reflate" their economies so as to pull the world economy out of recession, even though they no longer want to apply these policies in their own countries, since they now know that using the old, Keynesian remedies of pumping up total demand simply increases inflation rates.[5]

The only set of policy tools available to world leaders are those handed to them by economists and which cover a very narrow spectrum of options. This type of "flat-earth economics" has produced in most of the industrial democracies two major political parties, regardless of their designation as "Left," "Center," or "Right," or as Liberal, Conservative, Labor, Socialist, Republican, or Democratic. The party occupying the "Right" is usually sympathetic to investors and business and hews to what I call the Golden Goose model (i.e., the belief that all wealth is produced in the "private" sector and then is taxed to provide "public"-sector goods and services). All these parties accordingly believe in the "old-time religion" of the monetarists: squeeze the inflationary pressure out by cutting public spending; raise interest rates and reduce money-supply growth, accepting that unemployment will increase; then deregulate business and provide more investment tax credits on the trickle-down theory that this will create more jobs. On the opposite end of the spectrum is another party, which occupies the "Left" position. This party has its Keynesian economists (who are also in tune with the Golden Goose model), and all they know how to do is to print money and hope that the inevitable creeping rates of inflation won't become too noticeable. To further confuse the picture, in 1978's mid-term and in the 1980 elections the U. S. Republican Party campaigned nationally on a "tax cut" platform, a Keynesian approach for which it has always excoriated Democrats as "fiscally irresponsible"! The common sense of voters rejected the tax-cut panacea in 1978. Republicans touted their dubious plans to "cap" the federal budget and the panaceas of supply-side economics and re-floated the tax cut for the Reagan presidential campaign.

The tragedy is that under today's drastically changed conditions, none of these strategies will work. Totally new approaches are needed, which may not even be "economic." Indeed, macroeconomic-management approaches to national policy are failing precisely because the narrow economic data on which they are based are becoming ever more disordered and illusory. GNP does not differentiate between "goods" and "bads," wealth and "illth," and includes all the social and environmental costs of our disruptive, capital-intensive technologies, whether cleaning up the mess after production and consumption or coping with rising levels of carcinogens and cancer, hypertension (now a $2 billion new "industry"), the dislocation of workers, families, communities, and whole towns (such as Youngstown, Ohio) due to plant shutdowns and relocations, and of course, the rising tax burden of necessary efforts to coordinate and regulate private sector activities, mediate the conflicts and ameliorate all these social costs. Today, the only part of the GNP that is growing may well be this "social-costs fraction." Thus "progress" is little more than inflation, and "economic downturn" overlooks and cannot measure the real health and strength of the society. Other new sources of inflation are best understood from beyond the view of economics, for example the inflation due to the declining quality as well as quantity of our resources, which requires us to invest ever more capital and energy in extracting raw materials and energy from ever more degraded and inaccessible deposits. Thus, much of our "declining productivity" is not due to less diligent or greedy workers, but to the declining productivity of our *capital*.

Meanwhile, OPEC has understandably exploited our "petroleum habit" (a large cause of the sharp decline of the U.S. dollar) and is becoming more wary of accepting our inflating currency. The United States finds itself caught between its allies' demands that we reduce our energy consumption and oil imports so as to bolster the dollar and the OPEC position that if we try to accomplish this demand reduction (using price rises via oil-import fees or domestic wellhead oil taxes), that this will only demonstrate that energy prices are still too low. Thus OPEC has stated that it would prefer to "oblige" us with the price increase rather than allow the U. S. Government to reap the increased tax revenues on imported oil, which OPEC claims is unfairly diverting income from developing countries. At least the United States is now facing up to the role of oil imports in our balance of payments and weak dollar problems, although economists still decry attempts at mandatory allocation, rationing gasoline, or any other conscious policy choices, such as legislating smaller, more energy-efficient cars. Economists still prefer the two rather unpleasant choices their macro-model yields: 1) a recession, which drives down demand for energy at the excessive cost of depressing the whole economy; and 2) continuing to ration energy by price, thus adding to inflation and exacerbating the squeeze on citizens who are on low or fixed incomes. Thus, in 1979 and 1980 we saw the spectacle of still-deadlocked energy policies. At least we prevented poor people from dying of the cold in the interim, as fuel oil and gasoline hovered above a dollar a gallon. The U. S. Government meted out assistance funds of $1.3 billion, hoping to recoup these general-revenue funds from the windfall profits tax on oil companies—a case of first freezing Paul in order to pay Peter (by the deregulation of oil prices), then subsidizing Paul's fuel bills in the hope of recouping the money from Peter! Meanwhile, the much vilified OPEC cartel broke down, not as the United States had hoped, but on the upward price side, as member states such as Libya, Algeria, and others broke ranks and began selling their oil on the Rotterdam "spot market" at almost double the OPEC cartel's official price. Ironically, "spot market" prices were bid up by jittery consuming nations in the face of cutoffs of Iraqi and Iranian production. The Arabs' view of the situation is that if the United States had not been inflating its currency for a decade prior to the 1974 formation of the OPEC cartel, they would not have needed to take those steps to keep their oil's real price stable. They cite the dollar inflation caused by the Vietnam War and the flooding of world currency markets with dollars, as well as our increasingly huge military expenditures. At the Belgrade meeting of the International Monetary Fund, in late 1979, the Arabs asked U. S. Treasury Secretary William Miller why they should continue to believe in the dollar at all and accept payments for their oil in dollars, since the United States had not taken steps to bolster it by reducing imports of oil or instituted any serious measures to conserve oil, let alone ration or allocate energy use, prevent waste, or fund solar and other alternative, renewable energy resources at realistic levels.[6] As the Arabs converted their dollar holdings into gold and our European allies continued their plans to float a currency of their own (the European Monetary System, EMS), U.S. financial leaders at the Belgrade meeting were trying the "hard sell": promoting the soundness of the dollar, the government's competence to check the dollar's decline, and its political will to maintain economic stability. One doubts, however, that the advertising approach will substitute for domestic leadership and realistic policy.

It is now becoming clear that, despite the OPEC price increases, it is useless to make scapegoats of the Arabs. They deserve credit for the thankless job of teaching Americans some of the global realities of the declining age of petroleum. New oilfields, whether in Mexico or China, do not belong to us. We must *pay* for our oil imports. Each year, we will pay billions more, with the usual effects of worsening our balance-of-payments deficit and depressing the dollar while increasing domestic inflation. This, in turn, further depresses the dollar and the cycle begins anew, as OPEC says it must raise prices again to correct for the fallen value of our dollar payments.

However, short-term macroeconomic policies will still be checkmated if they remain blind to the full range of socioeconomic strategies available. The narrow monetary, fiscal, and price-system choices reduce to either accepting more inflation in order to keep the debt economy rolling over, or engineering ever more *inflationary recessions*. Economists need to accept the reality of the non-GNP socioeconomy and recognize the potential of consciously and democratically determined conservation and selective demand-reduction, instead of the "buckshot" of the old monetary, fiscal, and price levers that work only on across-the-board demand and supply.[7]

Selective, *non*monetary, demand management would target *only* the bottleneck areas, such as legislating better mileage and smaller cars and instituting equitable, white-market rationing of gasoline to reduce nonessential driving, so as to address the immediate problem of oil imports. We might target the energy overuse with full-scale public-service campaigns for conservation (as in Britain) and ban advertising that encourages energy waste (as in France). Varying the total volume of product commercials on radio and TV up or downward could provide another noninflationary demand-management tool. Another step toward political demand management has been taken in voluntary wage/price guidelines (proving that our economy needs cooperation now, as well as competition).[8]

The failure of Keynesianism can also be seen in our overreliance on the institutionalized, "formal," cash economy to provide for all our needs, goods, services, and jobs. Instead, it is bogging down in debt and inflation. We are perilously dependent on the now bankrupt economists' policy tools of centrally manipulating an abstraction called "total economic demand." A few simple levers are relied on: either continuing to print more money or administering the "old-time religion" of arranging a recession by squeezing credit and hiking interest rates, slashing the federal budget, or deregulating already high gas and oil prices. It is now obvious that all these policies are inflationary, just as the new Reagan supply-side economics will prove to be.

The old Keynesian tool kit worked with much smaller rates of inflation back when we could expand the economic pie using cheap inputs of energy and resources, fuel our consumption with credit, and provide larger slices to all the competing groups in society. Now these ineffective Keynesian Band-Aids are being peeled away to reveal underlying social conflicts about how to slice the now-constant GNP section of the pie. Some conflicts are intensified but familiar battles between special interests, as when older energy industries struggle to retain their subsidies in the face of newer upstarts like the burgeoning solar-energy industry. Others involve ominous new clashes that politicize credit, investment, and debt, such as conflicts between city workers in Cleveland and New York and those cities' bondholders and banks; labor and business arguing over wage/price guidelines; and Wall Streeters fighting over capital availability and possible credit controls, over which investments are "productive" and which are "unproductive," and over whether credit for homeowners must be squeezed as competing financial interests lobby to divert mortgage funds and force the housing sector to bear the brunt of recession. Similar crunches appear in the conflicting requirements between needed rates of saving, investment, and consumption, where equally painful trade-offs will have to be negotiated. Yet we must avoid the "easy" route of allowing the poorest, the powerless groups to bear the brunt of stabilization efforts, through job layoffs and rationing by price.

As the failures of Keynesianism institutionalize inflation, the linear, either/or logic of economists has again swung toward the monetarists in Britain, Germany, and France, as well as briefly, with the Joe Clark Conservatives in Canada and in the draconian policies instituted by Paul Volcker of the U. S. Federal Reserve Board. None of these new doses of the monetarists' "old-time religion" will do much more than increase unemployment. However, they have touched off another arcane debate among economists as to whether to attack the new outbreaks of double-digit inflation with high interest rates or, as the monetarists prefer, by trying to squeeze the money supply (very difficult in practice, since it has become almost impossible to measure, as we shall discuss further in Chapter 8). Another fashionable monetarist theory, promoted by none other than Milton Friedman (winner of the Nobel Memorial Prize, set up by the Central Bank of Sweden as an addition to the prizes set up by Alfred Nobel himself), is that of indexation.

To a linear-modeled discipline such as economics, indexation sounds like a sensible idea: simply deal with inflation by allowing for automatic increases in everyone's wages, pensions, rents, and interest payments, i.e., linking or indexing them to the current inflation rate. Any systems analyst could have predicted that such an add-on approach would simply institutionalize inflation by exerting a multiplier effect that would keep triggering price increases that would feed back instantly into ratcheting wage increases. In *Business Week* (November 12, 1979), Milton Friedman had to admit, in the face of growing evidence that indexation *increases* inflation, that he was wrong. This is a key admission, since the monetarists, whose dean is Dr. Friedman, are now back in the policy saddle, while Friedman himself is still an active policy maker, intervening with advice on how to manage the economies of countries who negotiate loans from the International Monetary Fund (IMF), such as Chile, and many others with negative trade balances and heavy foreign debt. The International Monetary Fund's advice, proffered with its loans, is often worse than the disease, and the IMF's draconian economic austerity has been accused of destabilizing the governments it seeks to help by causing high unemployment and hardship for the poor (thus contributing to riots and social unrest) and has earned the nickname "the roughest bank in town."[9] Meanwhile, companies and individuals try to live with inflation: companies by reporting "inflation-adjusted" profits and applying new accounting rules that are admittedly inexact, promulgated by the National Accounting Standards Board; individuals by continuing to increase their debt (which they figure can be paid back in cheaper dollars) and by buying everything from houses, consumer durables, diamonds, art, antiques, and gold, silver, and other strategic metals in an understandable flight from money.[10] Thus the economy kept "growing" through the spring of 1980 in spite of recessionary policies.

At the same time that such domestic economic instabilities are increasing, the hiking of domestic interest rates sets off another vicious spiral, as other countries begin bidding in the interest-rate sweepstakes to attract foreign deposits from each other. Thus another ratchet in general inflation rates is set in motion and the basic interest rates become increasingly difficult to lower, surging through the domestic economies and raising the costs of housing mortgages and car payments. In economies like that of the United States, which runs on massive borrowing and credit, they exert another multiplier effect on all prices. So we see the emerging double bind for economists: raising interest rates *used* to be effective in squeezing credit and reducing economic activity and employment, thereby reducing inflation, but now is effective only at reducing production and employment, which *no longer* decreases inflation but actually increases it! One suspects that the reason the U.S. economy failed to respond in the usual way to the watershed increases in interest rates of October 1979 is related to the fact that with inflation then running at approximately 15 percent, interest rates of the same rate are effectively zero, and the wealthy take greater risks to use money to buy real assets even though they must borrow it at high interest rates.[11] This effect has been visible in the housing market, where in spite of sky-high mortgages, people take risks to get on the inflation escalator by buying a house, and if mortgage money is unavailable, sellers can use their inflated equity in their house to offer a second mortgage to a buyer (yet another monetary illusion, or "funny-money" effect).

We can expect unprecedented situations of economic policy instabilities in all aging industrial societies in the 1980s. Their economic policy levers have stripped their gears and are beginning to swing wildly, less than ever related to the real world of production, consumption, and maintenance, to geographical and sectoral differences, climate zones, political power blocs, and shifting alliances. For example, the Federal Reserve's actions of October 1979 were, according to Paul Volcker, to curb speculation in financial and commodity markets. But this is a too highly aggregated view and cannot weed out the speculators from worried average citizens afraid that if they don't overextend their credit now, at even usurious mortgage rates, they will never be able to afford a home. Similarly, as Jonathan Gray, of Sanford C. Berstein & Co., a Wall Street investment firm, noted: "Every financial institution speculates on interest rates, and if we have violent swings in interest rates within a short period of time, these institutions simply will not commit funds long-term" (*Newsweek*, October 29, 1979, p. 71). Another example of growing economic-management instabilities involves the estimates of the federal budget deficit for fiscal 1980, which at first were for some $30 billion on expenditures of $547 billion. Significantly, the Office of Management and Budget (OMB), recalculating the effects of increases in the Defense Department's budget and, even more, the higher interest rates on the national debt that must now be paid due to the Federal Reserve Board's policies, revised its estimate of the deficit upward by *$10 billion*, to $40 billion (*Newsweek*, October 29, 1979, p. 31). Other examples include the incredibly large errors (in the billions-of-dollars range) uncovered by Congressman Henry Reuss, chairman of the House Banking Committee, in reporting by Manufacturers Hanover Trust Co., of New York, of money-supply figures, as well as the wholly new factor introduced by electronic funds transfer systems (EFTS) in the major multinational banks.

As Martin Mayer pointed out in "The Incredible Shrinking Dollar" (*Atlantic Monthly*, August 1978, pp. 59–65), because of the speedup of transactions allowed by the faster information provided by these electronic systems (such as S.W.I.F.T., the computerized system used by international banks), the same amount of money now supports five times as many transactions as previously. Thus the speedup of the *information* about money also increases the velocity of its circulation, adding another new factor to inflation. This added layer of complexity is almost impossibly difficult to plot and analyze in an already Byzantine global banking and financial system. I first raised this issue several years ago with some of the participants in a seminar held by the Institute for the Future (on whose advisory board I served) on the implications of computer modeling and its impacts on public policy (published in May 1974 in *Toward Understanding the Social Impact of Computers*, edited by Roy Amara). However, my fears that EFTS would further disorder economic decisions by the increasing speed of computerizing information on money flows were difficult to communicate, since they required acceptance of my view that economics' fatal flaw is precisely its hypnotism by money and that it mistakes the whole real-world production, consumption, and maintenance system for the half that is monetized. Thus, my argument was based on the problem of order-of-magnitude errors involved in the successive levels of abstraction involved; i.e., money is an abstract symbol system supposedly representing real-world transactions, production, and resources.

Money is already evidently diverging from that reality, as I showed in *Creating Alternative Futures;* capital investment decisions based on money, for example in energy-extraction processes, could result in *no net energy yield* as measured by thermodynamic units, i.e., kilocalories or BTUs (British thermal units), even though such useless, wheel-spinning activity might well yield "profits" to the investor. Another problem with money as an accurate tracking system for modeling real-world production and resource-allocation activities is that the money unit can be distorted almost at will by the power of large institutions, whether corporations' "creative accounting," which externalizes real costs to other systems, or worse, when governments simply create too many money units by increasing the money supply, manipulating bank reserve requirements, or allowing banks to create money by monetizing their loans, as well as proliferating credit cards, speculation in commodity "futures," and all the elaborate pathways to self-delusion that characterize today's financial halls of mirrors.

What happens when one takes the already disordered information about the real-world transactions—i.e., money—and adds *another* level of abstraction with an EFTS system composed of *units of information about the units of money?* This creates another order-of-magnitude error, even further divorcing the tracking system from what is actually occurring. Now, imagine the effect created when the new tracking system of *information about money* is speeded up with the much faster, more "efficient" operation of the computer as compared to the manual and less-automated auditing and tracking systems of paper check clearing, written statement of account, and the like. Another order-of-magnitude error occurs, as has now been proved by the fact that the S.W.I.F.T. system has increased by five-fold the velocity of money, as it supports ever more transactions at each point in the system. To illustrate the danger of these kinds of distortions, let us imagine that the EFTS systems were speeded up to what computer people call "real time," i.e., with almost no delays in transmission of information. At this point, in the case of EFTS, the velocity of the information-about-money flows would lose all relationship to the thermodynamic realities of the actual system (subject to lags, friction, inertia, and human pace), and the amount of "money" or "capital" available at any point in the banking system would tend toward infinity! As the information-about-money system became decoupled from actual events, all manner of new ventures and schemes might be initiated by false "promissory notes" signaling capital availability with nothing more than an electronic impulse at a computer terminal.

No wonder people are taking refuge in real goods, houses, and land, as well as in withdrawing from the cash economy whenever they can and entering countereconomies of more production for use, sharing, and reciprocity.[12] Taxing policies now embody the same kind of Byzantine system of confusion, misdirected policies, and distortions that President Jimmy Carter called "a national disgrace" in his 1976 election campaign and vowed to reform. No wonder, too, that there is increasing evidence of both overt tax revolts and also of taxpayers' simply "going underground" into mutual aid and self-help alternatives, "sweat-equity" restoring of old houses, and increased bartering and cooperative enterprises. Ordinary citizens who do not own capital simply see the tax code as an instrument of the rich and the large corporations, which can afford tax lawyers to find or lobby for new loopholes for them. Meanwhile, the Internal Revenue Service is trying to nip the ingenuity of the desperate citizens with no capital and incomes below twenty thousand dollars a year. For example, a 1979 ruling in Milwaukee may put dozens of food co-ops out of business by forcing the members of the co-op who put in free labor at the store in order to reduce their food bills to pay thirty-five hundred dollars in back taxes and Social Security on their discounts, which the IRS says are "income"! Each round of changes, or "reforms," in the tax code is skewed even further in favor of the investor and corporations by massive lobbying, and even the tax cuts manage to favor the already affluent. For example, the last time Congress cut taxes, through the Revenue Act of 1978, it managed to change the administration bill, which would have favored those earning less than thirty thousand dollars a year, to a version that shifted the benefits to those above that figure, so that those making more than one hundred thousand dollars a year received better tax breaks than those making between ten and fifteen thousand dollars. Congress justified this as its response to the tax revolt by the "middle class." As Tom Fields, director of the public-interest group, Taxation

With Representation noted, the middle fifth of all tax filers had had incomes (unrealistically expanded by inflation) of only $8,300 to $13,900 in 1978, whereas the lion's share of the tax bonuses went to people with incomes far above them. To add to all this, increased Social Security taxes are borne by the least affluent, and the new talk in Washington of imposing a European-style value-added tax (VAT) is even more ominous, since VAT is purely a tax on consumption and would hit hardest those least able to pay, as well as being highly inflationary, as the European experience has shown. In fact, a British observer noted in late 1979 that British Prime Minister Thatcher's increase in the VAT would make it more difficult to control Britain's inflation.

Such policy confusion reigns that, not surprisingly, public disbelief rises, and the suspicion and frustration erupts in tax revolts, such as the political tantrum of Proposition 13, in California. Sadly, voters as well as government leaders and planners are faced with such complexities in late-stage industrial societies that they can no longer model even their own self-interest. California's voters (40 percent of whom did not believe that Proposition 13 would result in any cuts in government services) believed in the imaginary pot of gold at the end of the rainbow promised by Milton Friedman (who ought to know better) and Arthur Laffer's "Laffer Curve," which proposes that reduced taxes will hype consumer spending and business activity in the private sector and thereby create jobs and prosperity (the Golden Goose model again, with a liberal dose of Keynesian trickle-down stimulation). Yet they must acknowledge that the very growth of government that has triggered the tax revolt has been our chief job-creation mechanism, as people have become disemployed in more automated production. For example, if government were to be cut back to its 1950 size, we would be experiencing today unemployment rates of approximately 17 percent.

None of the tax-revolt agitators, those behind the risky drive for a constitutional amendment to limit the federal budget, or those corporations and investors lobbying for further tax cuts for investments and on dividend incomes point out that across-the-board tax cuts and federal budget ceilings are very blunt instruments and suffer from the same overaggregated approach and statistical abstraction that have led to the denouement of macroeconomic management itself.[18] We now see the results of California's famous Proposition 13, which

axed local police, fire, and sanitation services, laid off low-paid service workers, forfeited billions in matching federal funds, while leaving untouched all the tax boondoggles about which all taxpayers
should get really mad: high-level administration costs, bureaucratic
featherbedding, bloated budgets for military and space hardware,
cost-plus contracts, cost overruns, and (by far the largest segment)
special tax subsidies to business, including the shipping, oil, nuclear-
energy, trucking, and construction industries, not to mention foreign
investments, commodity-price supports, investment-tax credits, accelerated depreciation allowances, and state tax incentives wastefully
competing to lure companies to relocate with "tax holidays." In addition we must note that 65 percent of California's Proposition
13 tax relief was a windfall to business and real estate investors,
rather than small homeowners. The investment tax credit, which
business incessantly lobbies to increase, has proved a costly, ineffective way to create jobs. Senator Edward Kennedy cited at the March
1978 Joint Economic Committee Hearing on Creating Jobs Through
Energy Policy that, between 1969 and 1976, the one thousand largest companies in the United States used 80 percent of the investment
tax credit and 50 percent of all the industrial energy and created
only seventy-five thousand new jobs, while the nation's 6 million
small businesses in the same period created 9 million new jobs. A
1978 Organization for Economic Cooperation and Development
(OECD) study prepared for the Bonn economic summit meeting
warned that participant countries may face an automation-created
era in the 1980s of virtually "jobless economic growth." A final
irony is the news that post-Maoist China, having started down the industrialization track of Deng Xiaoping, her Western-oriented Vice-
Chairman, now boasts that it is finally rich enough to have inflation.
After thirty years of steady prices, China, according to economist
Bryan Johnson, reporting from Peking in *The Christian Science
Monitor,* November 6, 1979, boosted prices of eight basic food items
by 30 percent, claiming that this was proof positive of their rising
standard of living.

The message seems clear: mature industrial societies must now
reconceptualize their situations and recognize that, in many cases, the
Golden Goose is no longer laying golden eggs. In fact, it has been on
a life-support system ever since industrial societies adopted the
Keynesian demand-stimulating and pump-priming adrenalin in

search of taxpayer-supported subsidies, tax credits, and government-rigged "markets," while its health becomes ever more delicate and it threatens localities that if the business climate isn't just right, it will pull up stakes, shed its payroll, and move to the Sunbelt or offshore, wherever it can find a cheaper labor force and virgin resources and environments to exploit. Our U.S. economy today is no longer a "free-market-system," but one of rigged, legislated, and monopolized markets, with a crazy quilt of taxes, transfers, and subsidies to large, well-organized economic interests. If tax revolters get to work repealing these kinds of waste and inefficiencies, they might live to see their individual tax rates really reduced.[14]

The anomalies have reached above the threshold of public apathy, however, as demonstrations have occurred outside the headquarters of oil companies and of their Washington voice, the American Petroleum Institute, leading to passage of Jimmy Carter's windfall profits tax. Meanwhile, as the oil companies' earnings continued their dizzy climb, in more than a hundred cities in thirty-five states, thousands of consumers demonstrated with signs saying TAX BIG OIL and passed out "Big Oil Discredit Cards" (*Time,* October 29, 1979, p. 70). Thus a *populist* tax revolt, quite different from that portrayed by Republicans and business investors, may be brewing. Such a populist revolt might be led by small savers, renters, and those with incomes below fifteen thousand dollars a year, and joined by the some 50 percent of all American voters who no longer think it is worthwhile voting at all. Independent candidacies at all levels, including the 1980 presidential bid of John Anderson, now express these frustrations and may lead to a third party by the mid-1980s.

Much erudite hand-wringing is devoted to analyzing why Americans have the lowest participation rate in elections of any democracy, but the simple explanation of Ralph Nader is his "Golden Rule of Politics": those who have gold, rule! The new coalitions around the populist view of tax reform include the new Citizens Party, the Progressive Alliance, and hundreds of local coalitions such as the Ohio Public Interest Campaign, Massachusetts Fair Share, and the Association of Community Organizations for Reform Now (ACORN), as well as many environmental and consumer groups. The neo-populist direction of their anti-inflation platforms is clear in "A Progressive Anti-Inflation Program" (Fig. 4), and as outlined in *Saturday Review*

SOME SPECIFICS OF A PROGRESSIVE ANTI-INFLATION PROGRAM

	Government Actions Advocated or Initiated			Citizens' Actions (In addition to Inquiry, Advocacy, and Electoral Politics)
	Federal	State	Local	
1. Price Controls	Permanent, selective controls, starting with Fortune 500. Opposing Carter on decontrol and anti-control	Tough utility regulation. Public interest advertising	City price commissions. Extensive consumer education, dealing with product quality as well as prices and helping in comparison shopping	Monitor sellers. Monitor public officials at all levels in consumer initiatives. Widespread initiatives in consumer self-education and protest
2. Spending	Cut war budget and corporate subsidies. Regulate capital flight and foreign investment	Reduce business subsidies. More efficient public administration	Reduce tax abatements. More efficient public administration	Protest welfare handouts to corporate rich and local speculators. Support public officials who resist private pressures
3. Taxing	Close major loopholes. General-fund financing of social security	Anti-speculation taxes. Progressive income taxes. Lower taxes for homeowners	Anti-speculation taxes. Less reliance on sales taxes. Lower taxes for homeowners	Tax-reform initiatives. Tax withholding under certain circumstances
4. Credit	Reduction of interest rates. Credit allocation. Activation of new national co-op bank. Promote credit unions	State banks. Productive use of pension funds	Municipal banks. Productive use of pension funds	Protest redlining. Withdraw deposits from certain banks. Withhold mortgage payments under certain circumstances
5. Energy	Public corporations for oil imports and energy development. Major shift to solar, biomass and neglected hydro energy, as well as hard and soft coal	Anti-trust actions against oil companies. State power generation and distribution	Municipal public power, with elected consumer representation on boards	Anti-nuke demonstrations. Install solar devices. Voluntary conservation and recycling

6. Transport	Massive mass transport expansion and improvement. Require more efficient, safer cars	Improved land-use planning to minimize trips to work	Bicycle lanes Free downtown buses Mini-buses	More bicycling and walking More car pools
7. Housing	Revive public housing for both rehabilitation and new construction	Low-interest loans for rehabilitation and new housing Public land purchase and maintenance	Rent control More public rehabilitation, construction and co-op promotion Public land purchase Zoning to prevent, not promote, speculation	Oppose condominium movements, gentrification, planned shrinkage and shortage-creating demolitions Tenant unions and rent strikes
8. Food	Less price support for agri-business and factory farms Family farm aid Promote direct marketing More anti-trust action	Promotion of co-ops Consumers on marketing boards Consumer protection laws and enforcement	Promotion of co-ops Sponsor farmers markets Consumer protection laws and enforcement	Consumer co-ops, food gardens Boycotts Protest automatic checkout scanners
9. Medical Care	National health service Regulation of medical-drug-hospital costs	Support health co-ops Cost regulation	City-owned health maintenance organizations, health promotion education in schools	Self-care in home; improved diet; more physical exercise and activity; client-controlled health clinics
10. Insurance	Public insurance companies Clear-wording laws	State insurance companies Better consumer information	City insurance Better consumer information	Buy term life insurance only Protest discriminatory auto rates

Prepared by Derek Shearer and Bertram M. Gross, with the help of Hunter College undergraduates

Fig. 4

by David Osborne in "Renegade Tax Reform: Turning Prop. 13 on Its Head" (May 12, 1979).

In fact, this more sophisticated understanding of how taxes are used to subsidize large corporations and often increase their control over our economy was summed up in a letter to the editor of a major newspaper, which I quoted in my article on the problems of corporate accountability in *The Nation* of December 14, 1970:

Dear Sirs:

I am disturbed and offended by your editorial statement that the Capitol Building belongs to the nation. This is a good example of the kind of muddled thinking that has led to the student revolt, riots, and crime in the streets. In the interests of peace, plenty, and the American Way of Life, you must print the truth:

The Capitol Building belongs to Senator Eastland.

The air belongs to General Motors.

The mountains belong to Con Edison.

The water belongs to U. S. Steel.

The oil belongs to Secretary Hickel.

The airwaves belong to NBC, CBS, and ABC.

The courts belong to the rich.

The taxes belong to the working man, etc., etc.

Corporate power is encountered daily by millions of citizens who attempt to fight polluted air, oil-smeared beaches, plagues of nonreturnable cans and bottles, supersonic transports, rampant freeways, deceptive advertising, racial discrimination in employment, exploitation of natural resources, mushrooming shopping centers and housing developments, as well as huge military appropriations. In all such battles, sooner or later, they come up against some corporate Goliath, and find their slings unavailing. Newly radicalized, they learn that the 500 largest corporations not only control more than two thirds of the country's manufacturing assets but also influence elections by carefully channeled campaign contributions that avoid legal restrictions. In Chapters 4 and 5 I shall explore the role of giant corporations and how they distort our democratic processes.

NOTES — CHAPTER 2

[1] An example of the new consciousness in the Third World was the organization of the First World Congress of the Association Mondiale de Prospective Sociale (AMPS) World Social Prospects Study Association, which I attended in Dakar, Senegal, in January 1980. Approximately two hundred participants from many African countries, India, Pakistan, Bangladesh, Indonesia, Brazil, Venezuela and other South American countries, Mexico, Eastern and Western Europe, and North America discussed the various crises of the northern hemisphere with much clarity and perception. They noted these northern crises as those of industrialism itself, its technologies, its monetary systems, its resource and energy overdependence and growing pollution, waste, and social disorder. Some feared that the countries of the northern hemisphere would redouble their efforts to "export their crises" to the southern hemisphere. There was much discussion of how the countries of the southern hemisphere might decouple themselves, their trade and economic patterns, and their technological choices, and develop their own regional trading, bartering, and monetary systems and restore vitality to their own culture and indigenous agriculture and resources. The Association is headed by Albert Tévoédjrè, of Benin, Africa, author of *Poverty: The Wealth of Mankind* (sic), Pergamon Press, London, 1979.

[2] Some fresh perceptions on Islam are disclosed in Pakistani physicist Dr. Ziauddin Sardar's two recent books *Science, Technology and Development in the Muslim World* (Croom Helm, London, 1977) and *The Future of Muslim Civilisation* (Croom Helm, London, 1979).

[3] For example, in the United States, where equal-pay-for-equal-work laws have been in effect for sixteen years, women still have average wages only 60 percent of those of men, and this wage inequality has not changed in the past forty years (New York *Times*, October 26, 1979). Only recently have mainstream labor unions begun to deal with such inequities suffered by women and minorities.

[4] At the February 11–15, 1980, session of the Parliament of Europe, Britain's Roy Jenkins, president of the European Commission, told the parliamentarians that "we face no less than the breakup of the established economic and social order on which postwar Europe was built. If we do not change our ways while there is still time, our society will risk dislocation and eventual collapse" (*The Christian Science Monitor*, February 19, 1980). Mr. Jenkins then compounded the error in perception that contributes to these crises by adding the familiar economic cant: economic growth rates would drop to 2 percent in 1980, unemployment would grow to 6 percent, and inflation would rise from 9 percent in 1979 to 11.5 percent in 1980, and that the European Community's external trade deficit would more than double. As long as Europe's "problems" are so hopelessly misstated as "economic," there will be little hope of addressing them. Similarly, the "economic crises" of inflation and the much-planned recession that refused to materialize until summer of 1980 in the United States requires a complete reconceptualization in noneconomic terms.

[5] An example of this bankruptcy of ideas was the lecture former Federal Reserve Board Chairman Arthur Burns gave at the Belgrade monetary meetings in

September 1979. He noted the new "intellectual ferment" in the world's democracies and the new "understanding" that inflation had become the number-one problem. Back home, his four-point program to combat it, however, was more of the same old economic nostrums: laws to make federal government deficits more difficult to run, leading to a constitutional amendment directed to the same end; a comprehensive plan to deregulate the economy to improve competition; a binding endorsement of restrictive monetary policies until the rate of inflation has been lowered; and scheduled tax cuts for business to release powerful forces to increase the nation's productivity (*U.S. News & World Report,* October 15, 1979, p. 102). In my view, it is precisely these four "remedies" that are contributing to inflation, and the new bout of Hooverism that Arthur Burns advocates would be as wrong as Herbert Hoover's policies were in trying to combat the Depression.

6 The OPEC nations finally realized that until some real signs of political will to conserve petroleum were evident in the United States and other consuming countries, they would have to pursue strategies of reducing their output so that their economies would not be overwhelmed with paper money and pressures to invest it abroad or in the now-suspect process of "modernization" of their own lands. At OPEC's London meeting in February 1980, the group split over the issue of maintaining current production in the face of growing opposition from Algeria, Venezuela, Kuwait, Iraq, Saudi Arabia, and Iran, which argued that only by reducing their output would they protect their finite reserves, maintain prices, and encourage consumer conservation. Kuwait accordingly announced it would cut back its production by 25 percent (*The Guardian* [U.K.], February 20, 1980). To escape the continual depreciation of the dollar, OPEC was advised by the Venezuelan oil minister, Dr. Humberto Calderón Berti, to return to the concept first proposed in Geneva in 1972 of requiring payment for their oil not in shrinking dollars but in a "market basket" of the world's stronger currencies.

7 Another example of the short-circuiting of the "funny-money" policy prescriptions was the announcement by the federal Energy Regulatory Commission that with a stroke of its pen it had doubled U.S. reserves of "economically" recoverable natural gas locked in geological formations called "tight sands." Chairman of the Commission Charles B. Curtis announced a "pricing decision" to put a 50 percent premium on new gas produced from such unconventional wells. The problem is that, in thermodynamic terms, there may be little net energy in extracting gas from tight sands, and worse, the process requires injecting large quantities of water into the rock formations in the Rocky Mountains, water that is already scarce for farmers. Water, in economists' models, is "free." Thus, I have predicted that the next crisis to "surprise" industrial societies will be the water crisis, and we can expect a Water Minister and proliferating Water agencies. Reality dealt the market model another blow as the Federal Aviation Administration warned that airspace over major airports would have to be rationed (*The Guardian* [U.K.], February 21, 1980).

8 I discussed a broad range of such policy alternatives for the economic transition to a more energy-efficient, renewable-resource society in Chapter 8 of *Creating Alternative Futures* (G. P. Putnam's Perigee Books, 1978).

9 Ron Chernow, "The IMF: The Roughest Bank in Town," *Saturday Review,* February 3, 1979, pp. 17–20.

10 The U.S. rate of savings is now the lowest of the Western democracies. By January 1980 it had fallen to 3 percent of disposable income, while installment

debt had risen to an unprecedented 18.4 percent of disposable income (*Business Week*, January 28, 1980).

11 The Federal Reserve Board, its last revision of money-supply indicators having been widely judged a failure, tried to revamp these indicators to give more reality to the "funny-money" world of banking and global investment flooded with computer-speeded Eurodollars. Board member Henry Wallich noted that the Fed now publishes a do-it-yourself list of money stock components (broader than the old M 1 + indicator of cash in circulation, savings and checking accounts + the check-drawing savings accounts), so that money-market analysts "will be able to roll your own" indicator. In addition, the Fed has composed a new indicator: L (= liquidity), which will include every liquid asset you ever heard of—from treasury bills to money funds—which will account for several trillion dollars, while a broader, M 2 indicator now includes overnight Eurodollars in offshore banks in places such as the Cayman Islands, plus money-market mutual-fund shares and savings and time deposits at thrift institutions as well as banks, plus about half of the $70 billion sold to bank-corporations through repurchase agreements (*Business Week*, January 28, 1980, pp. 37–38). Thus, the desperate chase to track the proliferation of "funny-money" becomes ever more arcane, lagging ever farther behind the imaginative games of financiers, bankers, and investors.

12 Consumers were correct in their disenchantment with the funny-money rat race. For most wage earners, real disposable income had been barely holding its own against inflation. In 1979, the nation's real discretionary income actually fell 4.6 percent below the level it had reached in 1973 (*Business Week*, January 28, 1980, p. 73). Even two-earner families could no longer keep pace with inflation's erosion of their purchasing power. Economists were shocked by the unexpected "strength" of the U.S. economy as more consumers rushed to spend their eroding paychecks. At the same time, Wall Streeters were hailing the "surge of strength" that the huge new Defense budget would provide—"giving the economy a $16 billion shot in the arm," even though this is estimated to increase the total federal budget for the 1981 fiscal year to $616 billion. These same analysts, investors, and other members of the financial community are the very same people who are demanding massive budget-cutting for domestic programs and a ceiling on federal spending (*Business Week*, February 4, 1980, p. 24).

13 Another baroque twist to the confused debate over the federal budget, whether and how to cut it, and at whose expense is the old trick of administration budget makers: the off-budget items, i.e., "back-door financing" of government projects. In the fiscal year 1981 budget, for the first time in our history, these "off-budget" items that are omitted from the formal budget were in excess of the initial deficit it projected! They will rise to $18.1 billion (a 68 percent increase in two years). This "off-budget borrowing" is conducted by the little-known Federal Financing Bank, a bookkeeping operation set up to control Washington's spiraling credit programs. The Bank borrows money from the Treasury to buy the loans that various federal agencies extend to finance everything from student loans to space communication systems. Once an agency has "sold" such a loan to the Federal Financing Bank, presto! it can treat it as "repaid" for its budget purposes. Creative accounting like this is catching on among government agencies, following the ingenious deals pulled off in this way by the Farmers Home Administration, whose far-flung lending activities extend beyond farms to loans for distressed steel companies and motel-chain operators. The Carter administration was moving to control such back-door credit proliferation

but caved in to pressure from various interest groups who receive such special credit largesse, whether from the Export-Import Bank or the some forty other agencies allocating credit. Congressman Norman Y. Mineta, of California, notes, "Credit subsidies are driving up interest rates for everyone. A control system is needed that will bring these programs into the budget process" (*Business Week,* February 11, 1980, p. 31).

[14] Even *Business Week* editorialized that "now that most economists and econometric models are changing their forecasts, it is relevant to ask whether the new predictions are worth any more than those being discarded. . . . Neither economists nor their elaborate models are able to anticipate consumer behavior, the vigor (or weakness) of the dollar, Federal Reserve policy, administration tax policy, prospects for controls and overseas events that will influence the speed of increases in defense outlays" (February 25, 1980, p. 45).

CHAPTER 3

The Bankruptcy of Macroeconomics

As the latest doses of monetarists' medicine failed to reduce inflation rates in most of the aging and maturing industrial economies, the spate of gloomy analyses of the failures of economists and governments grew.[1] By late 1979 dark predictions of another worldwide depression coincided with the fiftieth anniversary of the great Wall Street crash of 1929. Henry Kaufman, of Salomon Brothers, a respected Wall Streeter, publicly called for government "to declare a national emergency" to deal with inflation.[2] While the 1980 situation was a far cry from that of the 1920s, and the intervening period saw the institution of basic, built-in stabilizers such as unemployment insurance, Social Security, and federal financial guarantees and regulations on speculation and fraud, there is reason for sober evaluation of new factors.[3] The world oil situation is still creating the greatest vulnerabilities.[4] Managing the excessive oil-consumption habits of industrial countries as they live through the decline of the age of petroleum is the greatest challenge.[5] Transition strategies prepared by keen observers outside the halls of government, such as Amory Lovins' *Soft Energy Paths* (Friends of the Earth, 1977), Denis Hayes's *Rays of Hope* (W. W. Norton, 1977), and Dennis Pirages' *Global Ecopolitics* (Wadsworth, 1978) were beginning to be taken seriously, as were plans such as the Swedish Secretariat for the Future's *Solar Sweden* (1978). British parliamentarians, through the leadership of author Renee-Marie Croose-Parry's Parliamentary Li-

Parts of this chapter are excerpted from my editorial in *The Christian Science Monitor,* October 10, 1979. Used with permission.

aison Group on Alternative Energy Strategies (29 Woodberry Avenue, London N21 3LE), now meet regularly in the House of Commons to discuss these and other studies such as Gerald Leach's *A Low Energy Strategy for the United Kingdom* (1979). In Denmark a book coauthored by K. Helveg Peterson, former Education Minister, Dr. Niels Meyer, a physicist, and Villy Sorenson, a journalist, put energy and alternative technology plans together with a political strategy in the runaway best seller *Revolt from the Middle* (Marion Boyars, 1980).

The Carter administration in the United States capitulated to oil lobbying for multibillion-dollar plans reminiscent of the ill-fated Project Independence, of Nixon and Ford. President Carter hailed such supply programs while understanding that none of these highly dubious, inflationary schemes could produce much actual energy for a decade. Even though conservation was the only feasible plan in the short run,[6] Congress, stampeded by energy-company lobbyists in 1980, opted for the $20-billion Synfuels Corporation, more costly than the Space Program and little more than a boondoggle. Canada's Science Council's work on the "Conserver Society" and the comprehensive GAMMA Report from the University of Montreal entitled *A Conserver Society: Blueprint for Canada's Future* (1977), laid the groundwork for the debate in that country. The Club of Rome reports *Mankind at the Turning Point*, by M. D. Mesarović and E. C. Pestel (1975); *Reshaping the International Order*, by Jan Tinbergen (1976); and *Goals for Mankind*, by Ervin Laszlo (1977), were also being included in official discussions of the issues. France and Germany, then the strongest economies in the European Economic Community, still seemed self-assured with their existing heroic energy-supply plans based heavily on increased nuclear capacity in spite of mounting civic protests and the new worries caused by the accident in the United States at Three Mile Island. As energy loomed larger as the key variable in the new economic instabilities, some economists began to back away from their earlier predictions that the world market would adjust and that as long as the OPEC petrodollars were recycled through the world's financial system, there would be no irremediable harm. The economists' view that markets, if left to themselves, would adjust supply to demand, clouded the need for conscious policy choices in societies whose energy demand had been set in concrete, literally, with the automobilization and suburbanization of the population and where most of the domestic manufacturing base involved excessively wasteful production methods and end products.[7] In such a structural situation, it was unrealistic to expect that increasing inflation due to oil imports could be tackled by raising interest rates. So as Arabs began buying gold, investors in all major financial markets began scurrying for inflation hedges.[8] Short-term paper such as U.S. three-month treasury bills, and other securities offering greater liquidity than long-term investments, became the order of the day. Everyone looked for "safe" investments with good returns, trying to turn them over faster and faster to catch higher and higher interest rates. Multinational corporations in many cases could make more money more safely by trading and arbitraging currencies and international accounting of profits, dividends, investments, and depreciation than they could by going to the bother of building a factory to produce anything anywhere. Multinational banks in New York, London, and Europe carried on their books enormous loans (that they had hoped would be profitable) to developing countries such as Turkey, Brazil, and many African states. As the oil bills of these weaker states mounted in step with OPEC increases,[9] these loans became shaky and had to be renegotiated. If any one of these states defaulted, it might create major bank failures or worse.[10]

Robert Lekachman, an economist who has been a consistent critic of economics, author of *Economists at Bay* (1977), voiced the doubts in the minds of many. He reminded us in an article in the *International Herald Tribune* on October 5, 1979, that although we need not fear a replay of the great crash of 1929, the later worldwide Depression of the 1930s might be reenacted—a much graver threat. What was lost in the crash was merely money, whereas the Depression cost millions of people on both sides of the Atlantic their jobs, homes, farms, and self-esteem. Lekachman pointed to economic historian Charles Kindelberger's view of the Depression in which "Germany owes reparations to Britain and France and commercial debts to the United States. Britain owes to the United States about what it receives from Germany, and is owed war debts from France; France is to receive the lion's share of reparations, well in excess of its war debts to Britain and the United States." Lekachman notes that "no U.S. leader admitted that Allied war debts owed to the United States were linked in fact, though not in law, to the continued flow of German reparations to France and England. Nor did the politicians and bankers concede that the only way the Germans could pay up was with funds borrowed from the United States. As soon as Americans tired of pumping funds into Germany the Germans were certain to default on their reparations, and the British and French shortly afterwards, on their obligations to the United States." Lekachman adds that when this happened, international financial stability broke down, and the mutual recriminations paved Hitler's path to power. Today, a soberingly similar set of linkages exists in our relationships to both OPEC and the new free-for-all world oil market as well as within the weaker debtor countries mentioned earlier, which are struggling to meet their own increased oil bills and repay their crushing burden of loans and debts, where even interest payments are now often significant portions of their total GNP. Merely renegotiating or rolling over these debts can no longer suffice. Only expanded outright aid funds, international financial agreements, and a New International Economic Order can bring stability. Up to now, the financing of oil imports has been facilitated by OPEC deposits of oil revenues in large banks including New York's Citibank and Chase Manhattan and other financial institutions, and they in turn have made loans to Third World borrowers. Some bankers admitted the situation was desperate, and by late 1980 the World Bank and other international lending agencies increased stop-gap funding.

Thus we see the same kinds of delicately balanced linkages in the world's financial system as we did after World War I. The question was whether we would repeat the mistakes of the past, when the victorious Allies did not permit Germany to use her exports to finance her reparation payments. In 1980, the same kind of short-range protectionism blocked access to industrial countries' markets for the export goods of Third World debtor countries, and cash-rich industrial nations were obliged to keep extending them loans, because they had placed these debtor countries in a Catch-22 situation. Ironically, this reflected their own Catch-22 situation domestically, in which "lame duck," obsolescent industries together with their "lame duck" unions had the political power to force politicians to raise protective trade barriers against Third World imports. Whenever OPEC redirected the flow of its oil revenues, it could cause banks financial distress or raise the possibility of banks becoming unable to continue to roll over their Third World loans, raising the specter of national bankruptcy for some overextended countries.[11] Anytime some of these debtor countries themselves can decide that they are in a no-win international money game, in which they will always have to run harder to stay in the same place, and that, rather than pursue their current efforts via the United Nations, UNCTAD, and the Group of 77 to renegotiate their loans and terms of trade, it might be preferable simply to confront the international bankers with the facts. Many of these countries will never be able to repay their loans, and under the current rules of the game, industrial countries have left them no way to do this. Understandably, the talk at international conferences on the shape of the New International Economic Order is exasperated, as leaders of debtor countries exclaim "off the record," "What do the bankers expect us to do? Perhaps we should just be honest and say, 'Okay if you won't take ten cents on the dollar, why don't you just come over here and repossess our whole country!' " They might add, "Remember the case of Germany in the 1920s." Rising interest rates and the new "interest rate war" these policies have triggered not only pushed the United States, Japan, and European countries into recession but meant a staggering increase in the cost of balance-of-payment loans for the "less-developed" countries. Brazil, for example, was forced to pay an additional $1 billion in interest on its $22 billion of foreign debt during 1980. A global credit crunch loomed while the worldwide recessions caused by higher oil prices created horrendous political problems of unemployment *and continuing inflation.* Paul Horne, European analyst with the investment firm of Smith Barney in Paris, saw the picture clearly: "That would prompt governments to cancel efforts to follow conservative economic policies. The result would be for the global economy to emerge from recession in 1981 with inflation running an even higher fever than it is now" (*Business Week,* October 29, 1979, p. 174).

So, even in 1980, as monetarism was being retried, we could see that it will be abandoned again in a return to Keynesian-type, "reflating" or tax cuts as governments decide that this option is the only one that will keep them from being thrown out by voters or civil unrest. How much farther into the future will the economists' two bankrupt either/or remedies of deflating and inflating go unchallenged? These narrow policy options will continue to hold sway over more realistic policy analyses and choices as long as the economics paradigm remains central. Part of the problem is the virtual equating of the money view with the full reality. This money fetishism may yet prove the downfall of the current international order, as politicians, bankers, investors, and leaders of socialist and "less developed" countries mistake money and paper for real wealth, real production, and real resources.

Another key aspect involves a crisis of income and wealth distribution not only between nations but within the industrial nations. The huge imbalances in income and wealth have always been justified by economists in market-oriented nations as necessary for the accumulation of capital investment so that some citizens could use their surplus income and capital to invest in companies so as to create jobs for the rest—the Golden Goose model viewed in yet another aspect. However, if the impoundments of capital, money, or real resources such as land, minerals, and energy deposits become too large and too centralized in too few hands, bottlenecks of the kind just described between debtor and creditor countries are created, as well as investment and consumption bottlenecks visible today in the aging industrial societies. People with real unmet needs for basic food, clothing, and shelter who would pay for these necessities and keep up total demand in the noninflationary domestic sectors, buying basics from small, local, and regional farmers and producers, are read out of the system. Meanwhile, tax cuts go to the middle class and investors to encourage additional plant and equipment to supply the global and upper-middle-class domestic market with highly advertised, energy-intensive nonessentials—even while existing capacity remains idle and new taxes such as the VAT are proposed to discourage consumption. Here again we see the tragedy of the too highly-aggregated view, which cannot target *which kinds of consumption* are inflationary or involve balance-of-payments problems (as with oil) and which kinds are noninflationary and will merely keep local farmers and domestic industries humming at full production. Thus blockages in the production-to-consumption channels not only lead to hardship, but because money is "rolled over" within too narrow a segment of the population domestically and between nations, the economic instabilities will grow. Elmer G. Doernhoefer, an economic analyst in St. Louis, reiterates many of my arguments on the crisis of income and wealth imbalances and states in one of his many memos to Congress, "The situation stems from the fact that fully 25% of personal income in the U.S. consists of dividends, interest, and rentals," and he cites studies by the Wharton School of the University of Pennsylvania, that "1% of U.S. families with the largest income accounted for 47% of all dividend income and 52% of the market value of stock owned by all families, and that 10% of the families with the largest income accounted for 71% of the dividend income and 74% of the market value of stocks." In the Carnegie Council on Children's report *All Our Children* (Harcourt Brace Jovanovich, 1977), psychologist Kenneth Keniston cited research that "the top 2% of American families holds over 37% of total wealth. The top fifth holds over 60%. The typical family in the bottom fifth has no net worth. We believe this condition to be patently unjust. It alone is a compelling reason to change the overall distribution of material well-being."

But there is an even more compelling reason than justice or charity, and that is that such dangerous imbalances may shake both domestic economies and international financial structures to pieces. The lonely voice of one courageous economist, Herman Daly, has consistently pointed to these dangerous wealth imbalances. In *Steady State Economics* (1977), Daly held that societies passing out of the unsustainable industrial phase to steady-state, renewable resource use will need to legislate not only guaranteed minimum income levels to maintain purchasing power in noninflationary ways, but also to grasp the nettle of overconcentration of economic wealth and power and set maximum levels for individual ownership of resources and capital assets. Already many church groups in the United States and in Europe have been addressing the moral implications of the current huge disparities in wealth and income and challenging their congregations to face up to the anomalies. However, as long as the economists' model of the "invisible hand" remains unchallenged, the question of distribution and equity will remain a moral issue, rather than one pertaining to the very stability of the world's financial system and domestic economic management, as it is in reality. While the pragmatic case for redistribution is now overwhelming, as we shall explore in detail, the ecological case was made by biologist Barry Commoner in *The Closing Circle*. The scientific case for redistribution confirms the ethical case. Citizens are now suspecting that their societies may be basically healthy but suffering from overmedication by economists. Thus, the underlying theme of the politics of the 1980s will involve a shift of focus from macroeconomic management and its abstractions to the real world of local "micro-futures." Millions of Americans are developing thousands of real plans and action for energy conservation, less wasteful land use, urban rehabilitation, recycling for fun and profit, food co-ops, farmers' markets, do-it-yourself housing construction, health care, and generally bringing more of our lives back under community control.

This emerging scenario of rapid growth of countereconomies is a spontaneous devolution as citizens bypass paralyzed institutions and their bottlenecks, and simply begin recalling the power they once delegated to the state and to the executives of giant corporations to make all kinds of technological and investment decisions with enormous hidden social and environmental costs—now visible and coming due. Most of all, citizens are no longer buying the definitions of the "experts" and crisis managers as to what is happening.

This new energizing of civic action, local self-help, and the "town meeting" design of *practical* "Alternative Futures for Our Town" is a healthy response to the doubts that the government dinosaurs and the corporate paper tigers and their bureaucracies can continue managing our affairs from such rarefied heights of abstraction: centrally manipulating such statistical illusions as "aggregate demand," "average productivity," "per capita income," rates of "innovation" and "inflation," and "levels of unemployment." Citizens see clearly that such excessively abstract governance and corporate management have lost touch with reality. The legions of "experts" no longer know just where the rubber hits the road. For example, in the United States a "national energy plan" is a vast abstraction, and while the Department of Energy fiddles and Congress has check-mated itself, yet in spite of this, we see a thousand local and county energy plans blooming, as spelled out in *Energy Efficient Community Planning* (1979), by James Ridgeway, and the *County Energy Plan Guidebook* (1979), by Alan Okagaki and Jim Benson. Real people designing and building solar-based new towns, such as San Francisco's Solar Village, designed by former state architect Sim Van der Ryn, reusing an old military base. As in any society undergoing rapid transition (as are all mature industrial societies as they shift to new productive systems based on renewable resources and energy), the old center—the dinosaur's brain—is always the last to understand the change and get the feedback messages.

The unreality and loss of feedback on what is happening in the real grass roots is most evident in the stress on small business of Washington's credit squeeze and recession policies.[12] Economists, it seems, can keep only two ideas in their heads simultaneously: *either* supply *or* demand; *either* inflation *or* unemployment; *either* boom *or* recession. We should not be too hard on them—since this "either/or" logic is the heritage of simple, linear thinking and education that worked in our simpler, slower-moving, agrarian past but is now a major block to understanding today's complex, nonlinear, inter-woven, industrial societies. Not only are economists still cursed with this type of linear thinking, but we now have agencies and data collection frozen into these same patterns and unable to imagine alternative ways of collecting different data so as to point their statistical cameras at emerging social activity, which always begins locally.

Thus, the heroic macroeconomic conceptualizers in Washington miss important trends and huge geographical differences in the real functioning of the economy as well as the larger society. For example, they do not measure the growth of the countereconomy, because they cannot conceive of its existence. Similarly, a "national level of unemployment" of, say, 6 percent conceals enormous geographical and group differences, so that a "national," buckshot approach, such as an across-the-board tax cut, will miss most of its targets and simply increase general demand and inflation. Some recognition of these absurd levels of abstraction and data averaging has led to some more targeted, efficient approaches to actual pockets of unemployment although President Reagan's supply-side economists have fallen into the same traps.

These conceptual problems persist and are growing worse. Many economists, now recognizing the illusory efforts of Keynesians attempting to manage by manipulating levels of "aggregate demand" in their usual either/or fashion, are shifting their attention to another statistical illusion: "aggregate supply." Now that stimulating *demand* has failed, they seek to stimulate *supply,* worrying about increasing "productivity" and "innovation," giving more tax cuts to investment and business, and repealing regulation of working conditions and environmental protection. I will examine this fashionable "supply-side" economics in Chapters 4 and 10. For most economists, it seems, training in the simple, linear, logical world of supply and demand prevents them from doing anything but oscillate back and forth between the two concepts, rather than look for a third way: a shift of direction in technology and a redefinition of "growth" and goals. Economists will not find their way out of the either/or box until they can see that today the problems must be redefined as circular, chicken-and-egg problems, in which cause and effect are indistinguishable and not even sequential.

In fact, proliferation of these kinds of paradoxical "problems" usually signals the need to redefine the issues they raise in broader terms. For example, the moment we add conservation to the choices for increasing the availability of energy, we can no longer use narrow policies, say, of comparing coal with oil or nuclear energy. We must widen the whole range of choices and include questions of *why* we waste so much of what we produce and what we will use the energy *for*. We see the same sort of confusion in current debates in Washington (as well as London, Brussels, Paris, and Bonn) concerning whether "declining productivity" is the cause of "inflation" or "inflation" is the cause of "declining productivity"; whether "levels of savings" affect "levels of investment" or vice versa; whether "inflation" is adversely affecting research and development budgets or the reverse! In truth, when stated in these obsolete terms, these are fruitless, circular debates. Yet the old power centers now being undermined by the new global scarcities of energy and materials, on which their dominance was built, will not gracefully move aside for the new adaptations. The wounded dinosaurs may trample many of the weak and unorganized in their scramble to stay on top.[13]

Another ominous problem may be the search for scapegoats as we face these paradoxes and our energy and economic restructuring. Some of the advisers to President Carter had urged that he make the OPEC nations a scapegoat for our energy and economic challenges, rather than facing up to the conflict between our political-economic system and the need to reduce our waste and overuse of world resources and change our life-styles. All this seems obvious to those not trapped in the economists' either/or boxes. In fact, we now see how policies of raising interest rates in economies that run on credit and debt instruments will of course also *increase* inflation, by sharply raising the cost of borrowing—*and* probably also precipitate worse recessions, credit-crunch fears and widening bankruptcies. However, the economists' response to such fears led to passage of the Monetary Control Act of 1980, which may exacerbate our monetary illusions. While it does treat small savers more fairly by phasing out regulation and fixed interest rates, brings all banks under Federal Reserve rules, and insures deposits up to $100,000, the new law also gives the Fed sweeping new power to inflate the money supply and credit, and *lowers* bank reserve requirements. Thus, any administration faced with recession now can create runaway inflation.

Similar anomalies exist, though less frankly discussed, in Poland and other countries of the Eastern European bloc, amid the same exhortations to their people to greater "productivity." Among the market-oriented countries of the Organization for Economic Cooperation and Development (OECD), unemployment is again reaching politically untenable levels. Even though the OECD countries' leaders no longer whistle in the dark about stimulating demand and "consuming their way back to prosperity," their pronouncements are still confused by their economic advisers' either/or prescriptions. In most of these countries, old political labels are now inoperative. President Carter was right in his July 1979 speeches in facing up to the crises not only of energy but of the spirit and values that America is undergoing. This cultural confusion is all part of the hangover from past excesses and the need to adapt to the great transition. But Jimmy Carter avoided the crux of these issues: that we have a democratic *political* system, which relies on individual responsibility and maturity, cooperation, and a well-informed citizenry, while at the same time we still have an outdated *economic* system, inherited from eighteenth-century England, which is based on the theory that all people pursuing their selfish interests will somehow add up to the best of all possible worlds. We now know that this "invisible hand" does not work, as described further in Chapter 8, and that in today's world it is excessively competitive and overrewards greed, selfishness, pride, and aggressive, irresponsible behavior. Economic theory does not acknowledge that people are also generous, cooperative, and altruistic, since this behavior is unpaid and omitted from the GNP. Furthermore, the flywheel of mass-consumption industrial countries is mass-media advertising, which delivers sales and profits to its advertisers by huckstering children, glorifying violence, oversimplifying complex issues, and pandering to the most infantile fantasies of wish fulfillment. It plays on all our fears, sexual needs, and frustrations and tries to convince us that all our problems of life can be solved by buying something. In addition, its large, powerful corporate advertisers often misinform us on the most important issues, such as energy, if it is in their stockholders' short-term interests to distort them.[14] Thus we have a very serious paradox to face as a nation, one that Alexis de Tocqueville saw better than our founders when he warned (in his book *Democracy in America*) that America's noble experiment might lead to "a manufacturing aristocracy" as our economic system became more concentrated. The paradox, in a nutshell, is that our powerful economic system operates by regressing us to more infantile states, while our political system requires us to grow to fuller human maturity.

Perhaps the last, most dangerous expression of the old, either/or thinking is the growing sense of despair and loss of confidence of leaders who see that they are losing control of *that* part of the system *they* created and the dreams of technological glory slipping from their grasp. As they feel *themselves* toppling from their pinnacles, they cry, "Apocalypse," "Armageddon," generalizing their personal panic to the whole society. They rigidify their grasp on the wildly gyrating "controls" and redouble their efforts, not seeing that it is only they who are falling from their collapsing hierarchies, *not the sky* that is falling. They cannot see *what is growing* in their societies: the cooperative, localized countereconomy, our safety net and bridge to the dawning solar age. While they scare themselves with talk only of depression and gold bars, still mistaking money for wealth, the rest of us must continue in our communities to redesign saner, real-world alternative futures.[15]

NOTES – CHAPTER 3

[1] The Carter administration's 1980 budget accepted that a decline of 1 percent in the GNP (after inflation) and an increase by the end of 1980 of joblessness to 7.5 percent was in the cards, as well as at least 9 percent annual inflation. Such a budget message from a President seeking reelection showed unprecedented honesty and helped elect President Reagan. Usually, Presidents seeking reelection inflate the money supply or give tax cuts, so as to ensure that the good times roll until they are reinstalled, as Edward Tufte has documented in his book *Political Control of the Economy*, Princeton University Press, 1978.

[2] *The Christian Science Monitor*, February 25, 1980: "Kaufman Spoke and the Market Went Full-steam for Cover." After Kaufman's speech, the Dow Jones average dropped eighteen points.

[3] Social Security revenues are falling behind what will be needed to keep the system solvent over the next five years. The report of the Advisory Council on

Social Security released in December 1979 suggested as options for Congress either finding new sources of revenue or reducing the level of benefits. The Council also recommended raising the age of retirement above sixty-five and rolling back the recent increases in the payroll tax Congress enacted, which impose an additional inequitable tax burden on lower-income Americans. The entire Social Security system will have to be rethought in the light of new demographic realities. The number of Americans sixty-five years old and over is growing—up nearly five million since 1970—while there are six million fewer children under thirteen now than in 1970. This means that fewer and fewer citizens of working age will have to support more and more retirees as we approach the turn of the century. The massive proportion of the federal budget that goes in such mandated payments and income transfers as Social Security, Medicare, Medicaid, and supplemental security income puts the lie to glib calls for slashing the federal budget's social programs. Even though President Ronald Reagan is fond of pointing out that the lion's share of the budget goes to "social welfare" programs (one third of the 1981 budget of $616 billion) and and that this is more than the Department of Defense gets ($146.2 billion), he does not clarify that 95 percent of this "social welfare" goes to Social Security, Medicare, and Medicaid ($186 billion), while only $13 billion goes to Aid to Families with Dependent Children (AFDC) and supplemental Security Income, which is what is usually termed "welfare" (*The Christian Science Monitor,* February 12, 1980, "Social Welfare Still Getting Biggest Federal Budget Slice" and "More Retirees Expected and Fewer Working Americans to Support Them").

4 By spring 1980, governments and their economic advisers began to identify clearly the role of energy over consumption in the woes of *all* maturing industrial countries. Energy was the focus of the Venice economic summit for Western leaders in June 1980. Concern over oil price increases was expanded to the more crucial issue of future production cutbacks, now inevitable since oil in the ground is now preferable as the best future option for producers. Consumer countries will continue bidding up whatever supplies are available. For example, in 1980, the United States paid OPEC $80 billion for oil, up from $60 billion in 1979—but for less oil (*The Christian Science Monitor,* July 24, 1980). In Eastern European nations, cutbacks in energy use for industry and private consumption are a fact of life, and the U.S.S.R. has warned its satellites that it can no longer assure them supplies in the 1980s, since it will soon be transformed from an energy-exporting to an importing nation, as its own oil production falls short of its domestic needs. Also, conservation for these Eastern European economies really cuts to the bone, since their use is already so frugal. Similarly, although in absolute terms the United States could boast that it had reduced its petroleum consumption more than its friends in Western Europe, its overall per-capita consumption was almost twice as high, thus comprising much more waste in the first place. The thrifty Europeans simply have fewer places to cut.

5 Governments were shaken and toppled in 1979 and 1980 over the issues of rising energy prices, inflation, unemployment, and the other general symptoms of aging industrialism. Prime Minister Joe Clark, of Canada, was thrown out in February 1980, after only nine months in office, largely because he promised to solve these problems with the standard economic remedies and, not surprisingly, failed. Gasoline prices were the last straw for Canadian voters, although they are still low by world standards. Prime Minister Pierre Elliott Trudeau, who took back the reins reluctantly, had the unenviable task of persuading Canadians that

the party was over. France's political malaise has continued. Its Constitutional Council (Supreme Court) declared Prime Minister Barre's unpopular budget (decreeing austerity for workers and the same "old-time religion," tight-money policies now back in favor) illegal. The Socialists and the Communists took full advantage of the situation, hoping that a new coalition could emerge to break the current deadlock before the 1981 elections. In Britain, the honeymoon with Prime Minister Margaret Thatcher was also over, as unions resisted her turn-back-the-clock brand of old-time monetarism, rising unemployment (5.6 percent by February 1980), high interest rates, and an inflation rate hovering at 25 percent. It is increasingly clear that the latest retreat to monetarism and the rigors of "laissez-faire," deregulated, "free market" orthodoxy in North America and Western Europe, as well as in Japan, only result in various backlashes from the ranks of unions, consumers, operators of small businesses, and farmers, which then produce new coalitions with the old Left parties, the Socialists, and Communists. Britain's Mrs. Thatcher faced her first vote of "no confidence," in February 1980, and the stringent "free market" policies of her Industry Secretary, Sir Keith Joseph, were criticized even within her Cabinet as not truly conservative but heading down a radical path too steeply veering to the Right, and the Lord Privy Seal, Sir Ian Gilmour, declared that such a "hawkish" policy on the economy was dangerous "because of its starkness and its failure to create a sense of community" (*The Christian Science Monitor,* February 19, 1980).

[6] While urging Americans to make 1980 the year of energy conservation, President Carter in spring 1980 was still shying away from realistic measures to curb energy consumption. While deciding against a fifty cent tax on gasoline (which would have driven the price up to two dollars a gallon, thus unfairly rationing it by price), he stopped short of advocating coupon rationing, instead proposing a lame "standby" rationing plan, to be effected only if petroleum supply shortfalls approached 20 percent—a catastrophic level that would by then have produced economic chaos! Furthermore, the Administration gas-rationing plan would allocate the valuable gas coupons only to car owners (the six-car owner being entitled to six ration books!) while the almost 19 percent of Americans too poor, too old, too young, or too liberated to own even one car would receive none. A coalition of inner-city residents, low-income people, environmentalists, labor unions, consumers, and minority groups pushed for an equitable, fair-share coupon system (since the devaluing of the dollar and the erosion of our foreign-policy options affect all Americans), with coupons for all registered voters, and "a white market" so that those who already ride mass transit and conserve gasoline could sell their coupons to those who still guzzle gas. I had advocated such a plan in 1975, in a paper presented to the Joint Economic Committee of Congress (see *Creating Alternative Futures,* pp. 113–35), and in 1978 I and my former partner, Carter Henderson, self-published his monograph *The Inevitability of Petroleum Rationing in the United States* (available at $3.95 from P.O. Box 448, Gainesville, FL 32602), advocating such an alternative rationing plan, which was supported by Senator Edward Kennedy as the best way to save oil (1.7 million barrels a day over a three-year period).

With similar lack of resolve, President Carter set an oil import ceiling so high (8.2 million barrels a day) as to have no real effect and caved in to energy-supply interest groups in favor of price deregulation and the massive new government subsidies for synthetic fuels. But the realization grew that conservation was the only new "energy supply" available in the short run (due to long lead times for all the alternatives except local solar, wind, hydro, and gasohol for on-site farm use). Even the National Academy of Sciences, long a hold-out of the "productivist," "supply-side" philosophy, affirmed in its CONAES (Committee on Nuclear and Alternative Energy Systems) report *Energy in Transition 1985–2010* (1980) that the highest priority among all energy options must be conservation and its huge potential for increasing end-use energy efficiency, i.e., squeezing more real economic benefit out of every BTU. The report added that a national commitment to renewable energy sources could produce 30 quads (quadrillion BTUs) by 2010, which could be as much as 40 percent of total demand. Widespread misreading of the study in the press reports focused on its less central conclusions that nuclear power and coal would continue to be important in bridging to the future and that research should continue on breeder-reactor technology (*The Christian Science Monitor*, January 30, 1980).

7 Economist Anne P. Carter, of Brandeis University, noted that "a great deal of domestic R and D is being diverted into making the nation's transportation systems, heating units, and industrial plant and equipment less energy-intensive. Thus, precious resources are being taken out of more-productive uses, such as creating new products, and are being transferred into retooling and reshaping a capital stock that has become inefficient at high energy prices—just to get back to where industrial production was before" (*Business Week*, January 28, 1980, p. 74). This explanation is half right but makes the familiar economist's error in definition of what is and is not a "productive" investment, as we shall explore further in Chapter 10.

8 The gold boom (from thirty-five dollars an ounce only a few years ago to well over eight hundred dollars an ounce in January 1980) created many instabilities and odd effects, one of the strangest of which was pointed out by a London market analyst: at such astronomical levels, the United States' current gold stocks (268 million ounces) were within a range of money value as to almost match its foreign liabilities (then at $242 billion), and that if the price of gold reached exactly $903 an ounce, theoretically the U.S. could pay off all its foreign debt or reestablish a limited convertibility of the dollar into gold! (*International Herald Tribune*, January 21, 1980, p. 7.) That such a move was not attempted is another indication of the extent of the international monetary fairyland of the times. Similarly, banks were awash with all the new funny-money liquidity, with the soaring paper value of their gold stocks and with petro-dollars, while sensible investment opportunities eluded their rearview-mirror vision, and loan money was increasingly harder to place "prudently." Other effects of the gold boom included: gold profits that fed inflation, making not only governments and banks but individuals feel rich enough to spend profligately; a bonanza for the world's two major gold producers, the Soviet Union and South Africa (for example, at current gold prices the Soviet Union will have sufficient hard currency to offset U.S. grain and technology embargoes and buy elsewhere); and the higher price of gold, which is now driving up the price of oil in a "leapfrog" game, as Arabs' gold-buying drives up the price of gold, reduces the value of the dollar, and leads them to a new oil-price increase. Other such circular effects, all pointing to the as yet unacknowledged end of a long era of reliance on economics-based policies, are documented in "How the Gold Boom Is Escalating Instability," *Business Week*, February 11, 1980.

9 In 1979, the oil import bills of the less developed countries were $43.5 billion and for 1980 were some $60 billion (*Business Week*, February 4, 1980).

[10] An insightful study of international banking and financial dealings between industrial and developing countries is *Debt and the Developed Countries,* edited by Jonathan David Aronson, Westview Press, Boulder, Colorado, 1979.

[11] In any event, trust in private banks and their syndicated international loans and other dealings was dealt a heavy blow by President Carter's freezing of Iranian assets in U.S. banks in retaliation for the hostage-taking of November 1979. European bankers were aghast, since it called into question in the minds of all OPEC leaders the safety of their funds in any Western banks, since Iranian funds were also frozen in U.S. branch banks in London. Since then, predictably, OPEC countries are simply getting into banking themselves, as described in "Bankers in Burnooses," *Business Week,* July 14, 1980, p. 45. Many financial observers in Europe predicted then, as a result, the efforts to avoid or replace the dollar as the world's chief reserve currency would be stepped up and that the importance of New York and London as banking centers would decline, as OPEC nations would seek safer havens for their petro-receipts in Singapore, Hong Kong, the Bahamas, and Switzerland. The most precipitous action was that of Morgan Guaranty Trust, which, as a result of the freeze, obtained a court order in West Germany declaring Iran "in default," allowing it to seize Iran's 25 percent ownership of Krupp Industries. In all, seven U.S. banks, led by Chase Manhattan, were involved in the seizure moves (New York *Times,* December 10, 1979).

[12] The Independent Business Federation released a report in February 1980 documenting the plight of small businesses, many driven to the financial wall between high interest rates, inflation, and less credit availability (as I pointed out in *Creating Alternative Futures,* banks prefer to lend to large companies, which can afford high interest rates and are cheaper to service). One measure of the problems of small business was a downturn in their employment for the first time in three years, and the Small Business Administration braced itself for a new wave of bankruptcies (*The Christian Science Monitor,* February 27, 1980).

[13] For example, the U.S. nuclear industry, unable to build new power plants in the United States, has focused its marketing efforts in the Third World. Similarly, the military "hawks" in both the U.S.S.R. and the U.S.A. campaigned further costly arms spending. In an editorial in the New York *Times* on February 10, 1980, Sid Taylor, of the National Taxpayers Union, called the Pentagon's request for $158 billion for the MX missile, the MX 1 battle tank, the nuclear aircraft carrier, the new jet CX cargo plane, and pay raises "a new wave of election-year patriotism, in which the hawks never had it so good."

He added, "Our real war for survival now is not with Moscow or the Middle East but with inflation. America's first line of defense is not in the Persian Gulf, but right here in Washington in front of the Treasury. We are being invaded by deflated dollars in the hands of foreign nations. Our farm lands, real estate, businesses, industries and resources are being captured or bought by alien interests. No wonder the Doves are getting mad. The birds who will pay for all this military/economic blundering are the pigeons [American taxpayers]." I would add, "And our young people, facing registration and a new draft."

[14] Examples of this type of advertising are legion, from Mobil Oil's 1980 series of simplistic animal fairy tales to many corporations' trying to pin the entire problem of inflation on "big government" running the printing presses to pay for bureaucracy. (For example, Amway Corporation's ads portray a

bloated bureaucrat running a dollar-printing press and claim that energy and imported oil have little to do with inflation.)

15 Signs of the booming countereconomy in the United States and elsewhere were the subject of syndicated columnist John Chamberlain in mid-1979 in "The Other Economy Is Booming Now." Chamberlain noted that in Italy, the "other economy" is what keeps the nation from stumbling into communism, and in France and Belgium a growing number of transactions are kept off the books in barter deals, cash, and gold coins. Conservative Chamberlain notes that the "other economy" may be reprehensible but so is the value-added tax (VAT) and the bureaucratic tax collectors and that in fact the "other economy" may save bureaucratized industrial countries from depression.

CHAPTER 4

Post-Keynesians—Not Much Better

As the fatal flaws of both monetarism and Keynesianism became more apparent in the aging industrial societies and press criticism of economic remedies in general grew, still no clear new policy directives emerged. Everyone from columnists to former members of Carter's Cabinet began making their own analyses of the economic situation, and the general lament was "Where is the new Keynes?" Bill Neikirk's article "American Economists Have Run Out of Band-Aids," in the Chicago *Tribune* of October 21, 1979, was typical. "We are in trouble," he wrote, and wondered "if the nation is politically and socially ready for the tough choices that have to be made to prevent the decline and fall of the American economy in the 1980s and 1990s. . . . In the last decade America lost something it once regarded as very precious—its confidence that prosperity could be guaranteed far into the future."[1] Even more on target was the interview in the Washington *Post,* November 4, 1979, with Dr. Juanita Kreps, outgoing Secretary of Commerce, who noted that economics was no longer working and that she found it impossible to go back to her old job as professor of economics at Duke University, because "I would not know what to teach."

However, such humility is lacking in the new breed of economists, both the supply-siders and what they bill as new breakthroughs as well as those who call themselves "post-Keynesians." Since we shall discuss the supply-side school in Chapter 10, a review of the post-

Parts of this chapter are excerpted from my book reviews in *Business and Society Review* in 1973, 1976, and 1979. Used with permission.

Keynesian brand of economics is also in order. Perhaps the best wrap-up of the range of post-Keynesian thinking is *A Guide to Post-Keynesian Economics,* edited by Alfred S. Eichner (M. E. Sharpe, Inc., 1978). In the Foreword, British economist Joan Robinson sums up the general thesis of post-Keynesians as based on the recognition of uncertainty, which undermines the traditional economic concept of equilibrium. This sounds encouraging, but on perusing Alfred Eichner's introductory remarks one has misgivings. "Late in the day," he begins, "after they have had two or three drinks, many economics professors will begin to admit to their own reservations about the theory which forms the core of the economics curriculum. The theory, they will acknowledge, is at odds with much that is known about the behavior of economic institutions. 'But what else is there to teach our students?' they will ask. This question, it turns out, can readily be answered. There does exist an alternative . . . the post-Keynesian theory which is the subject of this book." With hopes thus raised, it is therefore a disappointment to sample the analyses and prescriptions in everything from macrodynamics, pricing, income distribution, and tax policy, to production theory, labor markets, monetary factors, natural resources, and general prognoses that are covered by, presumably, the best team that the post-Keynesians can field. A key problem lies in their inability to transcend the economic paradigm and method itself, thus limiting their prescriptions to tinkering with problems stated from the now-bankrupt economic perspective. Examples include Paul Davidson's useful chapter on resources, which critiques current energy policies and identifies corporate market power but still does not acknowledge the primacy of thermodynamics and is firmly trapped in the supply-demand-price model; Eileen Applebaum's critiques of the absurdities of traditional labor-market theory, highlighting the realities of discrimination and structural unemployment; and J. A. Kregel's chapter on income distribution, which shows that in complex industrial societies it is no longer possible to justify unequal incomes on the basis of differential productivity, since this is the result of social and political decisions, not economic laws. Although the post-Keynesians are still steeped in the orthodoxy of the overall economic paradigm, they often make very useful critiques of its more obvious absurdities. Their usefulness lies in the area of helping the rest of us to understand the shortcomings of traditional nostrums under which the body politic still labors.[2]

However, we should not put too much trust in their ability to help us see beyond economics or to help develop the newer, more systemic, interdisciplinary policy tools and methods we will need to map the future.

Before we proceed to examine some of the post-Keynesians' prescriptions in more detail, we can at least give them credit for vigorous rejection of the "invisible hand" model of the "free market." However, they are still hung up on prices and supply-and-demand albeit viewed more realistically in an oligopolistic setting of large corporations and government-agency policies that, in effect, now create markets, administer prices, and engender business cycles while still mystifying the results as the workings of "market forces."[3]

Post-Keynesians do recognize that our economy is now dominated by massive institutions—corporations and the government agencies that too often cater to them—and that much inflation is caused by big corporations' market power to impose "markup prices" on consumers.[4] But they, too, suffer from the abstract view and their own either/or mental traps. They talk of increasing "levels of investment" to increase "labor productivity," to ensure "economic growth" that will be "noninflationary," but few of them disaggregate these heroic abstractions and specify what *kind* of investments: whether more fast-food franchises and "research and development" of more patent pill remedies and cigarette brands, or the crying need for investments to capitalize a whole new type of economy based on renewable resources and energy and managed for permanent productivity.[5] Thus they also do not bother to redefine what *kind* of economic growth, presumably still accepting the now-suspect GNP measure. Thus they still would obscure the need to shift direction from quantitative to qualitative growth. Lastly, they fall into the same trap of defining "productivity" as only "labor productivity" (i.e., per-capita productivity), which is really an "automation index." Thus we raise the productivity of some fortunate workers at the expense of disemploying many others, who are read out of the production process as "hard-core unemployables," without ever dealing with the question of whether such an economy can produce enough jobs to go around.

Only a redefinition of "productivity" that recognizes that energy and raw materials have been taken for granted and undervalued can help us see that what economists deplore as "declining labor productivity" is balanced by *other gains* we have made: in *energy* productivity (via conservation); in *socio*productivity (i.e., growth in total employment, which has absorbed millions of new labor-force entrants, via more flex-time, part-time, job-sharing, self-employment, and cooperative, smaller enterprises); and *bio*productivity (i.e., investments in restoration of agricultural fertility, reforestation, and employment of human resources in reviving inner-city housing, restoring neighborhoods and railroads, and all forms of recycling).[6] In contrast to this type of smaller-scale, humbler, real-world view, all the post-Keynesians offer is "national indicative planning" and an "incomes policy" (one at least to limit dividend income as well as wages). Or, as one post-Keynesian, Basil J. Moore, put it, there are only three alternatives: continuing inflation, a slump and massive rise in unemployment, or some form of incomes policy, adding categorically, "There are no other games in town" (*A Guide to Post-Keynesian Economics*, p. 138). We will review in greater detail the approaches and policies of the post-Keynesian economists, zeroing in on two who have held high government posts in previous administrations: Walt W. Rostow, as outlined in his *Getting from Here to There* (1979), and John Kenneth Galbraith, from the perspective of his *Economics and the Public Purpose* (1973).[7]

Walt W. Rostow, an economist, economic historian, and high-level adviser in the Kennedy-Johnson administrations, is in the delicate position of having to incorporate new insights while avoiding a too-devastating critique of past U.S. economic policies during the generally expansionist period of the 1960s. During that time, many believed we could iron out business cycles, "fine-tune" the U.S. economy to produce steady GNP growth, high levels of employment, and increasing labor productivity, continue the march of technological innovation, expand social services, and withal, fight a costly war in Vietnam.

In the first seven chapters of his book, Rostow explains some but not all of the reasons why this pipe dream fell apart and the Soaring Sixties gave way to the Stagflation Seventies. More forthrightly than most of his colleagues in the economics profession, he discusses the set of intractable variables that economists prefer to consider external to their models: population pressures, dwindling energy supplies, shrinking arable land, raw-materials bottlenecks, OPEC's oil-price hikes, the new militancy of resource-rich developing countries and their demands for a New International Economic Order, and the new constraints on industrial production posed by a mounting social bill of environmental degradation. His notable omissions from this list include the inflationary role of the Vietnam War; the rapidly rising social and transaction costs of complex, mature industrial societies, which are now clearly beginning to saturate productivity gains in what I described as the syndrome of "The Entropy State" (*Planning Review*, April 1974); the now diminishing returns to excessive capital- and energy-intensive production; and the growing diseconomies of scale.

Much of this has become conventional wisdom to interdisciplinary policy analysts. However, some of it is still news to economists, who have the largest intellectual investments in traditional paradigms, which they are understandably reluctant to write off. Therefore Rostow's discussion in Chapter 3, "The Bankruptcy of Neo-Keynesian Economics," is very useful, since he may persuade his fellow economists, where the rest of us fail, to let go of their excessively simplistic notions, particularly that today's complex, structurally transformed, interlinked industrial societies can be "managed" by the old hydraulics of macroeconomic manipulation of aggregate demand. The chief problem of the neoclassical Keynesian "synthesis" is that no synthesis actually occurred. Keynes's essentially disequilibrium view never modified the basic, equilibrium models of the neoclassicists. Yet, at the same time, they adopted Keynes's policies and applied them *as if* industrial economies were still equilibrium systems, rather than having evolved in their institutional and technological structures into systems in chronic *dis*equilibrium.

Rostow touches on these and other problems of the neo-Keynesian paradigm, including time lags involved in large-scale industrial investments and technological development. In a similar vein to Jay Forrester, whose *World Dynamics* (1971) informed Meadows' *Limits to Growth* (1972), Rostow revives interest in the Long Wave, the approximately fifty-year-cyclic economic theory of Russian economist Nicholas Kondratieff. Rostow shares Forrester's view of the Kondratieff explanation for these long boom-and-deflation cycles: that boom periods have been based on the phases of development of whole new industrial sectors exploiting new technologies such as the era of the railroad and the later boom period sustained by the automobile/highway/suburb industrial complex. However,

The Children's Television Game
SWITCH

How to change the TV set from often ON to ON 'N' OFF
Players will embark on a mock journey through a
typical session of children's television viewing.
The object of the game is to reach the
"Sign Off Zone" free from cumbersome commercials
which obscure the view
of TV programs designed
for young
people.

act /46 AUSTIN STREET/NEWTONVILLE, MASSACHUSETTS 02160

enclosed is my check for $_____ for _____ copies of SWITCH
at $1.50 each.

Name: _____
Please Print

Street: _____

City: _____ State: _____ Zip: _____

Action for Children's Television is a national organization of parents and professionals,
dedicated to child-oriented quality television without commercialism. ACT began in
1968 in Newton, Massachusetts, and now has thousands of members across the
country and the support of major institutions concerned with children.

Plate 1

Plate 2

Wooden bowls and cornhusk dolls
Crafts build self-respect

Through Mennonite Central Committee's Self-Help Program, families in less developed areas can make local crafts to earn their living. This non-profit program is set up to aid the handicapped, the refugee and the person who because of a political or economic situation, is not able to earn. The program fosters self-respect by offering individuals a chance to become self-supporting. Special attention is given to make sure the artisans receive the maximum benefit from their work. Though the program demands high standards of quality in the product, it expresses more concern for the well-being of the craftsperson than for the product.

The Self-Help Program features crafts from over twenty domestic and overseas projects. Stuffed toys from Appalachia, wooden bowls from Haiti and jute hangers from Bangladesh are just a few of the handcrafts available.

Mennonite Central Committee is the joint relief and service agency of North American Mennonites and Brethren in Christ.

Catalogues available from:
Self-Help Program
Box M
21 South 12th Street
Akron, PA 17051

Plate 3

Plate 4

Rostow seems to part company with Forrester as to how the Kondratieff wave will unfold over the next two decades. Where Forrester sees technological doldrums and a shake-out of excessive capital investment, and a period of economic decline and stagnation ahead, Rostow sees what he terms a Kondratieff "upswing" (seemingly referring to an upswing in prices). My own view falls in between: I expect both periods of recession and continued rising prices, i.e., a series of *clearly inflationary recessions,* which will force economists to admit publicly their conceptual confusion. Rostow sees the same energy and resource pressures Forrester and others predict but finds plausible the familiar scenario of substitutions and the prospect of imminent technological "breakthroughs" fueled by impending achievements in the basic sciences: i.e., developmental biology, astronomy, astrophysics, brain research, and computerized measurement and control leading to new technologies of energy and materials, birth control, agriculture, and environmental protection. Rostow espouses the conventional view that the industrialized countries of the OECD must accelerate their GNP growth in order to permit the developing nations to "move forward" more rapidly in the future. He denies the thesis that the gap between rich and poor nations has widened as these "trickle-down growth" policies were pursued in the past.

All this onward-and-upward prescription belies his initial, more realistic analysis of the intractable problems of mature industrial societies, not the least of which is the overarching conceptual disarray he points out is implicit in the failure of neo-Keynesianism and the lack of any new theory; the stale conventional debates over "bigger or smaller Federal deficits: more or less welfare spending; lower or higher central bank interest rates" (p. 215). Paradoxically, part of Rostow's prescription is more economic planning and more attention to sectoral data, which he correctly points out the Council of Economic Advisers is not intellectually equipped or structured to provide. It will certainly be necessary over the next decade to reconceptualize our mixed economy of rigged markets, incentives, taxes, subsidies, rebates, etc., and admit that we have been planning our economy and our investments in an *ad hoc* and informal way for decades, for example, the fact that we have subsidized our oil, gas, coal, and nuclear-energy industries to the tune of approximately $130 billion (according to the March 1978 Battelle Institute study *Federal*

Incentives to Stimulate Energy Production). We need to accept the fact that we have been creating markets for a long time, and can no longer blame God or an unseen hand.[8]

Therefore, it would seem that in the face of such conceptual confusion among policy makers, it is extremely doubtful whether their planning efforts would improve matters or make them worse. One of the chief reasons for today's problems, as I and Rostow emphasize, is the excessive aggregation of data, leading to many statistical illusions (as Oskar Morgenstern used to say) managed by bureaucrats in Washington, London, Paris, and other capital cities. Yet Rostow's prescriptions continually fall back into the language of the heroic conceptualizer: "rates" of investment, "levels" of productivity (a concept also needing redefinition), "flows" of more modern technology, and the continually evoked economic "growth."

Thus Rostow's book, while containing much useful discussion and many interesting insights, is essentially a portrayal of his own mind in transition as he tries to reconceptualize his own thinking. This leads to the characteristic vacillations: the old onward-and-upward-linear-extrapolation-of-economic-progress typical of his earlier book *The Stages of Economic Growth,* interspersed with some genuine rethinking vis-à-vis the need to address real-world problems, such as consciously directing investments to specific sectors, the imperative of vigorous energy conservation, and stepped-up efforts to develop alternate sources such as solar, wind, geothermal, biomass, and fusion. But, as a policy guide, his book must be judged too little and too late. Proposals are timid or fail to address some of the more glaring structural problems Rostow chooses to ignore, such as which group shall bear the brunt of controlling inflation—labor or business. In prescribing for today's higher rates of inflation, he falls back on another admittedly Keynesian remedy: that of attempting to fix wages in the hope that prices will fall if productivity gains are achieved. In the light of experience, it is inconceivable that labor unions would submit to such an inequitable arrangement, which would have to be justified by the equally incredible assumption that in an economy characterized by large corporations and their market power, productivity gains would be passed through to consumers in lower prices. Here Rostow resorts to anecdote. During Rostow's service in the Kennedy administration, the President did jawbone the steel industry into holding prices down for four years, in spite of Walter Reuther's fears that only wages would be controlled, not profits or prices. History has overtaken another plank in Rostow's platform: voluntary wage-price guidelines, already in effect, and for which we are now subject to the worst of both worlds—a costly new burgeoning of red tape and "monitoring" without even the assurance of social results of enforcement!

Rostow's diagnosis and remedies are about the best that could be expected in the old, paternalistic, elitist mode. Hierarchical, patriarchal structures, policies, and leadership styles have simply run out of steam and new ideas. This form of centrist, top-down, technocratic decision making is now itself causing the bottlenecks in information that prevent new formulations and discussion of genuinely alternative approaches now rapidly cohering in the current flood of counterculture journals and books published outside the mainstream media and "straight" publishing industry. As a result, we are seeing the age of Keynes disintegrate while the grass roots flourish unobserved in the emerging countereconomy, and a scenario of spontaneous devolution of now unsustainable institutions as citizens simply recall power previously delegated to politicians and bureaucrats and to captains of industry in making far-reaching technological and product decisions.

Rostow's is a nostalgic view from the top, proving again that what you see depends on where you stand. We see this in his justification of the growth of the Sunbelt and his new home state of Texas (while overlooking the role of the skewed federal tax system, which encourages wasteful relocation via accelerated depreciation); his timid urging of higher energy prices, rather than mandatory import controls, curbing advertising promoting energy waste (as in France), or even an equitable, white-market gasoline-rationing program (all of which flow logically from his assessment of the severity of the energy crunch); and his call for labor self-discipline in wage demands. Rostow is clearly an establishment figure who still plays a team game, rather than breaking ranks to propose anything too new. Indeed, one gets the impression that Rostow himself would like another chance at bat. In his discussion of the need for more economic planning, where business, labor, and government (what ever happened to consumers?) could collaborate as in the French planning-commission style of Jean Monnet in 1946, he notes that it was a small team that labored with *esprit* in a congenially baroque townhouse in Paris. He then envisions how a similar intimate planning team could function in Washington, headed by "a person of obvious distinction who clearly had direct access to the President," and suggests "a town house of appropriate modesty on Lafayette Square might be found." These happy few planners would then set about resuscitating the old Reconstruction Finance Corporation from the 1930s and might create some regional development banks, "perhaps four," that would have "a good chance of operating free of those short term political pressures that afflict the process when public development funds are generated . . . in legislative bodies."

It is, of course, precisely this kind of high-handed, bureaucratic arrogance that has helped to produce voters' mistrust, the demands for decentralization, "small is beautiful"-type technologies, worker self-management, expanded employee ownership, citizen participation, consumer and environmental protection, corporate accountability, human-rights legislation, and the spreading tax revolts of today. The basic problem is that there is as yet no political consensus to mandate Rostow's planning. This time, lobbying the old-boy network will not suffice; like other would-be leaders, Rostow will have to go to the electorate. We sorely need a broadened public debate and multiple leaders from all quarters to interpret the massive transition now underway in all late-stage industrial societies. There is no shortcut in democracies, yet time *is* perilously short. The explanatory politics of reconceptualization must continue.

John Kenneth Galbraith's view is much more egalitarian, and he consistently has reached out to the electorate in his many readable and intelligent critiques of traditional economics. Perhaps his *Economics and the Public Purpose* (1973) best sums up his view of what must be done in maturing industrial democracies. While the public embraces such synthesizers as Galbraith, the organized profession of economics provides them less recognition than it does the armies of mathematical modelers and econometricians. These quantitative analyzers, with their slide rules and computers, seek to turn economics, once a broad sociopolitical avenue of inquiry, into a "value-free" science, as if its subject matter conformed to the immutable laws of the universe, rather than embodying the manifestations of all the messy, unpredictable behavior of the human species.

Galbraith, in his most blistering attack on such reductionist economics, still hews to the great tradition of those giants who founded the discipline—Adam Smith, Ricardo, Malthus, Marx, and John Stuart Mill—and the tradition of embracing within economic thought all those troublesome social, political, psychological, and ecological variables that are too often excluded from the analyses of the quantitative school. In their fruitless efforts to turn economics into a "hard science," this quantitative school has done much harm, both to economics as a useful discipline and to society, by tending to mask social and value conflicts and choices as if they were technical or economic issues. Only when economists confess publicly that economics is not a science will these social choices be revealed as embodying conflicts over goals and values that can be resolved only by political processes.

Not only is economics a normative discipline embodying the value preferences of economists—for example, unwarranted assumptions regarding human motivations that are hotly contested by psychologists, and the feeling that more is necessarily better, which is challenged by ecologists—but, as Galbraith points out, its predominant neoclassical school still largely ignores the role of power in altering economic outcomes. Economists, Galbraith claims, have too often become advocates and apologists for the existing economic arrangements that sustain and employ them. In recognition of this unconscious advocacy role performed by economists, the Public Interest Economics Foundation, of San Francisco, and Accountants in the Public Interest recruit and coordinate volunteer economists to perform analysis and representation for those public-interest and citizens groups that cannot afford to deploy their own economists. Often, the proponents of private or public works projects such as downtown redevelopments, highways, and sports arenas employ economists to prepare cost/benefit analyses that inevitably tend to justify their plans. One of the flaws of cost/benefit analyses is that they are blind as to how the costs and benefits of a given project will be shared. Who will get the benefits—the contracts, the bond-issue business, or other proceeds of the contracts—and which groups will have their oxen gored, such as low-income homeowners in the path of the bulldozers or trapped in new pollution, noise, and congestion zones, or the taxpayers who will bear various scantily documented social costs?

These new public-interest activities by both accountants and economists suggest the extent to which social and environmental externalities have been ignored by traditional economic analyses. Even stalwart establishmentarian Paul A. Samuelson has at last recognized —in the latest editions of his textbook *Economics*—the need for documenting all the diseconomies, disservices, and disamenities produced by economic activities, and the need to subtract them from national income accounts. Accordingly, he now proposes a life-quality indicator he calls "net economic welfare" instead of GNP.

Galbraith bores in on all the erroneous assumptions of neoclassical economics that still parade as revealed scientific wisdom in too many classrooms and textbooks. These assumptions—that 1) most economic tasks will be accomplished in response to instructions of the market, that 2) the firm is subordinate to the state, and that 3) the consumer is the final arbiter of the nature and flow of goods and services provided—are all cheerfully attacked and often convincingly relegated to the realm of ancient mythology. Galbraith documents how the sovereignty of the consumer has been steamrollered by the power of corporate producers and their ability to control markets by advertising and their dominant role in supporting our system of commercial mass media. Even in the capacity of voter, the consumer has now also succumbed to the influence of corporate power wielded through election-campaign contributions, lobbying, and the growing interlock (first predicted in President Eisenhower's warning of the growth of "the military-industrial complex") between large corporations and the mushrooming bureaucracies of the federal government.

This pervasive détente between our major corporations and the U. S. Government, which Galbraith explored in *The New Industrial State,* has, he believes, generated nothing less than a planned economy, albeit informal in nature and even hotly denied by those within its orbit. In *Economics and the Public Purpose,* he refined this earlier analysis and proposes that alongside this planned economy there is still a vestigial market economy whose development is thwarted and twisted by the power of the dominant planning sector. He stressed— rightly, I believe—that this unequal development and the income inequalities resulting from it bear no relation to need, productivity, or efficiency; rather, they are the results of the unequal deployment of power. Left to themselves, Galbraith concluded, economic forces do not work out for the best, except for the powerful. The phenomenal growth of the social movements for consumer and environmental protection, racial and economic justice, and corporate accountability seem to have validated Galbraith's thesis. If the consumer really were king, if markets could rationally allocate resources with a minimum of government interference, and if the external costs of private economic activities were merely aberrations that could be internalized according to economic theory, then such social movements would not exist and we would already have achieved the economist's optimal promised land.

One is forced to conclude either that consumers and citizens (whose estimations of corporate probity, according to Opinion Research Corporation and other pollsters, have plummeted to new lows) are all mad or that, indeed, corporations do exert a growing and often detrimental effect on our lives and our society. Moreover, as the Council on Economic Priorities has demonstrated in many of its comparative studies of corporate social performance, the influence over our lives and our social system of various corporations is highly arbitrary and capricious, not necessarily correlated with social priorities, needs, or the public welfare, but, rather, reflects the subjective corporate concerns for growth and profit-maximizing goals. It becomes increasingly obvious that Milton Friedman's argument that corporations should just stick to maximizing profits and have no business trying to affect society is schizophrenic. For, in fact, just by doing their profit-making thing, corporations impose enormous social and environmental costs and other effects on society. Or, as futurist Willis Harman, of the Stanford Research Institute, notes, corporate goals are increasingly perceived by the public as misaligned with social goals. One might even say that the beneficent "invisible hand" envisioned by Adam Smith has become for increasing numbers of Americans a clumsy, heedless "invisible foot," which tramples on social, human, and environmental values, rather than responding to them.

Galbraith defined the dominant "planning system" as composed of the some one thousand manufacturing, merchandising, transportation, and financial corporations producing approximately half of all goods and services not provided by the state. These corporations, he claims, represent a high degree of concentrated economic and political power: the 333 largest industrial corporations, for example, account for 70 percent of all assets employed in manufacturing. An assembly of the heads of such firms doing half of all the business in the United States would, Galbraith remarks, be unimpressive in a university auditorium. These giant corporations of the planning system, he adds, have successfully accomplished the "euthanasia of stockholder power" and pursue their goals through ever-more-sophisticated organization, group decision making, and information control, in such a way that it is difficult if not impossible for individual managers, technicians, stockholders, directors, government agencies, or anyone else to piece together all the relevant facts needed to critique the corporation's policies or challenge its actions. However, Galbraith also points out, such corporations are careful not to flaunt the collective power of their managers. They present painstakingly structured information to their boards of directors with such deference that ratification is assured.

Of course, all organizations, whether the great bureaucracies of the public or of the private sector, are "defense mechanisms" for controlling or screening information to pursue their internal goals. In all such bureaucracies, top managers are adept, when put on the spot, at imputing authority to others in the endless, convoluted buck-passing with which corporate activities are rife: "We must answer to the stockholders," "My board would never go along," and the like. As Galbraith notes, power is not diminished by being attributed to someone else; rather, it is usually enhanced and made easier to exercise. While not doubting the importance of greed in human affairs, he also believes that the neoclassical assumption that corporations maximize profits is now only partially true, and that the security and growth of the enterprise are now important goals, not only because large-scale technological operations require such size and security, but also because the managers themselves are more protected from the vagaries of rugged competition and their stock options and bonuses are more assured by continuous corporate growth.

In short, says Galbraith, these large corporations have the power to extensively force their will on society, to fix prices and costs, to influence consumer behavior, to organize their supplies or materials and components, to mobilize and internally generate savings and capital, to manage labor relations by "buying off" their fortunate workers with wage increases whose costs are passed along to the consumer or the taxpayer, and to influence the attitudes of voters and the state. Lastly, he observes that the corporation in its multinational form is a logical extension of all these properties.

Qualifying his sweeping conclusions, Galbraith noted some exceptional cases in which stockholders have become aroused by losses, as well as the potential, not often utilized, of aggregating stockholder power by institutional investors such as insurance companies, foundations, and mutual and pension funds, most of which, however, are still passive with respect to management policies. Galbraith dismissed the activities over the past few years of those hundreds of groups, both large and small, that have sought to influence corporate social policies through proxy power and the politicizing of the annual meeting, and the institutional investors. These have been documented since 1970 by the Council on Economic Priorities in its many detailed reports on corporations' social impacts in many areas including consumer and environmental issues, minority rights, military contracting, and the social effects of U.S.-based corporations in other countries, and in annual surveys of stockholders' actions in its newsletter, "Minding the Corporate Conscience."

Galbraith dismissed the efforts of Campaign GM, which served as a model for most subsequent corporate campaigns, claiming that such efforts are naïve. Yet one wonders if Galbraith's tone would be as fearless and forthright had not the corporate-accountability movement so ably tilled and fertilized the field of public opinion.

Unfortunately, there appears to be a cultural lag visible in most of the social-science disciplines. At best, they may trace some of what has happened in recent social interactions, but they are of no help in "real time," let alone providing any predictive power in the unfolding of events. Such a cultural lag is now glaringly evident in economics. Just when economists finally embraced Keynesianism *en masse*, the social and environmental costs of such policies as pumping up the whole economy to address specific problems of structural unemployment and income distribution are becoming apparent in worsening inflation, social disruption, resource shortages, and pollution.

How does Galbraith seek to address the problems he so persuasively documents? His prescriptions range from the exhortatory to the pragmatic. First: He calls for the emancipation of belief, to which many would heartily subscribe. Economists and their outworn theories have become a major stumbling block to the rational public discussion of resource allocation and social choices. Somehow their stranglehold on the metaphors and rhetoric of such urgent public debates must be broken, and they must be made to acknowledge the limits of their professional competence, beyond which their own value preferences can be legitimately asserted only as individual citizens. Here the movement to politicize economists and accountants can provide significant ripple effects within the walls of academia. Economics is one of the last disciplinary strongholds of unalloyed academic pretension, and in the U.S.A. it has been additionally shielded from reality by the taboo status of Marxian and other competing paradigms. This has made clearheaded critiques of market economics almost impossible for lack of any firm intellectual terrain from which to view the market system and realistically assess its strengths and weaknesses. Now such academic groups as the Union for Radical Political Economics, active in Galbraith's backyard at Harvard and M.I.T., as well as at dozens of other prestigious universities, are beginning to exert a beneficial effect in freeing economists from past conceptions, and, one suspects, encouraging such free thinkers as Galbraith to greater endeavor.

Second: Galbraith calls for emancipating the state from corporate control by the forces of the planning system. He advocates taming the corporation to make its goals serve, not define, the public interest. But Galbraith disagrees with many activists who would like to see antitrust laws enforced so as to break up corporations exerting monopolistic or oligopolistic power. Such strategies, he believes, misinterpret the workings of antitrust laws, which he sees as tolerated by the planning system as a convenient decoy whereby the public is lulled into believing the economic myth of rugged competition in a free-enterprise system. Most students might agree that doctrinaire assertions of the merits of competition are little more than cant. But, unlike Galbraith, who seems neither to fear organizational size nor to question supposed economies of scale, activists and millions of citizens now distrust bigness itself, not only because they suspect that there are significant diseconomies of scale (probably both internal and external) but because of the demonstrated imperviousness to citizen/consumer feedback of all large, bureaucratic structures. The environmentalists also distrust bigness and centralization because of their understanding of the principles of ecological system behavior: that diversity and responsiveness to feedback characterize all efficient, stable ecosystems. Nor is Galbraith interested, it seems, in decentralizing or diffusing capital ownership, through such means as the Employee Stock Ownership Plans proposed by Louis O. Kelso, which are now being set up in many corporations and offer another route to democratizing corporations by broadening stockholder ranks with millions of workers.

Third: Galbraith urges, as he did so brilliantly in *The Affluent Society,* the restructuring of resource use away from overdeveloped sections of the economy (for example, in the private consumption of frivolous goods) and redeploying resources into areas of unmet needs, as well as making technology serve public, not technocratic, interest. There is already widespread recognition among large corporations of the approaching saturation of some private-sector consumption and of the need to regroup themselves to serve the unmet needs of the public sector. After the embarrassing "boosterism" of the late sixties, when some corporations claimed that they could educate our children, rehabilitate the slums, and rebuild our cities, there is now, fortunately, a chastened self-image of corporate capabilities. But if corporations wish to serve such new "markets" as mass transit, housing, health care, pollution control, and recycling, they must understand the social movements pushing for these priorities. These movements represent nothing more terrifying than potential consumers who have been forced to aggregate their demand politically since such societal needs apparently cannot be signaled or fulfilled in the traditional marketplace.

So how shall we produce such new social goods and capitalize such production? Certainly, many existing corporations will be in line for the new contracts. It will therefore be necessary to carefully define by political processes the performance criteria, costs, and other conditions of such contracts if we are to avoid the current debacles of military contracting: poor performance, cost overruns, and the unhealthy influence of the corporate contractors themselves over procurement and even the processes of defining Defense Department needs. Here Galbraith is vindicated by the 1979 recommendation by the Brookings Institution: that about $500 million could be saved annually if the Pentagon stopped paying civilian, blue-collar employees more than comparable nongovernment workers earn. The General Accounting Office discovered another $300 million waste in the Defense Department's overly complex phone system (*Time,* October 29, 1979, p. 33).

Galbraith suggests that we might beef up the weaker market system to fill many of the new needs by encouraging small firms to organize more effectively and see that they receive the comparable government largesse accorded to firms of the planning system. Many small firms, he believes, may be more intrinsically suited to providing some of the needed public-sector services than larger corporations whose forte is churning out cascades of identical goods. He also advocates a major increase in the minimum wage to protect the workers in the weaker market system, which President Reagan seeks to reverse, and applauds the current drive to unionize workers in the services sector, municipal governments, and agriculture.

Galbraith asserts that for the performance of some essential services such as public transportation, low-cost housing, and medical care, state-operated corporations are the only answer, since these services do not lend themselves to the capabilities of the planning system and may preclude, by their very nature, profit margins of interest to large corporations. Therefore, what Galbraith refers to as "the new socialism" would not seek out centers of power in the economy, but centers of weakness to undergird. In addition, he calls for public ownership and control of corporations such as Lockheed and General Dynamics, or any others that do more than half their business with the government, in the belief that one clearly visible bureaucracy would be preferable to the mutual conniving and lobbying. In newly conservative Canada, such a conversion is planned to turn Petro-Canada into a quasi-publicly owned corporation (*The Christian Science Monitor,* October 17, 1979). Such proposals for turning major government contractors into such quasi-public corporations merits consideration, since, more often than not, it is the taxpayer that is asked to bail their stockholders out in hard times.

To combat structural inequality of income, Galbraith espouses the various guaranteed-income proposals put forward over the past decade, which culminated in the defeat of the Family Assistance Plan in 1972, chronicled by Daniel P. Moynihan in his book *The Politics of a Guaranteed Income.* It is hoped that we can go back to the drawing board and come up with another version of guaranteed income that will meet with more success, for it is one of the necessary strategies to cushion the individual hardships wrought by technological change in advanced industrial economies. Galbraith urges a familiar list of tax reforms to narrow inequalities in income distribution, and he calls for labor unions to demand a narrowing of the huge differentials in salary scales between workers and managers.

For the environment, he proposes more outright prohibition in more cases of environmental abuse than most of his fellow economists. But in considering the possible limits to currently defined economic growth (see, for example, *Limits to Growth,* by Donella Meadows et al.), Galbraith falls into the trap in which most of his fellow economists have already landed. Here he reveals a deterministic view not at all consistent with the massive social reforms called for elsewhere in his book. A reduction in growth, he asserts, becomes a decent remedy only when distribution of incomes becomes more equal. He seems to have swallowed the current corporate propaganda, typified by Mobil Oil's recent advertisements, that economic growth is the only means of achieving greater income equality. This, of course, contradicts the main thrust of his other social remedies, which embrace income *re*distribution. He cannot maintain, as he does, that limiting economic growth would cause people and groups to be frozen at their present levels of consumption without admitting the impossibility, and therefore naïveté, of his programs to fight income inequality. The recently perceived realities of energy and resource shortages may bring Galbraith to recast his position on economic growth, as defined in this book. It is clear even to economists that resource shortages will result in curbing economic growth in any event; in which case, the past calls of environmentalists for strategies of redistribution, sharing, communal life-styles, and less emphasis on material consumption may be taken more seriously.

In all, Galbraith espouses standard liberal, humanistic proposals and the assumption that somehow a package of needed reforms can be won by our processes of democracy. But how? By influencing the Congress, says Galbraith, to take up the concerns of individual constituents and the public interest, rather than its all-too-organized special-interest constituents. But this is where we came in. The preemption of such political processes by corporate power, which he so convincingly documents, is what originally drove political dissent into other, *ad hoc* channels such as that of the annual meeting and the proxy machinery, and occasioned the recent politicizing of the mass media by the movement for citizen access. Even if one is persuaded by Galbraith that bigness is inevitable and that all we can look forward to is the computerized Leviathan state as the only answer to the Gordian knot of problems generated by industrial power and economic growth, one wonders how he can display such confidence in human ability to manage such labyrinthine complexity. The truth is that we do not know how to model the complex sociophysical systems we have created, let alone manage them.

Neither do the post-Keynesians.[9] However, if there is to be a form of economics that survives the current crucible of aging industrialism, we must leave no stone unturned in our search within the discipline of economics. Thus, we turn now to a brief review of the clearest statement of policy issues from the United States' most respected Marxian scholar, Michael Harrington, in his *Twilight of Capitalism* (Simon & Schuster, 1976).

Quoting Jürgen Habermas' predictions, Harrington notes that macroeconomic mismanagement has led us to today's crisis of political legitimacy. All market-oriented democracies are now caught in the crunch created by their citizens' expectations of smoothly managed economic growth, high levels of employment, and rising individual incomes and opportunities—the beguiling promises of overconfident economists. As a result, government's old medicines for managing inflation, i.e., clamping down on credit and money supply with high interest rates and thereby creating unemployment, are now politically unacceptable. Indeed, this was the message of Jimmy Carter's ousting of President Ford—who was powerfully backed by business—by means of a coalition of the less affluent. It seems that there are now so many Americans for whom our economy is not working that they can constitute an electoral majority. Carter himself was ousted by Reagan's similar promises to cut taxes and curb inflation and unemployment.

Harrington soberly explores our economic system, now structurally transformed from the simple model described by Adam Smith. He uses the first half of the book to measure our economic woes against Marx's predictions about the internal contradictions of capitalism, and finds that declining profits and the need for a periodic recession—"the pause that refreshes" the private sector—still prevail. On the other hand, Marx's predictions of the workers' revolution have proved wrong, since so many workers still believe in Horatio Alger and identify with those who have made it. In fact, Harrington generally exposes the latter-day cant surrounding Marx, shows how some Marxists misunderstand their guru, and demonstrates that many U.S. sociologists have relied on Marx for insights while loudly denying his contribution to their analyses. Harrington emphasizes that Marx never asserted production relationships to be the basic factor in determining the cultural superstructure, but that he was much more subtle in portraying their interrelationships. He also points to Marx's belief in the progressive role of greed in contributing to the evolution of an economy.

Perhaps most important, Harrington reminds us that the "free market system" was not derived from God or any natural laws, a point to which Marx, Polanyi, Max Weber, and legions of anthropologists have attested. Karl Polanyi examined this point more closely in his book *The Great Transformation* (1944) and illuminated the delicious paradox that, in fact, the laissez-faire, free-market system was actually a package of social legislation! This embarrassing truth is becoming more visible as the age of industrialism, to which this social legislation gave rise (essentially land enclosure and the commoditizing of labor) draws to a close. As Harrington notes, "Minerva's owl flies at dusk," and only at its twilight can we begin to see the era in which we have lived.

Harrington also points out that Marx never believed in economics as a discipline, but considered economists as simply professional apologists for capitalism. Indeed, the recent rise of "public interest" economics has helped to smoke out economists from their pretensions of objectivity and reveal their almost religious commitment to the now fading free-market ideal and their simplistic macroeconomic policies based on the hydraulics of supply and demand. Harrington also slashes away at the weakest conceptual link of economics: marginal analysis. He explains how in the war of the two Cambridges (between economic theorists at Cambridge, England, and those at Cambridge, Massachusetts), Joan Robinson and the British team almost destroyed Samuelson and the U.S. team by showing that, despite all their fancy math, the Americans couldn't define capital, the linchpin of their conceptual model! Beyond recounting such intellectual hijinks, Harrington reviews the 1960s and characterizes the Johnson administration's war on poverty as doomed from the start, since its attempts to alleviate poverty by the trickle-down method of bolstering the business and investment sector were bound to end in halfhearted, spasmodic programs.

Harrington also attacks the prevailing view that wealth is produced only in the private sector, from which it is siphoned off to "unproductive" government programs. Instead, he proposes a revised model based on an analysis of how the system has changed into "The

State As Milch Cow." This new model stands market theory on its head and portrays the government and taxpayers as a stupid and patient milk cow, asked to underwrite investments, risks, and the social costs incurred by the private sector, while having no voice in these investment decisions and being denied a share in the profits. His analysis resembles Galbraith's in that both writers describe an economy in which risks and costs are socialized, but profits remain privatized.

Harrington makes a very useful contribution to the current debate over the changing structure of our economic system. First, it would seem, we must lay to rest an imaginary dilemma: the so-called trade-off between unemployment and inflation, which economists chart as the "Phillips Curve." Ironically, Phillips himself never postulated a Phillips Curve, but merely described a hypothetical relationship based on scanty data in his 1958 study in England. It is now possible to prove that the Phillips Curve is inoperative in the test tube of the British economy. Britain's woes may be giving us a preview of the last act of the drama of industrialism. As in all mature industrial economies, not only has unemployment become structural, but so has inflation. Both are related to excessive capital intensity and resource dependence, and both may be alleviated only by running such economies on a mix lean in capital, energy, and materials, and rich in labor, the more plentiful and underutilized resource.

In fact, in such economies as Britain and the U.S.A., labor is now the more efficient factor of production in many processes, while two new sources of inflation cannot even be modeled within the paradigm of economics: inflation caused by internal, systemic complexity and its resulting soaring social costs and growing public sector; and inflation caused by the external effects of a declining resource base, which requires that more and more investment capital be employed to extract resources from ever more degraded and inaccessible deposits, with steeply declining net yields. Harrington draws attention to the first new systemic source of rising inflation rates—the soaring social costs incurred by private profits—but he misses the second. However, one hopes that he has helped open up the debate about inflation to more precise formulations and given courage to economic journalists, who at last are relying more on their own common sense and observations of reality, rather than slavishly interviewing economists. Some business journals now sense that a major paradigm change is under way, as evidenced by a polemical and confused essay in *Forbes* a few years ago entitled "Inflation Is Now Too Serious a Matter to Leave to the Economists." The key factor, overlooked by *Forbes,* Harrington, and most economists, is that advancing technology creates interdependencies that systematically destroy free-market conditions. There is therefore a monumental paradox facing all mature industrial societies: laissez-faire policies have become progressively less workable, yet we do not know very much about how to plan such sociotechnical complexities. We must now face this paradox if we are ever to begin mapping out a "third way."

NOTES – CHAPTER 4

1 And with good reason. By the spring of 1980, every citizen, worker, and consumer in the United States was feeling the pinch somewhere, and the experience was equally bleak in other industrial countries. In Europe, hardships caused by rising prices and unemployment triggered increased social welfare programs from 41 percent of gross domestic product in the European Community countries in 1973, to 47 percent in 1979. In Germany, workers had to accept wage gains of only 4 percent while inflation reached 6 percent, leading to the first major steel strike in fifty-one years. In France, prices rose 11 percent while wages lagged, increasing less than 10 percent, and taxes zoomed; in Japan, the rise in real income slowed from 4.8 percent in 1979 to approximately 2 percent in 1980. When their energy and resource overdependence hit home, all maturing industrial countries saw sharper conflicts between consumers, labor, and capital investors and business, as the no-longer-growing pie had to be shared by increasingly vociferous political tussles. A useful roundup of these new conflicts was "The Shrinking Standard of Living," *Business Week,* January 28, 1980, pp. 72–78.

2 By spring 1980, with inflation roaring at 18 percent annually and more unemployment, politicians were caught between the demands of the military-industrial complex for ever-larger defense expenditures and slashes in social programs, and the increasingly restive urban areas, labor unions, low-income groups, and consumers. Meanwhile, our foreign oil bill in 1980, estimated at $83 billion, had required 40 percent of all U.S. exports to pay for—up from 33 percent in 1979. In 1979 alone, the world oil price had increased 98.1 percent, adding almost 4 percent to the inflation rate. Thus, fertile ground for post-Keynesian remedies brought more calls for mandatory wage-price controls. Most advocates of wage-price controls profess their abhorrence of them, but while the Carter administration stoutly refused consideration of controls, many joined former director of President Carter's Council on Wage and Price Stability Barry P. Bosworth in calling for such a freeze. Included were such unlikely converts as Wall Streeters Henry Kaufman, of Salomon Brothers, and Felix Rohatyn, of Lazard Frères & Co.; former president of the Minneapolis Fed-

eral Reserve Bank Bruce K. McLaury; Senator William Proxmire, chairman of the Senate Banking Committee, and Representative Henry Reuss, chairman of the House Banking Committee; and 1980 presidential candidates Edward Kennedy and Howard Baker. Even *Business Week* proposed a temporary six-month freeze. Events also forced another post-Keynesian remedy: credit controls, a short-lived, cosmetic attempt to target speculative buying by victimizing small borrowers, which utterly failed to prioritize investment-capital uses. Wage-price controls can do little to alter the many structural imbalances in the industrial economies; they are really a desperation measure that can, by the mere mention of them, send businesses rushing to anticipatory price increases to get in under the wire.

3 Perhaps post-Keynesians can also take credit for some more realistic addressing of structural problems in the U.S. economy—such as rationing gasoline—as a less-inflationary way of reducing oil imports, finally backed by conservatives such as Felix Rohatyn and New York *Times* analyst Leonard Silk. In 1980 polls, a majority of Americans already favored both wage-price controls and gas rationing. Another, more traditional remedy proposed was to increase banks' reserve ratios, addressing a problem of long-standing wherein banks are able to create money out of thin air by "monetizing" the loans they write for their customers by posting the loans in the customers' account balances as if they were no different from deposits. The late Ralph Borsodi devoted much of his life to exposing this inflationary role of banks' increasing the money supply. Forcing banks to adhere to higher reserve requirements to back their loans would reduce one source of inflation, as the late Irving Fisher (University of Chicago) and others proposed in the 1920s. New calls for draconian reserve requirements came from Donald R. Wells, of Georgia State University, in "Controlling Inflation with a 100% Reserve System" (*Business*, September–October 1979, pp. 26–28), and from the Committee for Orderly Financial Reform, of Canada, which proposed that Canada's chartered banks be prohibited by law from monetizing loans or performing any other functions with monetary-expansion effects, all of which powers would revert to the Bank of Canada alone ("The Economy; An Analysis and a Cure," by W. J. Blackman, professor of economics, University of Calgary, November 1979, sponsored by The Committee for Orderly Financial Reform, Niagara-on-the-Lake, Ontario, Canada). Ironically, as mentioned in Chapter 3, the Monetary Reform Act of 1980 made matters worse by *reducing* bank reserve requirements!

4 The issue of corporate concentration and power to mark up prices became unavoidable. Maverick economists Gar Alperovitz and Jeffrey Faux, of the National Center for Economic Alternatives, in Washington, launched a creative program, based on the work of Leslie Nulty, to highlight the extent of inflation in four basic necessities of life—food, energy, housing, and health care—which imposes a much higher effective inflation rate on those with moderate incomes, and pointed out that these four basics make up two thirds of the budget for 80 percent of American families. Many of these increases Alperovitz ties to corporate power to set prices and lobby for special favors, and he favors wage-price controls and a long-term program to decentralize the U.S. economy and shift its base to renewable energy resources. Similarly, the April 17, 1980, teach-in on Big Business Day was supported by post-Keynesian J. Kenneth Galbraith, as well as Alperovitz, and other economists including Robert Heilbroner, Robert Lekachman, and long-time advocate of democratic socialism and adviser to President John Kennedy, Michael Harrington, author of *The Twilight of Capi-*

talism and *The Vast Majority,* together with churches, labor unions, consumers, blacks, women's groups, and environmentalists.

5 A brief bright spot in reconceptualizing the new situation was the 1980 turnaround in U. S. Department of Agriculture policies under Carter's Secretary Bob Bergland, who infuriated large mechanized farmers and agribusiness interests by declaring his intention to stop subsidizing agricultural research that would increase farm automation and energy dependence at the same time that it disemployed more farm workers and continued to put small farms out of business. Even as he lobbied for greater research and development for farm research, Dr. Jarvis E. Miller, president of Texas A & M University, allowed that the agricultural sector could potentially reduce its energy consumption by 20 percent with effective conservation, as well as use less fertilizer and cut down on the use of tractors, heavy equipment, and water (*The Christian Science Monitor,* February 20, 1980).

6 The kind of productivity gains to be derived from giving workers a piece of the action, whether in Employee Stock Ownership Trusts (ESOTs) as mentioned earlier, or various worker-ownership and self-management schemes, more locally controlled, community-based businesses, cooperatives, land trusts, and other programs of land reform have enormous potential. Such model programs are now legion, as described by Daniel Zwerdling in *Democracy at Work* (1978) and promoted by several organizations including the Institute for Community Economics, Inc. (see Plate 4); National Land for People (see Plate 6); the Agricultural Marketing Project (Plate 5); the Campaign for Economic Democracy, of Los Angeles; the Peoples Business Commission, of Washington, D.C., which has published *Own Your Own Job,* by Jeremy Rifkin (Bantam Books, 1977); The Institute for Local Self-Reliance, of Washington, D.C., which publishes the monthly *Self-Reliance;* and the New School for Democratic Management (see Plate 24).

7 Walter Heller, economic adviser to President John Kennedy and to Senator Edward Kennedy, appears still to be a committed, old-style Keynesian. In a syndicated newspaper roundup, "America in the 80's," December 30, 1979, Heller saw greater strength and productivity in the economy. He still believes in further attempts to increase per-capita, labor productivity: "In the 70's we were substituting labor for capital. . . . In the 80's we'll be substituting capital for labor, putting more investment into both physical capital, such as plants, equipment, and machinery, and into research and technology."

Heller's statement neatly sums up the traditional economics, as well as the errors that the post-Keynesians have inherited from them: the too highly aggregated view of "productivity," "investment," "capital," "technology," etc.— rather than examining the vastly different realities underlying these unscientific abstractions, as we shall discuss in detail in Chapter 10.

Another well-known mainstream Keynesian began switching course. Paul Samuelson prognosticated in *Newsweek,* December 3, 1979, that "while planning for a minor recession, we should also prepare for the worst." His worst-case scenario (scenario building is a new tool that economists have picked up from futurism, which at least permits them to clearly state their assumptions) included: an inflation rate nearer 20 percent than 10 percent; a collapse in bond prices; prime interest rates going to 20 percent; and another run on the dollar, forcing import quotas, foreign-exchange controls, and international investment curbs. Since then, several of his "worst-case" events were in evidence: 18 percent inflation, interest rates at 20 percent and a collapse in the bond market. Curiously, however, Samuelson offered no prescriptions other than noting the need for "contingency safeguards" by the President and Congress; and he advised individual investors to "go with the odds that the recession would be mild and that inflation would drop to around 9% and not to pass up those money market funds yielding 13% interest."

[8] Robert Eisner, professor of economics at Northwestern University and author of *Factors in Business Investment*, spells out "An Economic Alternative to Slow Growth and Recession" (Gainesville *Sun*, December 23, 1979), embracing this view of inflation as the result of special favors to powerful interest groups: price supports for dairy products, "trigger prices" to protect our steel industry, sugar quotas, and other privileges of oligopolistic industries, as well as higher payroll taxes that raise the cost of labor and prices even as they decrease take-home pay, and the *costs* of higher interest rates. A graphic illustration of all this special-interest-group lobbying in the Congress is the recent ad to large corporate advertisers soliciting the purchase of space in *Roll Call,* the magazine of Capitol Hill, "where it counts" (Plate 8).

[9] Perhaps the most graphic illustration of the increasing unreality of the entire debate between economists of all persuasions is the debate over the issue of the declining rate of savings of the American people, generally viewed with alarm by all and used as the justification for switching concern from the "demand side" of the economy to the "supply side." This view is fashionable across the whole spectrum of economists, including the post-Keynesians, thus revealing their shared core of assumptions and the economics paradigm itself. It has led to the drive to increase interest rates for small savers and the phaseout of the infamous Regulation Q (limiting the interest rates unfairly for small savers) in the Monetary Control Act of 1980, mentioned earlier. While this is the place to target concern, the aggregate view of the economics paradigm has won out in considering *all* savings equally sacrosanct, in order to encourage *general* levels of investments, so that this may "trickle down" and create jobs, etc., in the Golden Goose model. Thus economists have generally pushed for less and less taxation of dividend income and capital gains; more tax credits for corporate investment; faster depreciation; and the whole "supply-side" shift of strategy, from Arthur Laffer's famous Curve, beloved of the conservatives, to the "productivists" on the Democratic side, who call for massive subsidies of dubious energy and for synfuels projects.

The current savings rate, calculated at about 3 percent on personal disposable income, is compared with alarm to the Japanese rate of 25 percent, fueling yet another economic tack: that of shifting taxes away from savings and investment by *increasing* taxes on consumption by enacting the value-added tax (VAT), (which would only increase the inequities in the tax system and in the distribution of wealth and income). My counterargument to all this unreal, circular economic theorizing is twofold: First, the statistics on the rate of savings are calculated on personal savings. However, as Jeffrey A. Nichols, of Argus Research, in New York, points out, ". . . the rate of total savings by households, businesses and government is a more valid indicator than the personal savings rate by itself." He adds that the ratio of *gross* savings to gross national product has in fact *risen*, from 12.8 percent in 1975 to an estimated 15.4 percent in 1979, and this ratio has been stable, ranging from 14 percent to 17 percent for all but three of the past 30 years (*The Christian Science Monitor,* February 25, 1980). Secondly, the informal, nonmonetized sector of the economy, although invisible to most economists, is growing, as noted earlier, and more and more Americans are saving *not money* but real tangible assets (which economists account for as "consumption"!) such as jewelry, paintings, antiques, and rare books, not to mention installing wood stoves and insulation, fixing up their homes, etc.—all of which are the most prudent form of saving and investment as the funny-money-economy bubble deflates and its "statistics" become ever more meaningless. Similar circularities of economic reasoning characterize the debate about whether or not to index an inflationary economy to deal with the new phenomenon of "taxflation," in which taxpayers are pushed into ever higher tax brackets by illusory wage increases. Republicans use these arguments for tax cuts and/or indexing, both of which, in my view, are inflationary. Another tack taken by some economists is to fiddle with the statistics used to compose the Consumer Price Index so as to "exclude" trouble areas such as rising interest costs of housing. These are all examples of the "end of economics" itself and a new era approaching of "posteconomic" decision making and policy analyses.

CHAPTER 5

The Changing Corporate-Social Contract

We now turn to one of the major features of industrialism, which is rapidly becoming its most dominant characteristic in both market-oriented societies and the centrally planned, socialist countries: the giant industrial corporation and its huge centralization of economic power and technological means.[1] The giant corporation was not anticipated by any of our Founding Fathers, even though Thomas Jefferson and Benjamin Franklin both saw the dangers of such concentration looming. Jefferson noted in 1814, "I hope we shall crush in its birth the aristocracy of our moneyed corporations, which dare already to challenge our government to a trial of strength and bid defiance to the laws of our country." Jefferson added that corporations, "penetrating every part of the union, acting by command and phalanx, may, in a critical moment upset the government."[2] Franklin understood the difference between the sanctity of private property for guaranteeing individual autonomy and self-reliance, and property endlessly accumulated to the point of oppressing others. Franklin stated in 1785, "Superfluous property is the creature of society. Simple and mild laws were sufficient to guard the property that was merely necessary. When, by virtue of the first laws, part of the society accumulated wealth and grew powerful, they enacted others more severe, and would protect their property at the expense of humanity.

Parts of this chapter are reprinted from *Human Resource Management,* Volume 17, #4, Winter 1978, University of Michigan, and excerpted from my paper before the North American Society for Corporate Planning, Oct. 12, 1977, Ottawa, Canada. Used with permission.

This was abusing their power and commencing a tyranny." As mentioned, even Alexis de Tocqueville noted, in 1835, the way in which our noble experiment in political freedom might lead to economic totalitarianism. A nonlinear thinker, Tocqueville saw relationships and feedback loops invisible to the linear mind: Equality of political condition would lead to increasing incomes, which would lead to greater demand for manufactured goods, which would require greater division of labor. This specialization would increase the relative differences in income and "mental alertness" between workers and owners, which, in turn, would lead to a manufacturing aristocracy.

Nor was the emergence of the giant corporation and its culmination in the multinational form we see in today's world trading and monetary system envisioned in Marxian theory, either. Marx did envision a stage of capitalist enterprise that would become global (i.e., imperialism) after absorbing all the smaller businesses and petit-bourgeois shopkeepers and merchants in its domestic milieu. However, Marx imagined this would be a *nationalistic* stage, i.e., a period of imperialistic domination and exploitation by capitalistic nation-states. Today, Marxian theorists are dealing with the fact that multi-national corporations are no longer creatures of the states that originally created them with royal grants and charters and legislation limiting their liabilities. Today's giant corporate enterprises (many heavily supported by both capitalist *and* socialist governments!) do not fit into the Marxian analysis any more than they do the market-economic or democratic-political analyses of the West. Transnational corporations have emerged as the most significant and anomalous institution on the planet. Well beyond the reach of the states that created them, but able to influence the policies of the nations in which they are domiciled, they are subject only to the criteria of economic success, maximizing of production, and the interests of their owners and investors (whether private stockholders or, increasingly, taxpayers and government/banking consortia).

For example, when attending the White House Conference on the Industrial World Ahead, in 1972, I heard the now-famous speech by Richard Gerstacker, president of Dow Chemical Company, in which he looked forward to the day when his corporation could elude the controls and laws of all nation-states, including the United States, and domicile Dow Chemical on an island that he hoped Dow would buy, somewhere in the middle of an ocean, out of reach of all states' regulations. Thus, today we see that multinational corporations take no special heed of the needs or policies of their own domestic governments, or domestic workers, and will pull up stakes and move offshore whenever wages and costs can be reduced by moving to less-developed countries with cheaper labor forces and unpolluted environments to exploit. Corporations are also free to dump dangerous or toxic products, such as drugs and pesticides that have been banned in the United States and European countries, in less-developed countries, where the dangers are not known and no regulations exist. An article in *Mother Jones* (October 1979) described some of the more horrifying aspects of this kind of corporate dumping of drugs and pesticides banned in the United States, with the collusion of the State Department and its Agency for International Development. It was to address these international issues of the social and environmental impacts of corporate global activities that I proposed to a meeting of the UN's Institute for Labor Studies, in Geneva, in 1976, the formation of an International Data Bank on Corporate Accountability.[3] The need for such a data-sharing institution is greater than ever; it might serve to give greater disclosure to the dealings of worldwide corporations and subject them to more sanctions of world public opinion until a body of international law can properly oversee their activities.

Therefore, before proceeding it might be useful to try to clear up the fuzziness that leads many business spokespersons to confuse the large modern corporation with "free enterprise" and even with private property. Luckily the American people, according to surveys, still can distinguish the crucial difference between genuine free enterprise and large, bureaucratized corporations operating beyond the classic checks and balances of Adam Smith's requirements for free markets to function as efficient resource allocators: i.e., that buyers and sellers meet each other in the marketplace with equal power and equal information and that no "spillover" nuisance effects be visited on innocent bystanders to the transactions. I need hardly add that such conditions are rarely met in today's industrial economies. Countless surveys have also shown that while public confidence in large corporations and their management has plummeted, there is as much support as ever for free enterprise and private property rights,[4] and indeed there is a resurgence of concern and sympathy with the smallest businesses, genuine entrepreneurship, and the growing num-

Plate 5

*La tierra pertenece al que la trabaja...
the land belongs to those who work it*

poster #1: green, olive & brown

land for the landless

National **Land** *for* **People**

stomach relief

This Land was made for you & me...

poster #2: blue, green & brown

You can help the landless get land and give your stomach a chance to escape corporate control — by supporting National Land for People (NLP).

For years, NLP has been fighting to give small farmers and would-be farmers (including farmworkers) their legal fair share of federally-subsidized irrigation water on Western farmland.

A 75-year-old federal law — the RECLAMATION ACT — supposedly guarantees small farmer preference in purchase of land in these subsidized irrigation projects which have made part of the West (particularly California) the richest farmland in human history.

BUT ... for years this law has been evaded by agribusiness. Evaded by giants like Southern Pacific R.R., Tenneco, Boswell (connected to Safeway), the Times-Mirror Corp., and several oil companies as well as socalled "family farms" that each operate thousands of acres.

NOW ... NLP has focused attention on these land-water scandals. Enforcement bills are being considered by Congress and new administrative policies are being developed by Carter's administration under legal pressure from NLP.

THE ASPIRATIONS OF THE AMERICAN LANDLESS ARE DIRECTLY TIED TO THE OUTCOME OF THIS STRUGGLE...SO IS YOUR STOMACH!

●Reclamation land is scattered throughout the 17 Western States
●30% of the US fresh & processed vegetables are grown on reclamation land
●most of the above in California
●much of this land is monopolized by big landowners
●your stomach will suffer the indignity of food more saturated with chemicals and your pocketbook will suffer the pangs of higher food prices which increasing monopoly will bring if the water law is weakened.

TO WIN FOR THE LANDLESS AND ALL OUR STOMACHS, WE ALL MUST PITCH-IN. The big land companies are spending over $50,000 monthly lobbying in Wash. D.C.

HERE'S WHAT WE CAN DO:
All the gift suggestions listed include a card expressing your concern for your friend's stomach & pocketbook.

For $5: one of the above 3 color, hand screened posters, an NLP newsletter and assorted fact sheets (including a Stomach Connection quiz).

For $10: both posters plus information packet (above)

For $15: the $5 material plus 1 year subscription to monthly newsletter

For $20: the $10 material plus 1 year subscription to monthly newsletter

All gifts listed are tax deductible, but NLP needs non-tax deductible lobbying $ too. Please designate donations for lobbying when you can.

National **Land** *for* **People**

2348 N Cornelia, Fresno, CA 93711
(209) 233-4727

Plate 6

The Corporate Giants have us in their clutches.

And as little people we haven't had the power to fight back effectively.

The Giants are too big.

Corporate ripoffs cost the public over $200 billion a year, a Senate subcommittee estimates. Even the U.S. Chamber of Commerce admits that every year American firms commit Crimes in their Suites totalling $40 billion. Investigations have uncovered union-busting efforts, increasing hazards on the job, inadequate toxic waste disposal and additional air and water pollution.

To protest these abuses we have named April 17, 1980 national **Big Business Day**.

Join the growing coalition of labor, church, consumer, women's, minority and environmental groups in turning around our country in the 80's. Join us in discussing the alternatives—
- consumer co-ops
- credit unions
- small businesses
- Corporate Democracy Act of 1980.

Help us organize teach-ins, film festivals, demonstrations and other activities in your area. Help us make April 17, 1980 the end to Business As Usual.

Big Business Day
1346 Connecticut Avenue, NW
Room 411
Washington, DC 20036
(202) 861-0456

Yes, I want to help fight Crime in the Suites.
☐ Let me know how I can participate in Big Business Day.
☐ I want to be a local coordinator.
☐ Here's my check for _____ to make Big Business Day a success.

Name _____ Phone _____

Address _____

City _____ State _____ Zip _____

BIG BUSINESS DAY April 17, 1980

Among the initiators of Big Business Day are: RALPH NADER (Consumer advocate), JOHN KENNETH GALBRAITH (Prof. Emeritus, Harvard University), WILLIAM H. WYNN(Pres., United Food and Commercial Workers), DOUGLAS A. FRASER (Pres., United Automobile, Aerospace, and Agricultural Implement Workers of America), PATSY J. MINK (Pres., Americans for Democratic Action), JAMES FARMER (Exec. Dir., Coalition of American Public Employees), ED ASNER (Actor), CAESAR CHAVEZ (Pres., United Farm Workers of America), BISHOP THOMAS GUMBLETON (Auxiliary Bishop, Archdiocese of Detroit), JOYCE MILLER (Pres., Coalition of Labor Union Women), ARTHUR SCHLESINGER, JR. (Albert Schweitzer Prof. of Humanities, City University of New York), RABBI MARC TANENBAUM (American Jewish Committee), WILLIAM W. WINPISINGER (Inter. Pres., International Association of Machinists and Aerospace Workers), JERRY WURF (Pres., American Federation of State, County, and Municipal Employees). Institutional affiliation for identification purposes only.

Plate 7

Use This Logo

To Project The Image Of

Your Logo

WHERE IT COUNTS
...in Congress
Advertising Rates and Information on Request

201 Mass. Ave., N.E.   **Washington D.C. 20002**   **202-546-3080**

Plate 8

bers of self-employed, self-reliant citizens.[5] This clarity on the part of voters is encouraging and bodes well for our form of political democracy, although, not surprisingly, it discomfits many corporate managers. Similarly, the U.S. public still appears to understand the difference between the inviolable sanctity of individual, personal property as a bastion of political liberty and an assurance of personal dignity and security, versus the license to hide behind property rights (encouraged by interpretations of the Fourteenth Amendment to the Constitution in ascribing "personhood" to corporations chartered for only limited financial, not social, purposes). As we know, today such agglomerations of property, divorced from the control of stockholder/owners (as Berle and Means established in the 1930s),[6] can often deny or conflict with the property rights of individuals. State-controlled companies such as Italy's Istituto per la Recostruzione Industriale (IRI) provide similar cases of centralized economic power. IRI controls major banks, half of Italy's steel output, state broadcasting, and most Italian highways, through a maze of six hundred subsidiaries—making it the European Common Market's largest employer.

This brief clarification is necessary because all late-stage industrial societies are going through an inevitable transition due to the very success in the past two hundred years of the industrial revolution in maximizing labor productivity and the GNP-measured growth of institutionalized, monetized economies (as opposed to total productive societies). This economic transition involves an inevitable shift from economies that maximize rates of production and consumption based on nonrenewable resources, to economies that minimize such wasteful rates of throughput of energy and materials and will be based on renewable resources and managed for sustained-yield, long-term productivity. The symptoms of this great transition, as mentioned, include increasing rates of inflation, structural unemployment, the failure of macroeconomic management, and growing tax revolts. The transition also marks the end of the age of Keynes and the reliance on theories of stimulating aggregate demand so that greater consumption by the more affluent would "trickle down" to benefit the poorer groups by increasing sales, profits, investment, and jobs in the private sector. The largest industrial enterprises are structured and dependent upon this type of "trickle-down," resource-intensive growth, substituting capital and large-scale technologies for labor.

This type of economic growth worked as long as cheap, abundant energy and resources were available or could be imported. However, in an age of higher global expectations and of understandable demands from resource-rich, less-developed countries (LDCs) for a New International Economic Order, cheap resources are now denied to the newly vulnerable industrial countries, which all face an unavoidable period of retrenchment.

During the transition, however, I expect that the emerging renewable-resources sectors of industrial economies will continue their rapid growth and provide both safety nets and bridges to the future (including their emerging solar energy, wind power, biomass, intermediate-scale hydropower, ocean thermal power, waste management, recycling, and electricity cogeneration). Similarly, the emerging countereconomy, based on the newly localized and regionalized efficiencies of scale dictated by higher energy prices will continue to grow. This countereconomy is the inevitable outgrowth of the exhaustion of the logic and possibilities of maximizing the institutionalized, monetized, GNP-measured economy at the expense of what Scott Burns calls *The Household Economy* (1977) and James Robertson, author of *The Sane Alternative* (1979), calls the informal economy, the more reciprocal, convivial, localized economy of use value, rather than exchange. In *Creating Alternative Futures,* I describe the limits of the attempts to maximize this institutionalized, monetized economy as involving a set of fundamental societal trade-offs: between the division of labor and specialization on the one hand, and the inevitable transaction costs in coordination, communication, centralized organization, and bureaucratization (both public and private) that would result, on the other hand. My thesis has now been given some unexpected support in a recent paper by Belgian information theorist Jean Voge, "Information and Information Technologies in Growth and the Economic Crisis" in *Technological Forecasting and Social Change* (1979). Voge verifies that the logic of efficiencies of scale in production are meeting diminishing returns and bogging down in the even larger information and coordination costs they incur, resulting in increasing bureaucratic sectors. An example of these costs involves the spiraling expenses of operating the European Common Market, where language translation alone costs $20 million annually.[7] Voge demonstrates what E. F. Schumacher and I had asserted: that when industrial economies reach a certain limit of centralized, capital-intensive production, they will have to shift direction to more-decentralized production technologies and decentralize economic activities and political configurations, using more laterally linked information networks, if they are to overcome the severe information bottlenecks in excessively hierarchical, bureaucratized institutions.

I referred to this change of direction as a scenario of "spontaneous devolution," in which citizens simply begin recalling the power formerly delegated to politicians, administrators, and bureaucrats, and the power they delegated to business leaders to make far-ranging economic, production, and technological decisions, as Charles Lindblom explores in *Politics and Markets: The World's Political and Economic Systems* (1977). The growth in all industrial countries of citizen movements, the gathering tax revolt, the drive for worker self-management, the growth of the human potential movement and holistic health, demands for humanly scaled, "small is beautiful" technologies, the resurgence of libertarianism, and the demands for autonomy being made by indigenous ethnic peoples in socialist and capitalist countries—all are part of this new "spontaneous devolution" of old, unsustainable structures. Decision makers in the old institutions lose their confidence, and bureaucracies become checkmated as the "dinosaur" stage is reached. Happily, unlike the hysterical Mobil Oil "image" advertising, it is never a matter of growth versus no-growth but, rather, what is growing, what is declining, and what must be maintained. Thus, neighborhood- and community-based enterprises are mushrooming as viable alternatives to both multinationals and large state-operated enterprises both in the Western countries and in the Eastern bloc.

The rise of worker-owned, self-managed enterprises, and of bartering, sharing, self-help, and mutual aid is documented by the Institute for Local Self-Reliance and the Cooperative League of the U.S.A., and in *Mother Earth News* and other journals. An economy based on renewable resources carefully managed for sustained yield and long-term productivity of all its resources can provide useful, satisfying work and richly rewarding life-styles for all its participants. However, it simply cannot provide support for enormous pyramided capital structures and huge overheads, large pay differentials, windfall returns on investments, and capital gains to investors, nor finance the overblown executive stock options and already extended pension liabilities, the monstrous office buildings, corporate jet aircraft, country-club memberships, art collections, and other "perks" subsidized in the receding era by cheap fossil fuels and resources. This overhead has been masked by the convenient economic theories of "externalized" costs, passed on to taxpayers, society at large, the environment, or future generations. Nor, I would add, can renewable resource-based economies sustain massive nuclear arsenals and permanent war economies or costly space adventures.[8]

In the transition, we will simply have to stretch capital, energy, and resources further, cutting the some 50 percent of waste from our energy system and combining our precious capital with more productive people in smaller, flatter-structured enterprises that liberate human initiative. The question of world competition arises. But we must remember that all industrial societies, both market-oriented and centrally managed, are experiencing similar stresses, and the United States, with the richest and most wasteful economy, is in the most advantageous position to cut out flab without cutting into muscle. Most of this flab is at the top, in organizational overhead—not at the bottom. Meanwhile, the demands of tomorrow's labor force for more opportunities for personal development and job satisfaction most often favor the small company, which can be more democratically managed and can release human potential and productivity through greater identification with enterprise and motivation. Indeed, our greatest future productivity gains will come from learning to trust people to do a good job.[9] Already, there is heightened interest (due to the worldwide interest in human rights, perhaps) in the question of civil liberties for employees, described by David Ewing in *Freedom Inside the Organization* (1978).[10]

All this is not to say that large corporations will fade away. They will not. We will still need to pour steel and aluminum and maintain telephone systems and electrical grids. But some of these systems may already have reached an optimal size, and the growth of newer sectors of the economy may better satisfy new needs in wholly new ways for which the old corporations and their existing technological configurations may be quite unsuited. Inevitably, therefore, the functions of public and consumer affairs will continue but in the larger, most obsolescent companies and industries they will increasingly involve educating *management,* rather than consumers and voters, and the usual lobbying of government officials.[11] Most corporate leaders were acculturated during the now receding age of petroleum, and they have not yet grasped the fact that the socioeconomic transition to the solar age will require an economic paradigm shift, i.e., a shift in their entire world view. This shift will involve replacing the linear, static logic of market-equilibrium economics with a much more realistic, general-systems view of the larger social and ecological *contexts* of management's decisions. The most fatal flaw of traditional, "flat-earth" economics involves its assumptions of an equilibrating economic system (such as actually existed during the eighteenth-to-nineteenth-century era of Adam Smith). Today, due to the "fine tuning" of a generation of activist economic policy makers and to the development of ever more complex, capital-intensive, socially and ecologically disruptive technologies, we now have created industrial systems that are in chronic states of *dis*equilibrium. In addition, they are linked globally, riding on the same international roller coaster of today's world trade and monetary systems. Yet business leaders, while dealing every day with these realities, still heed advice from economists who believe that such complex dynamic economies are still analogous to simple hydraulic systems of levels of total supply and demand and that "market forces" can equilibrate such nonlinear, interlinked societies with their thousands of interacting variables. The simple kit bag of economic tools offered by the profession, whether monetarists or Keynesians, econometric modelers, supply-siders or neoclassical traditionalists, is now inadequate to the task of managing mature industrial societies. Indeed, managing the decline of their unsustainable sectors may require specifically nonmonetary approaches, as discussed earlier.

Corporate leaders, still guided by such bankrupt economics, continue to exhort us to simply turn the clock back and deregulate the economy, without acknowledging that the economic processes, such as the two-hundred-year development of the industrial revolution, are not reversible but are evolutionary transformations, as Nicholas Georgescu-Roegen shows in his *The Entropy Law and the Economic Process* (1971). Massive corporations inevitably create equivalently large government infrastructures to coordinate and regulate them, not to mention huge disequilibrating global flows of capital between countries, as described in "Stateless Money" (*Business Week,* August 21, 1978). For example, can one imagine the auto industry having attained its preeminence without the interstate highway system, bridge building, or the provision of driver- and vehicle-licensing bureaucracies, not to mention the staggering costs of traffic-police systems? Likewise, how does one repeal the tax-supported airport and traffic-control system that underpins the airline industry, or the Federal Communications Commission, which must, whatever the outcome of current legislative debate, attempt to coordinate the use of the electromagnetic spectrum and communications satellites and deal with licensing CB radio, ham operators, etc.? It is also axiomatic in economic development theory that LDCs cannot hope to emulate Western-style industrial expansion without the necessary *government-created* infrastructure: roads, railways, telephones, trained bureaucrats, public education for basic literacy, sanitation, seaports, and airports. So the growth of government is always symbiotic with the growth and scale of private enterprise. Furthermore, a recent study, by a policy group at the Massachusetts Institute of Technology, of government regulations in five countries showed that these regulations had stimulated, rather than stifled economic growth.[12]

Another mistaken argument heard from corporate leaders is that the decline in our nation's and most other industrial societies' "productivity" is due, first, to greedy workers and, secondly, to environmental and OSHA regulations and consumerism. However, even *Business Week* editorialized (August 21, 1978), "Wage settlements are not the primary cause of inflation." In Chapters 9 and 10, I critique the January 1978 study for the Commerce Department by Edward Denison (which purported to show that due to "productivity declines" caused by environmental and OSHA regulations and rising crime, the U.S. GNP had lost some $40 billion), and point out that these are not the primary reasons for the "productivity" decline we are experiencing. First, we must distinguish that what is being equated with "productivity" is more accurately termed "labor productivity," since increasing labor productivity has been the main thrust of the industrial revolution, i.e., placing more machines and energy at the disposal of each worker, via automation. Now, as we undergo the transition phase in which capital, energy, and resources are scarce and we have automated large segments of our population in all industrial societies into the ranks of the structurally unemployed, we must change our "productivity" measures to reveal that it is the declining productivity of our *capital investments* and *energy/resource utilization* that is a new cause of our overall decline in productivity. Another problem inherent in Denison's definition of productivity as measured in terms of output per unit of input is that it is based on the historical ability of producers to externalize costs to consumers, taxpayers, and governments, or pass them on to future generations as with the costs of decommissioning nuclear reactors.[13] Therefore economists have simply overstated productivity for decades, due to their externalizing all these social costs, which are all coming due as social bills to be paid.

Corporate leaders follow the economists' faulty measures of productivity and its decline to argue that this is one of the basic causes of inflation. They are partly right, but for the wrong reason. Many business leaders proceed from this to assume that this productivity decline is also the culprit in the decline in our rates of technological innovation in the U.S. economy. As described earlier, today's inflation rates are best understood from beyond the vantage point of economics: from the general-systems-theory view of the systemic increase in social complexity, transaction, and information costs, and the unanticipated social-impact costs of private-sector activities, all added to the GNP (rather than subtracted), while additional inflation, viewed from the thermodynamic standpoint, is due to the declining quality of our energy-and-resource base. Raising prices and hurling ever more precious capital into "calling forth more supplies" will further reduce the net yields from extraction and bring further declines in the productivity of such capital investments, for example the Synfuels Corporation, set up in 1980, will sink $20 billion into dubious coal gasification and liquefaction research and other low net-energy-yielding oil shale projects. Similarly, some electric utilities are still constructing huge nuclear power facilities to meet demand that may never materialize, since in 1980 overall reserve capacity averaged 40 percent.[14]

Now let us turn to the corporate leaders' fears about "declining technological innovation" and the United States' losing its technological lead in the world. We cannot continue to lump together apples and oranges under the rubric of technological innovation, such as, for example, the real technological innovations being made in the electronics industry that are increasing the capacities of microprocessors, and the introduction of a new cigarette brand, patent medicine, cereal, or hair shampoo. We must also remember that some technological innovations have value and add to our wealth; others are better described as "illth," or disservices such as junk phone calls, oversugared cereals, and cigarettes. Many technological innovations create unanticipated social costs and impacts, creating or destroying jobs, causing population shifts, or creating ubiquitous health problems (such as the many industries based on highly carcinogenic compounds that erupt after many years in cancer and other pathologies). Technology assessments and values-clarification exercises will be necessary before the study of technological innovation can expect to shed any new light on these issues. A case in point was President Carter's Domestic Policy Review, on which I was asked to serve in 1978. I declined, based on the insuperable misdefinition of innovation issues in its mandate. The report was released October 31, 1979, and simply restated the old business and economic catechism about the need to increase productivity and give more tax breaks to business for more capital investment—also favored by Reagan. Not surprisingly, many corporate leaders, confused by such economic paradigm shifts, are now spearheading the charge for lowered taxes.[15] The Republican Party (in the past known for accusing Democrats of fiscal irresponsibility) based its 1980 election campaign on the tax-revolt issue, and its policies will be highly inflationary.

William Niskanen, the chief economist of the Ford Motor Company; the University of Chicago's Milton Friedman; and the University of Southern California's Arthur Laffer are leading economic propagandists behind this tax-cut drive and business-supported legislative proposals to lower capital-gains taxes. Corporate leaders, bankers, and investment advisers and their economists have stepped up their speaking, lobbying, and proselytizing on the tax-cut/government-spending issues, since so many of them support the Republican Party. Public vigilance of corporate advocacy advertising and grass-roots lobbying is essential.[16] We must hope that business spokespeople will take time to rethink their ideological positions on some of these issues, as well as reassess their own corporate future options. Meanwhile, corporate managers have the prerogatives of their corporate-image advertising budgets and lobbying capabilities to propagate their own political and economic views (not necessarily those of their stockholders). One can gauge the clout of corporations in this marketplace of political ideas from Mobil Oil's recent "editorial" ad campaign and its sponsorship on public television of the Ben Wattenberg series "In Search of the Real America" (which has over-simplified, slanted, and distorted many important public policy issues, including claiming that there was no energy crisis at a time when President Carter was trying to warn the country of the need to look ahead and grapple with it). By contrast, Mobil behaves very differently in France, where the government's Energy Conservation Agency, created in 1974, has vigorously intervened to purge oil-company ads of material urging greater energy consumption and is empowered "to prohibit all advertising which is of a nature to favor an increase in the consumption of energy" (*Fortune,* July 17, 1978). In France, Mobil emphasizes its Mobil Economy Run, and that country achieves its similar per capita GNP with only half as much energy as we do in the United States, where we import 50 percent of our petroleum each year.

Therefore, I believe that additional safeguards against such use of stockholders' assets for political advocacy and grass-roots corporate lobbying are needed. I commend the innovative testimony of Professor S. Prakash Sethi, of the University of Texas at Dallas, before the Sub-Committee on Commerce, Consumer and Monetary Affairs, of the House Government Operations Committee (July 18, 1978). Professor Sethi, who has authored several books on corporate/consumer media/government issues, suggests several ways in which the image- versus advocacy-advertising issue can be codified, how grass-roots lobbying can be regulated more efficiently, and how the right of access to media coverage and counteradvertising for consumer groups' views can be assured. Sethi underlines an important legal distinction between the corporation's right to free speech and the management's right to speak for the stockholders. He proposes that managers legitimate their views in behalf of the corporation by specifically soliciting the views of stockholders (not *stockholdings*) by means of proxies, since, as he adds, "unlike property rights, political rights are *not* subject to trade and transfer." The problem of access to media for consumers is exacerbated by the natural biases of the business press, mindful of their advertisers' interests. For example, *Fortune,* in an article on the "Backlash Against Business Advocacy" (August 28, 1978), while pointing out that there was a lack of symmetry in allowing corporations and their trade associations to deduct lobbying expenses while foundations and consumers groups have less deductibility and individuals none, nevertheless justified the situation thus: "But it can be reasonably argued that most corporate grass-roots lobbying is a legitimate business expense, since it usually has to do with issues directly affecting the profitability of the company." What *Fortune* failed to point out is that this assumes that free markets are functioning, while in truth a large, powerful corporation can increase its profitability *at the expense* of consumers and taxpayers.[17]

Some corporate leaders are honest enough to admit that corporate agendas may not be coterminous with the public interest. Others are honest enough to admit that their lobbying and calling for tax cuts without government spending reductions is sheer irresponsibility, as launched in the 1978 Kemp-Roth bill and supported by Ronald Reagan. Even Alan Greenspan, former member of the Nixon Council of Economic Advisers who supported Kemp-Roth as a last-resort measure to cut the growth of government, refutes the contention that large tax cuts would increase economic activity and eventually make up for the loss in higher revenues. This is the reasoning behind Arthur Laffer's now-famous Laffer Curve, which promised such an unlikely pot of gold at the end of the rainbow to the gullible California voters for Proposition 13. Harvard economist Martin Friedman dismisses the theoretical principle behind the Laffer Curve, i.e., that, at *some* point, reducing tax rates increases tax revenues, as "something we teach in the first week of the course on public finance." Most economists agree that the key question is *at what point* on the curve, and even Laffer can't say where we are on what can only be described as his Laffable Curve.

A constitutional convention to limit federal spending would be an even more disastrous blunt instrument than Proposition 13 (40 percent of those who voted for it did not expect that any public services would be cut, while 60 percent of the tax relief went to corporations, rather than individuals). Kemp-Roth seeks to reduce everyone's federal taxes by one third over three years, and after it was roundly rejected in 1978's midterm elections, even *Fortune* (December 18, 1978) described it as a major Republican blunder, since polls showed that Americans favor most social and environmental programs of government, as well as the fact that it would have cost some $124 billion in lost tax revenues.[18] Yet Kemp-Roth-type tax cuts are the keystone of Reaganite economics. The effects of Kemp-

Roth forecast by Data Resources, Inc., showed that if no compensatory spending cuts were made, the result would be to increase the federal budget deficit to $100 billion by 1983—a figure likely to be exceeded sooner.[19] The fallacy of simple tax-limitation and tax-cut proposals lies in the fact that they do not have the power to restrain local, state, or federal government from deficit spending. The federal government can deficit-finance, essentially using the money-printing presses or its new powers under the Monetary Control Act, and local and state governments can issue bonds and incur deficits in other ways. But the real danger and demagoguery are in the dishonesty in most of the debate so far about what programs are to be cut.

The hidden agendas, as always, involve whose ox is to be gored, and when corporate managers go public with their views on these issues, it usually depends on the business of their corporations. Construction companies don't want the cuts to be made in highway, bridge, and dam-building projects; oil and gas companies don't want to give up tax credits for intangible drilling costs; shipbuilding firms don't want to give up maritime subsidies; and most of all, aerospace and military-contracting firms don't want to see cuts in the Department of Defense or NASA's budget. We heard the howl that arose from such quarters when President Carter announced his veto of a $36-billion weapons-procurement bill due to the inclusion of the $2-billion nuclear aircraft carrier a Washington *Post* editorial called "an expensive pleasure, recreational vehicle for status-seeking admirals."[20] J. Fred Bucy, president of Texas Instruments, called in a typical speech for a return to "free enterprise,"[21] while his company is one of the country's largest government contractors. Mr. Bucy did support government spending cuts to compensate for tax cuts, and he was also honest about what he wanted to see cut: arguing to spare the Department of Defense and urging that the cuts be made in the budget of the then Department of Health, Education, and Welfare.

Thus, it will be necessary for honest politicians, business leaders, and those in labor and the voluntary sector to force this vital discussion of the priorities issue, and clarify exactly what is to be cut: whether the nuclear aircraft carrier, or education, day care, public-service jobs, food stamps, and school lunches. In this effort to make the debate about tax and budget cuts more honest, former Senator Edmund Muskie, of Maine, when he was chairman of the Congressional Budget Committee, fired the first legislative salvo to counter the tax and budget limiters. Muskie introduced the Sunset bill, which even *Forbes* of August 21, 1978, described as by far the most honest and realistic approach, which would mandate that all federal spending programs would automatically close down within ten years unless Congress specifically voted to extend them. It is illustrative of the "whose-ox-will-be-gored" nature of the whole tax-cut/government-spending issue to see those who first lined up in support of Sunset: the White House; the U. S. Chamber of Commerce; Senator Edward Kennedy, a Democrat; and Senator Robert Griffin, a Republican; and who was opposed: the National Association of Manufacturers; the Business Roundtable; and Senator Russell Long, chairman of the Finance Committee; who all avowed support for reduced federal spending but worried about their own special prerogatives, such as investment tax credits and deductions of intangible oil- and gas-drilling costs.

It is a familiar story of *priorities,* and as always, one interest group's tax "reform" is another's financial ruin. We must acknowledge that we no longer have much of a free market system left in the United States or in any other mature industrial society. They are now better described as mixed economies of legislated, rigged markets; vast taxing and transfer systems involving investment tax credits, incentives, depletion, and depreciation allowances; subsidies; rebates; research and development funding; demonstration grant programs; and price supports, as well as the more-often-singled-out welfare and food-stamp programs. Thus the oil, natural-gas, coal, and nuclear-energy industries, whose subsidies, as noted earlier, amount to approximately $134 billion, now fight similar subsidies to solar and renewable energy on the grounds that they should "compete in the free market"!

Now, creating markets via legislation is a perfectly sensible thing to do to accomplish all kinds of policy objectives in democratic societies. Indeed, the legislative action that created the first major "free market system" of resource allocation in Britain, some three hundred years ago, was a major social innovation. As discussed in *Creating Alternative Futures* and Chapter 7, Karl Polanyi showed in his *The Great Transformation* (1944) and *Primitive, Archaic and Modern Economies* (1968) that free market systems of resource allocation are a rare aberration in the history of human societies, which characteristically have used two other major resource-allocation systems: *reciprocity* and *redistribution*. We must now clarify the economists' obfuscations and admit that the "invisible hand" is our own and that markets are our servants, not our masters.

As the economic transition continues to engender all kinds of rolling readjustments throughout the 1980s, I expect the old "flat-earth" economic paradigms to adjust to the new, unavoidable, visible realities. Meanwhile, public debates among leaders will be very confused and confusing to the electorate. Leaders in all industrial societies will continue to sound very much the same, debating the sterile, "flat-earth" politics of Left and Right.

Today neither will work, and we will need to call in experts from many other disciplines: political science, psychology, sociology, biology, thermodynamics, physics, and general systems theory if we are to construct more realistic models of these societies in all their dimensions. We must also not expect the emergence of a new political consensus until we can call forth some more explanatory leadership, to describe the transition that is occurring, and to promulgate viable alternative futures based on the more realistic, diversified, decentralized economies of renewable resources and sustainable-yield productivity of the dawning solar age.

During the turbulent transition of the 1980s, all the issues I have mentioned are key to the survival of our democratic system and to assure that free enterprise remains a vital component. The bankruptcy of existing economic theory is now obscuring the public debate about *what is valuable* under changing conditions. The limits of macroeconomic management must become a national issue. The debates between corporate leaders, consumers, labor leaders, and government now involve the very rationality of corporate decision making, capital allocation, and technological innovation under changing conditions; the shouting must subside and the reasoning together must begin. For example, one corporate leader sets an example: Rudolf W. Knoepfel, president of Solvay American Corporation, stated recently, "Western-based companies can no longer wheel and deal on a global scale, setting prices and conditions to serve their own advantage, using the LDCs mainly for their raw materials and cheap labor." He goes on to say that he does not mean to be unduly critical of multinational corporations "but that *times have definitely changed*"[22] (italics added). On such reassessments of the real world, new dialogues can be built. A similarly frank statement was made by former Secretary of the Treasury Michael Blumenthal, himself a Ph.D. in economics: "I really think the economics profession is close to bankruptcy in understanding the present situation—before or after the fact."[23]

However, for decades, business has commissioned studies to "prove" that even very large corporations with oligopolistic control over their markets can still produce "efficiently." The idea has been to purvey the notion that great economic size, even extreme concentration, was still okay in spite of the theory of neoclassical economics that free, competitive markets were conditional on large numbers of buyers and sellers meeting each other with equal power and equal information and that only insignificant spillover effects were imposed on innocent bystanders.

Economists at the University of Chicago and elsewhere have asserted in numerous studies that big companies can be competitive—even if only with each other or where their opposing product divisions slug it out in the market. Such endless academic redefinitions of what constitutes a "market," what is "competitive," as well as "economies of scale" and "efficiency" will, no doubt, continue to provide professors with grants from corporations and pro-business groups for years to come. The basic business position on antitrust issues is already well known; it is summed up by University of Chicago economist Yale Brozen as follows: "In essence, any member of the top 200 [corporations] is there because it uses resources more productively" (*Fortune,* March 26, 1979).

One might ask, "So what about Chrysler?" Or one might ask a different kind of question, implicit in the antimerger bill introduced by Senator Kennedy in 1979. He explored the question of concentrated economic power from another and even more important viewpoint (overlooked by economists simply because economics is not the discipline that deals with it): "What is the social impact of economic concentration and what effects does it have on a democratic society that values decentralized decision making?" The Kennedy bill addressed these issues by calling for limits on mergers and conglomeration, some of which are not much more than pyramiding assets, where nothing productive is added—in fact, the "product" has simply become more money.

Thus, even if one *conceded* (which I don't) that economic concentration delivered the goods and services to consumers more *economi-*

cally and efficiently, this still would not address the *social* trade-offs and disruptions inherent in the decisions of very large corporations: to relocate, to shut down a plant, to develop a dangerous technology, or simply to make managerial mistakes on the scale that Chrysler did. The bigger the company the more people get hurt. Incidentally, this is the fallacy in the vaunted "efficiency" of the national decision-making system of business and government known as Japan, Inc. They can go down the right road very efficiently—and down the wrong road just as efficiently through what systems theorists call "loss of feedback." Decentralized decision systems may look untidy, but there is lots of feedback in all the diverse views and groups, which can help prevent spectacular disasters.

Thus it is vital, as the U.S. economy enters its mature stage, that we shift traditional antitrust concern to such larger, social-impact issues, just as Congress did in 1973 when it set up the Office of Technology Assessment to look at the social impacts of technology. We are at last beginning to see the subtle threats of creeping corporate bureaucratization and corporate welfare statism in loss of social efficiency and flexibility. We see the social costs, waste, and resource depletion that come with economic concentration. The very capital-intensive and energy-intensive technologies (whose scale, by definition, systematically destroys free markets' functioning) that have justified the expansion of corporate scale in order to manage them, now have created their own problems. We faced them all in dealing with Chrysler, which was overcommitted to producing large, materials-intensive, gas-guzzling cars, which *historically* had been very profitable. Chrysler even tried a crash program of *voluntary* divestiture, but it was too late. The society now faces the consequences: hundreds of thousands of workers, whole communities and economic subsectors too dangerously dependent on the decisions of petty, unelected managers, unaccountable to the taxpayers now forced to bail them and their stockholders out. The price of the bail-out was equally high for the workers who pledged their pension funds to Chrysler; it remains to be seen if their president, Douglas Fraser's, seat on the company's board will be worth its cost.

We can also learn from other mature industrial societies, notably Britain, which has been facing them for a decade in unsuccessful attempts to bail out overgrown, obsolete corporate lame ducks. Britain might have saved some of these industrial dinosaurs from their own follies and overexpansion, or even saved the hard-pressed British taxpayers from the supersonic Concorde, itself a typical product of overcentralized economic decision making. In fact, the European Economic Community has now begun to crack down on state-owned monopolistic companies on antitrust grounds.[24] Thus large corporations today, while seemingly impervious to political control by host nations or the reach of international law, are now changing from within in many new ways, including the increasing accommodation of workers and other outsiders on their boards. Mark Green, Ralph Nader, and Joel Seligman have outlined many internal reforms, such as national chartering of companies, in *Taming the Giant Corporation* (Norton, 1976).

Today we are seeing in the increasingly turbulent corporate environment nothing less than a challenge to the Divine Right of Management. We see the phenomenon not only in the external challenges to large, bureaucratized corporations and other institutions by consumers, environmentalists, civil-rights and women's organizations, and those demanding greater corporate accountability, but also from within these institutions themselves.

As more previously disadvantaged group members enter the structure of these institutions, large numbers bring their own group goals with them. Thus, for example, many women, having entered the corporate world, do not simply strive for the same upward career paths as their male counterparts have—but often sacrifice their own career goals for solidarity with their sisters and efforts to change the corporate structure from within. For example, they petition for day-care facilities, flexible work schedules, part time and job sharing, better health benefits, and education opportunities.

Another behavior pattern emerging from these newly incorporated disadvantaged groups is their "networking" activities with others, across corporate boundaries. I pointed to the likelihood of this phenomenon in the 1960s as a probable response to their psychological pain and alienation as "token" members of their group or "window dressing"—with no legitimate roles in the corporation other than to satisfy public opinion or legal statutes. For example, in the United States, the new black employees who were usually designated as either "managers of special markets" or as "urban affairs officers" quickly sought solidarity and psychic reassurance by forming *ad hoc* intercorporate associations (most often without the funding or the blessing of their institutions) and shared their frustrations at conferences and via newsletters.

The frustrations of such new personnel stemmed from this *ad hoc* response—in which they were hired to fend off the new social forces, rather than in recognition that nothing short of altering the corporation's own internal structure and behavior would address the grievances. People hired to deal with these emerging issues, therefore, were not only given no effective tools or mandate—they were often shunned by top management as the bearers of bad news—but were often treated as "pariahs" by their peers managing more traditional, "bottom-line" corporate functions.

Similar new networks were established by corporate women, environmental-control officers, consumer representatives, and, indeed, any new functional group set up by corporate management to respond to social pressures. In the 1970s, these networks began linking up with many citizen networks, working for corporate accountability, consumer and environmental protection, human rights, and economic justice. This "network model" of social change is exciting great curiosity today, and much superficial, faddish reporting. This new organizational form is already functioning widely in all maturing industrial countries and is visible to those whose perception is attuned. Networks are metaphysical organizations, and their participants describe themselves as "networkers." They have no headquarters, no leaders, no chains of command, but are free-form and self-organizing, composed of hundreds of autonomous, self-actualizing individuals who share similar world views and values. Political analysts, including Virginia Hines, Byron Kennard, and Jessica Lipnack, have described them, but no organizational theorists have yet captured their dynamics, because they ebb and flow around issues, ideas, and knowledge. Many hundreds of thousands of these networks exist today in and between media-rich, industrial societies, although quantification is impossible, because such spontaneous organic forms elude outside observers, who create "static" on the lines. This is instantly picked up by the participants, who then regroup, using alternative channels. Their chief product is information processing, pattern recognition, and societal learning.

Networks can now create a recognizable, media-reportable, national event expressing grass-roots interest in a political issue in a matter of hours, as the antinuclear movement has shown. Communications-rich industrial nations now require this kind of instant political signaling system to their decision centers in order to overcome bureaucratic inertia and hardening of political arteries. Networking crosshatches all existing structured institutions and links diverse participants who are in metaphysical harmony. It is a combination of invisible college and a modern version of the committees of correspondence, which our revolutionary forebears used as vehicles for political change. Luckily, networks are linked by the mimeograph machine, the mail, the telephone, and user-activated, small computer-based conferencing systems, all decentralized technologies accessible to individual users, with constitutional guarantees of privacy. This "network" model of self-organizing social-change activity provides a new model for understanding how institutional change is possible, even in the case of today's mammoth corporations and sluggish, Kafkaesque bureaucracies. Thus maturing and aging institutions, as well as industrial nations, may have produced internally the means for their own transformation.

Public relations, advertising, and lobbying are key information-structuring activities of organizations which are functional in the early, growth phase, but as size increases, the increased ability to distort information and screen out feedback eventually becomes dysfunctional. The transmission of information is brought to a fine art in public relations and advertising, but (as with anyone who is talking all the time), listening becomes more difficult and learning ability is impaired. Thus the laws of nature, which hold for all biological species, hold for humans: the very attempts to grow and dominate more variables in the immediate environment eventually become self-defeating, because this leads to loss of feedback and, consequently, maladaptation.[25]

This basic evolutionary law that "nothing fails like success" is the mechanism that keeps the total ecosystem or human society in homeostatic balance. It eventually checks overgrowth of subunits that have reached the dinosaur stage and prevents diseconomies of scale while encouraging diversity, experimentation, and continual learning and adaptation of the whole system to change. In fact, human societies and the subsystem we call our "economy" are continually evolving, operate within the basic laws of physics, and conform to the evolutionary processes of growth and decay, ordering and disordering; i.e., the syntropy/entropy cycles of all natural systems. Just as the decay of last year's leaves provides the humus for the new growth of the following spring, so if some parts of our evolving economy are to grow, other sectors must decline, releasing their components of capital, land, and human talent to the growing sectors. Thus networks, countercultures, and dissident views are always vital for societal and institutional renewal.

I suggest that all these new efforts to restructure large, bureaucratized institutions from both within and without are vital adaptive feedback mechanisms to the growing unmanageability of large institutions, about which I and many others have written, including the report of the Trilateral Commission in 1975 on the *Governability of Democracies*. Soviet dissident Vladimir Solovyov points to similar problems in the U.S.S.R., whose outward appearances of strength, he asserts, "are merely weaknesses in disguise" ("Who Sees Russia's Feet of Clay?", *The Christian Science Monitor,* October 26, 1979).

Another study of this crisis of unmanageability was published in 1977 by the Stanford Research Institute for the Office of the President's Science Adviser. Some models of the growth of social systems were compared, including a biological growth model (growth leading to stabilization) and an economic growth model (growth leading to diminishing returns). Interestingly, they did not include discontinuous, morphogenetic models of growth such as those designed by Magoroh Maruyama, or René Thom in his "catastrophe" theory, or Ilya Prigogine in his pulsating model of systems that achieve "order through fluctuation," discussed in Chapter 11.

Nevertheless the Stanford Research Institute study is useful, since it synthesizes four stages of growth of social systems or institutions:

STAGE I: *High Growth (Springtime) Era of Faith.*
In the "springtime" of growth, the relative level of systems comprehension is high, and the scale, complexity, and interdependence of the bureaucracies are low. There is a strong faith in the efficacy of shared values and goals. . . . The social leaders have considerable legitimacy, and the high performance of the system speaks of unbounded potentials.

STAGE II: *Greatest Efficiency (Summertime) Era of Reason.*
In the "summertime" of growth, the relative level of systems comprehension is moderate, and the scale, complexity, and interdependence of the bureaucracies have increased substantially relative to the ear-

lier period. . . . The level of systems performance is still increasing, but the bursts of vitality of Stage I have been replaced by a more methodical planning and implementation process.

STAGE III:　*Severe Diseconomies (Autumn) Era of Cynicism.*
In the "autumn" of growth, the relative level of systems comprehension is low and dropping rapidly as large, barely comprehensible bureaucracies have grown to largely incomprehensible supersystems. As leaders disavow their responsibility for error and maximize the visibility of their own increasingly modest achievements, the system's constituency becomes increasingly disillusioned, apathetic, and cynical. Both faith in the basic soundness of the system and trust in rationality to solve the mounting problems is virtually exhausted. Leaders are more tolerated than given active support and legitimacy. . . . Decision makers are increasingly unable to cope with complex problems that demand superhuman abilities. Costs and problems of coordination and control are mounting rapidly, and the benefit to the constituency seems to be declining with equal rapidity; consequently, people are less willing to support the actions of the bureaucracy.

STAGE IV:　*Systems Crisis (Winter) Era of Despair, then . . . ?*
In the "winter" of growth, the relative level of systems comprehension is minimal. The systems are on the verge of chaos and collapse. There is a rapid turnover of leaders, prevailing ideology, and policy solutions—yet nothing seems to work. Every attempt at creating order (short of a highly authoritarian structure) seems overwhelmed by growing levels of disorder. The level of systems cohesion is very low and, in turn, exacerbates the problem of the system's leaders, who govern virtually without support. The rigidified bureaucracy is made somewhat more resilient by the rapid turnover of personnel and policy, but the vulnerability of the system is so high, and mounting crises are of such seriousness, that whatever additional resiliency has been added to the system is quickly depleted in a grinding downward spiral into bureaucratic confusion and chaos.

The report then examines four likely responses:

1. *Muddling through.* The incrementalism and chronic inertia we see today in many oversized institutions (and mature industrial societies).

2. *Descent into chaos.* Increased employee turnover, with demoralized leaders frantically co-opting new ideas and creative people, thus aborting new, alternative institution-building.
3. *Authoritarian response.* Unsustainable rigidity—greater rationalization of human behavior—increasing vulnerability—withdrawal of loyalty and societal mandate.
4. *Transformational change.* The transforming of internal structure and goals.

My own model is the "Spontaneous Devolution Model," which involves some of the effects I described earlier, e.g., the spontaneous actions of external forces such as citizens' movements modifying the institution's behavior and changing its social mandate.[26] Simultaneously, the activities of employees and members of network organizations lead to internal restructuring, possibly aided by development of worker self-management programs or demands, and increased cross-institutional associations.

The underlying model of my "devolution scenario" is that of "self-organizing systems" (i.e., organic, and biological, systems incorporating both positive and negative feedbacks and displaying behavior modes both of deviation-damping morphostasis and deviation-amplifying morphogenesis). As the crisis of unmanageability unfolds, individual and group responses include:

1. Increasing demands for citizen participation at all levels of decision making.
2. Rapid societal learning via citizen movements (that are adult-education-based on the "each one teach one" model).
3. Proliferation of heterarchical (as opposed to hierarchical) communication in, for example, citizen-based media, newsletters, telephone trees, study groups, consciousness raising, and networking of new perceptions both within and across existing organizations.
4. Multiple-leadership model, i.e., heterarchy. As information-handling overwhelms hierarchical decision centers, they become bottlenecks. Normal channels for data input become overloaded. Similarly, there is insufficient capacity in existing political channels.
5. Retaking of individual responsibility. Public-interest-oriented individuals and public-issue-oriented groups appear at all levels of government. At the international level, for example, there are the

Club of Rome, Amnesty International, Friends of the Earth. At the national, similar groups proliferate, and the new phenomena of separatism and devolution appear, such as the demands of the Quebec separatists and of Native American peoples in Canada and the U.S.A.; the secessionist movements in Scotland and Wales, whose members have realized that London can't do much for them any more; the meeting held in Trieste in 1975 by the major ethnic groups of Europe demanding greater self-determination and later meetings in Lappland, the Netherlands, Australia, and elsewhere. We also see the new demands for community control and citizen participation of dissident minorities in the U.S.S.R. and Eastern Europe, as foreseen by Leopold Kohr in 1957 in *The Breakdown of Nations*.[27] Kohr made the case that nations are too big for the small problems, while today we find they are also too small for the big problems. At the corporate level, we see the new experiments at worker self-management, new forms of worker ownership, networking, and the whistle-blowing mentioned earlier. In government bureaucracies and professional societies, there are creative dropouts and radical caucuses, while the churches, too, are dealing with their growing ranks of dissidents and heretics. Even the Chinese are restoring small enterprises and collectives, rather than concentrating on big, state-owned factories exclusively.[28]

In all, what is occurring is the recalling of once-delegated power and the reasserting of leadership at more functional system levels. I have speculated as to whether this type of almost spontaneous response and self-organizing realignment is not a sort of "body wisdom." The genius for self-organization that all human bodies possess (e.g., one does not have to tell one's heart to beat) may in fact be encoded in our DNA. This devolution scenario that I have described is closely related to the ecological models of dynamic, interactive, co-evolving, steady-state systems that are intensely microdynamic so as to achieve long-term macrostability. A useful analogy from physics is to be found in the heating of a substance or structure, which increases the speed of motion of the molecules, and can lead to structural transformation.

Similarly, the theories of "state-specific" physics, which incorporate the observer's own consciousness or "vantage point," are illustrative. The best examples are Fritjof Capra's *The Tao of Physics,* Yztak Bentov's *Stalking the Wild Pendulum,* and Sarfatti and Wolf's *Space Time and Beyond.* We are beginning to realize that the most complex systems are self-organizing and that only the system can model the system or manage the system. Perhaps the most sweeping model of what may be occurring in this twilight of mature and industrial cultures and manifested in their institutional crises of ungovernability is that described by Harvard sociologist Pitirim Sorokin as the disintegration of sensate culture in his four great futurist volumes published between 1937 and 1941, *Social and Cultural Dynamics.* As we shall see in Chapter 7, this work, not incidentally, combines careful scholarship with an intuitive insight and creative, poetic vision.

Out of all our current social ferment, organizations are slowly learning that if they and our society are to survive, they will need to reformulate their goals and restructure themselves along less paternalistic, less hierarchical lines. Such participatory, flexible, organic, and cybernetic design is now mandatory in the face of cataclysmic changes. Articles unthinkable in the past, such as "The Androgynous Manager," *Planning Review,* November 1979, now appear in management journals. Organizational theorist Warren Bennis claims that "Democracy becomes a functional necessity whenever a social system is competing for survival under conditions of chronic change." This is the theme underlying the new clashes between the rights of capital and of management to manage, versus human rights. As one European worker put it, "We are going to repeal the Divine Right of Capital because it's just as arbitrary as was the Divine Right of Kings."

NOTES – CHAPTER 5

1 Useful clearinghouses for information on corporate power, abuses, and ways citizens can help in achieving accountability, include: The Council on Economic Priorities, 84 Fifth Avenue, New York, NY 10011; The Interfaith Center on Corporate Responsibility, 475 Riverside Drive, New York, NY 10027; INFORM, 25 Broad Street, New York, NY 10004; The Peoples Business Commission, 1346 Connecticut Avenue, N.W., Washington, D.C. 20036; Environmental Ac-

tion, Room 703, 1346 Connecticut Avenue, N.W., Washington, D.C. 20036; Foundation for National Progress, 625 Third Street, San Francisco, CA 94107; The Institute for Food and Development Policy, 2588 Mission Street, San Francisco, CA 94110; and The Institute for Community Organization, 628 Barrone Street, New Orleans, LA 70113. These organizations can help guide citizens to local resources and groups as well as provide reading lists for further study.

[2] *Voices of the American Revolution,* People's Bicentennial Commission, Bantam Books, 1975.

[3] The International Data Bank on Corporate Accountability is discussed more fully in *Creating Alternative Futures,* pp. 362–65.

[4] See, for example, *Study of U. S. Economic Knowledge and Attitudes Toward Business,* November 1975, p. 2, The Business Roundtable. Also *Gallup Poll Opinion Index Report 140,* March 1977, p. 16, American Institute of Public Opinion, Princeton, NJ.

[5] Large corporations, naturally alarmed at the prospect of losing their cloak of respectability if small businesses become effectively organized as a separate lobby, tried to muddy the waters again at the White House Conference on Small Business in January 1979. Paradoxically, the chairman chosen by the White House for its Small Business Conference was safely aligned with the bigger businesses (the Small Business Administration's definition of a "small" business is one with annual sales of below $50 million). He was Arthur Levitt, Jr., also chairman of the American Stock Exchange. Levitt, editorializing in *Business Week* on March 10, 1980, capitalized on the new cloak of affection and legitimacy small businesses enjoy as the backbone of American enterprise. He admitted that big business will have problems with the real entrepreneurs' demands for a more equitable share of federal procurement and contracts, and other issues in which big and small businesses have diametrically opposing interests. However, he pointed out that big business could try to make an ally of this feisty new movement for entrepreneurship and economic "freedom" and share the new public and legislative goodwill—thus thwarting the critics of concentrated corporate power and the demands for economic decentralization and democracy.

[6] A. A. Berle and G. C. Means, *The Modern Corporation and Private Property,* New York Commerce Clearinghouse, 1932.

[7] *International Herald Tribune,* October 25, 1979.

[8] Commenting on President Carter's "guns and butter" budget for fiscal 1981, Leonard Silk editorialized in the New York *Times* on January 25, 1980, warning of delayed inflationary pressures in this course, since many of the weapons would be procured over long periods, building future inflation into the economy. Silk cited the Congressional Budget Office's figure of a 1981 deficit of $30 billion, rather than President Carter's more optimistic $16 billion, even if there is no faster defense buildup. Reagan's promise to furthur increase military spending will increase inflation.

[9] In Germany, Japan, and Sweden, for example, dealing workers into management decisions and stock in their companies is established policy. Sweden's Volvo auto company gives workers voting representation on its board after successful experimentation in 1973 was made permanent in 1976. American unions, such as the United Auto Workers, are slowly rethinking their situations as a result of the Chrysler management failures (*The Christian Science Monitor,* February 1, 1980).

[10] One little-known area of employee civil liberties was explored by David F. Linowes as chairman of the U. S. Privacy Protection Commission, which pre-

sented its recommendations to President Carter in July 1977. In October 1979, when it became clear that business groups had lobbied for time to make voluntary compliance with the Commission's recommendations regarding their mishandling of employee records, Linowes went public at a Washington press conference. He noted that the single largest source holding comprehensive, sensitive personal data on citizens is the corporate employer. Regarding the right of employees to see, copy, and correct their records, Linowes reported that although 76 percent of companies comply in allowing employees to see their records, only 46 percent allow them the right to copy them. Although 79 percent of the companies comply with employees' right to correct misinformation in records, three out of four companies do not forward these corrections to those who had received the incorrect data from them. Two thirds of the companies do not inform personnel of the kinds of records they keep on them, how these records are used, and who is given access to them, while two thirds of the corporations do not inform the individual that they give personal credit information to credit grantors. Three out of four companies use medical information in their personnel decisions, yet 83 percent do not allow their personnel to see it. Linowes called for legislation to enforce these basic human rights of employees (*The Christian Science Monitor,* October 26, 1979).

11 Ideological support is still purveyed to corporate leaders, so they may gird up their loins against critics, in such books as *Capitalism and Sources of Hostility,* edited by Ernest Van den Haag, Heritage Foundation, 1980, which recites the familiar arguments relating "free markets" to individual liberty and economic growth, and citing as "success stories" Brazil, South Korea, and other countries whose governments have achieved these "successes" by often repressive means and exacerbating inequalities. Van den Haag notes that corporate critics are generally coddled, power-hungry malcontents, while public-interest activists are dubbed by big business apologists as "the new class" (*Fortune,* January 28, 1980, pp. 114–15). In similar vein is George Gilder's *Wealth and Poverty,* Basic Books, 1981, which is admired by the Reagan administration for its attack on the welfare state and liberalism and defense of conservatism.

Another think tank providing corporate leaders with ammunition against their critics is the Washington-based American Enterprise Institute, which has spearheaded the "counterreformation" in Washington in which business is promoting deregulation of the economy. Interestingly, the regulatory agencies that get all the flack are those with mandates crosscutting many industries, such as the Federal Trade Commission, the Occupational Safety and Health Administration, the Food and Drug Administration, and the Environmental Protection Agency. Rarely does industry try to deregulate an agency whose mission is to regulate its affairs exclusively, such as the Federal Communications Commission or the Federal Aviation Agency, since these agencies provide costly coordination services to their industries at taxpayer expense and are generally in a fairly cosy relationship with industries' leaders.

One of the more blatant corporate "education" efforts is that offered to judges on the federal bench at the University of Miami Law School's law and economics center. Directed by conservative economist Henry Manne, the center is funded by Procter & Gamble, IBM, General Electric, and others, who claim that they have no control over the thousands of dollars donated. A confidential Senate Judiciary Committee memo calls the seminars for judges "a brazen attempt by the Business Roundtable crowd to influence the enforcement of antitrust laws" (Washington *Post,* October 29, 1979). Most corporate funding of the biggest university-based business schools is more subtle, the favorite method of promoting "sound economic education" being to endow chairs of business and economics and to fund studies of the need to return to the "free-market," laissez-faire world of untrammeled business freedom. The views of Richard R. West, dean of Dartmouth's Amos Tuck School of Business are fairly typical. He deplores the lack of understanding among the general public of "how the economy works," and adds, "Young people should be taught basic economics, such as how government economic policy is formed and how demand and supply function in markets." One becomes more dubious as West praises the Joint Council for Economic Education, a nonprofit body "helping" some five hundred school districts provide "economic education" to their students (*The Christian Science Monitor*, November 5, 1979). The problem is, of course, that not only is economics bankrupt but it has always been nothing more than politics in disguise, as we shall show in Chapters 7 and 8.

William B. Cannon, of the University of Chicago and an adviser to the Progressive Alliance, called for government restraint on financial gifts to business schools and law schools as well as their contract research, since these departments of universities have been expanding at the expense of more humanistic education and promoting antidemocratic attitudes.

[12] New York *Times*, July 20, 1975.

[13] *Electrical World*, McGraw-Hill, N.Y., February 15, 1978, pp. 44–48.

[14] It is in this light of the poor investment decision that corporations are making today that we must view the enormous windfall profits of the large oil companies, the taxing of which was the central legislation passed in President Carter's energy program. Oil companies will reap huge sums from the subsidies to synthetic fuels. Reported 1979 profits were up by huge percentages over 1978. Standard Oil of Ohio's earnings were $1.186 billion—up 163 percent; Texaco's were $1,759 billion—up 106 percent; Mobil's were $2.01 billion—up 78 percent; Gulf's were $1.32 billion—up 68 percent; Standard of California's were $1.78 billion—up 64 percent; Exxon's were $4.29 billion—up 55 percent; Atlantic Richfield's were $1.16 billion—up 45 percent; and Standard Oil of Indiana's were $1.5 billion—up 40 percent. These colossal sums were retained to search for increasingly hard-to-find oil and to fund R&D schemes that favored wasteful, high-technology energy supply and various synthetic-fuels conversion processes with dubious net energy value. Much of this largesse went to buying up other companies: Shell purchased Belridge Oil for $3.6 billion (adding concentration to an already oligopolistic industry), Exxon spent $1.2 billion to buy Reliance Electric, and Mobil paid $800 million to acquire General Crude. Many bought small solar-energy firms and extended their holdings of coal and other energy resources, adding to their grasp of an ever wider spectrum of energy sources (*U.S. News & World Report*, February 11, 1980).

[15] As corporations step up their political activities, there will be an even more urgent need to bring them under democratic control. This might be achieved by passage of the corporate democracy bill, which is the chief goal of the Big Business Day teach-in coalition. The bill would extend public accountability of large corporations in the areas of shareholder and worker rights and would open up corporate boards of directors to broader membership than the current preponderance of insiders, including members of the public. The bill would apply only to corporations with assets of more than $250 million and five thousand or more employees and would also expand disclosure requirements of corporate employment practices, environmental pollution, job health and safety, foreign produc-

tion, performance of directors, shareholder ownership, actual tax rates, and expenses such as auditing fees. The bill would require a firm to give a community where it is a major employer two years' notification of plans to relocate or close down a plant. It would also expand workers' rights, prohibit anyone from being a director of more than two corporations, and increase civil and criminal penalties for corporate criminal acts. Such reforms, by and large, have been in effect in most of the European democracies, including Sweden and Germany, both of whose economies have been stronger, less inflationary, and with far less industrial strife and lost productivity, as well as higher economic standards of living, than the U.S.A. in the past few years.

[16] Recently, Ralph Nader released a report, by Mark Green and Andy Buchsbaum, entitled *The Corporate Lobbies; Profiles of the Business Roundtable and the U. S. Chamber of Commerce* (1980), on two such corporate lobbying groups, the U. S. Chamber of Commerce and the less known but more influential Business Roundtable. The styles of the two groups are complementary: the chief executive officers of the Roundtable see themselves as corporate statesmen and lunch with senators and congresspeople and have easy access to the White House, while the U. S. Chamber is more strident and can organize its members in the backyard of every member of the Congress. Both groups make heavy contributions to political campaigns through various channels. Typical of the Chamber of Commerce's style was its "red alert" sent out to members on the plans of the coalition for Big Business Day. Its Special Report of February 1, 1980, noted that these "self-proclaimed consumer advocates," "closet socialists in the labor movement," and other activists and "political gurus" are kicking off a decade-long drive to enact legislation that would end the private enterprise system as we know it in America today." Similar smear tactics were used publicly by Mobil Oil in an advertisement that ran in many media in September 1979 under the headline "The Commissioner Bares His Motives," which targeted New Jersey's crusading Energy Department Commissioner Joel Jacobson as advocating "some form of socialism" and noting that he is a former labor-union official. Apparently, it outraged Mobil's executives to find that there are some government posts not controlled by corporate interests or not filled by former business executives.

[17] Another approach to controlling the political debate about corporate power is the use of in-house analysts referred to as "public-affairs-management" or "public-issues-management" groups. This function grew out of the failures of the "hard-sell" public-relations approach. Most large corporations now have this type of in-house capability, ranging from defensive programs of monitoring "radical" groups and journals and general intelligence-gathering on activists, all the way to harassment and apparent attempted murder, as witness the suspicious circumstances surrounding the "near-miss" deadly accidents suffered after threats by some corporations, directed at anti-nuclear campaigners, including Dr. Rosalie Bertell, whose car was forced off the road near Rochester, N.Y., in a highly suspicious manner after an anti-nuclear speech documenting health effects of radiation on nuclear workers. A wrap-up of the more traditional and lawful activities of corporations is included in the article "Capitalizing on Social Change" (*Business Week,* October 29, 1979).

[18] *Forbes,* August 21, 1978, p. 35.

[19] *Business Week,* August 7, 1978, p. 62.

[20] Daniel Greenburg, Washington *Post,* August 22, 1978.

21 J. Fred Bucy, *Without Free Enterprise There Is No Freedom,* speech before the Odessa Country Club, Odessa, Texas, August 10, 1978.

22 *AMA International Forum,* Vol. 67, #6, June 1978, p. 29.

23 "I Don't Trust Any Economists Today," *Fortune,* September 11, 1978, p. 31.

24 The European Economic Community announced its new antitrust crackdown on state-owned enterprises in October 1979. During the past decade, the EEC's trust-busting activities have been directed mainly at private companies, including BASF, Continental Can, Phillips, General Motors, United Brands, Hoffmann-La Roche, and Kawasaki. Focusing on the state-owned giants, the EEC's first priority was disclosure, since these companies employ 16 percent of the European Community's 100 million civilian labor force and account for 25 percent of its capital formation. The Paris-based International Chamber of Commerce, representing the lobbying arm of the private multinationals, pronounced its delight in being able to help the EEC's antitrust enforcers for a change (*International Herald Tribune,* October 25, 1979).

Another area where state-owned companies are challenging private corporations is the oil industry. Whereas a decade ago the big private oil companies controlled 70 percent of the world's oil trade, in 1980 the percentage was below 50 percent and falling. Pushed by events since the formation of OPEC, in 1973, including rising nationalism, the Iranian revolution, and dwindling supplies, many countries share the view of Italy's state-owned oil company: "Oil is a political commodity now, not something to be left to markets and businessmen." The French and the British have controlling interest in their largest domestic oil companies, and West Germany and Canada have recently followed suit with newly formed state companies, Veba-AG and Petro-Canada. Many OPEC nations prefer selling oil in packages directly, government to government, where they have greater control (New York *Times,* December 30, 1979).

25 There are numbers of examples of the maladaptive syndrome of greater corporate/government dealings regarding plant-location decisions in the United States to which our kind of competing levels and geographically diverse political institutions lend themselves: the emergence of bureaucratic "entrepreneurship." The new breed of entrepreneurs are the state and local officials who put up taxpayers' money to woo businesses into their jurisdictions by underwriting their risks, giving them tax holidays, and bargaining away their citizens' rights to healthy environments and working conditions. These deals in our troubled economy become more and more "competitive," with each locality and state bidding against others in the race to subsidize corporate activities that, at least in the short run (during their officials' average term of office), seem attractive. The social costs and bills that must eventually be footed by the area's taxpayers come later—when the plants pull up stakes, throw workers onto local welfare rolls, and leave a backlog of dirty air and water, and waste dumps to be cleaned up. Robert Goodman describes this wasteful, maladaptive process in *The Last Entrepreneurs: America's Regional Wars for Jobs and Dollars* (1980).

26 This leads to the proliferation of third-party and quasi-political groups now visible in all bureaucratized industrial democracies: the Ecology Party in Britain, Les Vertes in France, the Citizens Party in the U.S.A., and many others. The new need for expanded political channels has also led to revivals of some older forms, such as the Libertarian Party in the United States, whose two hundred candidates for local offices in 1978 won 1.3 million votes and held their own in 1980. Libertarian thought is a strange amalgam of fundamentalist, rugged-

individualist, and laissez-faire economics riding a new wave of disenchantment with big government and big business, but its philosophy relies too heavily on the notion that the only bastion of political freedom is property ownership. Thus it is still confused over the role of corporate power parading under the guise of "private property" and "free enterprise." Nowhere was this confusion more evident than in the party's choice as its presidential candidate in 1980 of corporate lawyer Ed Clark, an employee of the Atlantic Richfield oil company. This blind spot regarding the proper role of property leads to the Libertarians' knee-jerk opposition to almost all government, a quasi-anarchist position against almost all the time-honored ways that civilized communities have assessed themselves to purchase public goods and services—whether park lands, libraries, schools, hospitals, fire protection, or waste collection. At the same time, they remain myopic about corporate abuse of the power of accumulated property and the Fourteenth Amendment-based interpretations of corporations as "persons," but with only rights—not the concomitant duties that real persons also are bound by: to pay taxes and fight in wars, etc. In addition, as the Libertarians have tried to update their philosophy, their individualism sometimes shades off into the competitive nastiness exemplified by the recent spate of books with titles like *Winning Through Intimidation* and *Looking Out for Number 1*, by Robert Ringer, who has just published a libertarian manifesto entitled *Restoring the American Dream*. And as they try to address issues of a newly interdependent planet, Libertarians must learn that two thirds of the human race do not and never will have any private property (their lands, houses, and tools being shared communally and reciprocally among villages). A useful article is "The New Libertarians," by Michael Nelson, *Saturday Review*, March 1, 1980.

[27] Leopold Kohr, *The Breakdown of Nations*, Dutton, 1978.

[28] New York *Times*, "China Restores Small Businesses to Provide Jobs," October 8, 1979, p. 1.

CHAPTER 6

The Transition to Renewable-Resource Societies: Nuclear Versus Solar Energy as Symptom of the Paradigm Shift

As aging industrial societies redoubled their efforts to continue on their now unsustainable, resource-intensive path, the clearest symptom of their pathologies of material abundance and of waste was the rising battles over nuclear energy versus a whole new approach to energy production and utilization embodied in solar technologies. Even before the shock waves of the accident in March 1979 at the Three Mile Island nuclear plant in Pennsylvania, governments in Austria and Sweden had toppled due to growing citizen opposition to nuclear energy and its admittedly unsolved problems of disposal of radioactive wastes. Even in France, where strong governmental authority had imposed its nuclear priorities with police power, labor unions revealed in late 1979 that cracks in both reactor vessels and steam-generator systems had appeared in two Westinghouse-designed, French-built nuclear plants. Worse, the state-owned utility, Électricité de France, had covered up these facts for over a year, then had admitted that similar faults might exist in six other operating reactors. Despite the utility's assurances—"We understand the cracks are small and do not constitute a safety concern"—unions representing more than 20 percent of the state utility's work force refused to refuel the reactors and threatened to strike nationwide if other technicians were brought in to do the job. The labor groups were joined by seventeen other organizations, including the Socialist

Party and French environmentalists, in denouncing the government's secrecy (*Business Week,* October 29, 1979). In the United States, the Kemeny Commission, appointed by President Carter to investigate the Three Mile Island accident, while deeply divided on the need for a moratorium on all new nuclear plants, did raise serious questions as to their safety, indicting the Nuclear Regulatory Commission's bungling and company operating errors. As an interim measure, it called for greatly increased safety vigilance, including recommending that operating licenses be denied in states where civilian emergency and evacuation plans had not been developed. Eight of the eleven commissioners favored a complete moratorium on nuclear-plant construction; only their disagreement on the wording prevented this becoming an official recommendation. In addition, the cost of the accident was summed as between $1.047 billion and $1.858 billion, demonstrating anew my contention (and that of many others) that if nuclear power were to be made safer, it would be uneconomical. This is a far cry from the early promoters who asserted that electricity from nuclear power would be too cheap to meter (*International Herald Tribune,* October 24, 1979).[1]

The waste-disposal aspects took on new safety dimensions as hazards of truck transportation and leaking burial sites were discovered. The seventy-two licensed reactors in the United States in 1979 produced between eight thousand and fourteen thousand cubic feet of wastes a year, some of which remain radioactive for several thousands of years, although the U. S. Department of Energy assured the public in 1979 hearings that it is "the first thousand years of the disposal period that are critical" (*The Christian Science Monitor,* October 16, 1979). Nevertheless, Washington's pro-nuclear former governor, Dixy Lee Ray, was forced to close the Hanford waste dump (one of three in the country) because of leaks from faulty containers and the dangers of trucking them on interstate highways. The voters of Washington had reacted to the news that hundreds of truckloads of radioactive debris would be shipped to their state from Three Mile Island. Similar anxieties emerged at the nuclear dump in Barnwell, South Carolina, as Governor Robert List, of Nevada, followed suit and ordered his state's nuclear-waste dump closed, leaving South Carolina for a while with the dubious distinction of being the nation's only radioactive-waste disposal site.[2] Similar protests against major facilities for storing and reprocessing spent nuclear fuels in Britain's

Windscale project and that proposed for Gorleben, West Germany, brought rising doubts as to the political, if not technical, viability of nuclear energy. Horrified voters in Britain learned that nuclear debris was to be shipped to the Windscale plant on cargo vessels from Japan, with all the attendant dangers of such long sea voyages. Meanwhile, the escalating costs of reactors' downtime for safety checks, the debacle at Three Mile Island, and the waste dilemma had already resulted in a *de facto* moratorium as investor financing dried up. Utilities' stocks plummeted, and some, particularly General Public Utilities, the holding company that owned Metropolitan Edison, the operator of Three Mile Island, faced imminent financial "meltdown."

Rather than face consumers' wrath, state utility commissions denied companies rate increases. In a landmark case, Missouri's Public Service Commission, in July 1980, denied Kansas City Power and Light Co. permission to add its $165-million new plant into its rate base as an investment. The Commission held that in light of reduced demand, the plant should never have been built.[8]

Thus, as the outlook for oil grew gloomier and the number of possible supply-disruption situations proliferated, the once bright hope of the nuclear alternative faded, leaving the industrial societies to face up to their energy addiction and to reconceptualize their situation. Like the behavior of rats in a maze, the first response was pure instinct: rather than thoughtfully reviewing goals and life-styles, they rushed in all directions looking for more supplies, as if energy were an end in itself, rather than a means to other ends. Recriminations and accusations by the Europeans that the United States, and to a lesser extent Japan, were to blame due to their energy gluttony were rife. Finally, even Western media began to come to the defense of OPEC, highlighting the fact that in terms of constant, non-inflated dollars, their price hikes had not been so great and that some OPEC members, notably Saudi Arabia, had been pumping far more oil than needed for her own "development" so as to appease the pain of conservation in the United States. In fact, as consuming nations hypocritically blamed OPEC, they contributed to the upside breakthrough of OPEC price levels on the soaring "spot market" in Rotterdam by bidding against each other in the scramble to hoard oil supplies (*The Christian Science Monitor,* November 9, 1979). Meanwhile, the Third World inspired by China's successes, forged ahead in renewable-energy technology, as the industrial world may learn at the UN Conference on New and Renewable Energy, to be held in September 1981 in Nairobi, Kenya.

As environmentalists had been warning for almost two decades, energy conservation was the only short-term option. At last, in late 1979, President Carter won a limited gasoline-rationing mandate from Congress and sent to the lawmakers an oil-import quota plan involving three alternatives: 1) an oil-import auction system, which would set import limits and allow importers to bid for licenses, 2) a license-fee system, with the government collecting a $2-or-more tariff on each barrel, and 3) an outright allocation program, in which government would apportion the right to import crude petroleum and refine it according to a formula based on past use. This led to a fifty-cent gasoline-tax proposal which was promptly rejected by Congress as unconstitutional. In the winter of 1979–80, oil imports were about four hundred thousand barrels a day below President Carter's 8.2-million-barrel-a-day ceiling, enabling him to tell the public that in 1979 an overall reduction in petroleum consumption of 2 percent had been achieved. This figure was the first evidence to the Europeans and the Japanese of U.S. conservation results.[4] Some analysts held that even the reduced *rate* of importation had not been due to reduced consumption of energy, but had been achieved by withholding from further stockpiling and by juggling the temporary glut of natural gas that resulted not from increased production but from increased conservation by industrial gas users. Thus robbing Peter in natural-gas use was a drastic short-term Band-Aid, reversing more-thermodynamically sound policies of conserving this high-quality, clean fuel for highest-priority use. Factories and other low-priority users were encouraged to switch back to gas. Although coal adds more carbon dioxide to the air, it is more plentiful and can be made much cleaner-burning with existing pollution-control technologies.[5]

The absurd Carter administration $88-billion synthetic-fuels bill was funded by Congress at $20 billion, but even though it was clear that such a massive, inflationary "investment" could not produce a drop of fuel until the mid-1990s, the bill was passed in 1980 due to a crescendo of energy-company lobbying. The reconceptualization failed to dawn on the industrial countries' leaders and their energy advisers: continuing on the path of increasing energy supply was rapidly becoming impossible as it pushed the limits of the laws of thermodynamics. President Carter, in mid-1980, backed off his commitment to conservation and solar and renewable energy in favor of greater funding of nuclear, coal, and synthetic-fuel programs. Worse, even coal, which is the last plentiful fossil fuel in the planet's crust, has its own inexorable limit: its combustion is now significantly raising the levels of carbon dioxide in the earth's atmosphere, which could trigger wholly unprecedented climate changes. Most scientists expect more of the recent extreme weather variability now affecting crops, whether via the "greenhouse effect"—trapping the sun's heat in the atmosphere (leading to a warmer climate and melting of polar ice caps, which would probably flood all the world's coastal cities), or an opposite effect—a cooling leading to an ice age. As policy makers were urged to triple world coal production, which would raise carbon-dioxide levels further, it became clear that energy policies in the future would have to factor in climate change. Accordingly, the U. S. Department of Energy now has an Office of Carbon Cycle Analysis.

In spite of this, the Department of Energy pushed ahead with its mandating of power plants still using oil to switch to coal, and the Administration introduced legislation to provide $12 billion in federal grants to help pay power companies' costs in switching to coal or alternative fuels. The tragedy is that so many alternatives exist from increasing the utilization of waste heat by installing co-generation systems in existing coal and gas fired plants, converting the solid waste stream to gas and boiler fuel, retrofitting small dams with efficient turbines, using wind power on the Great Plains, and since we are spending such colossal sums on palliatives, such as the $12 billion to switch to coal, it makes sense to leapfrog these expensive stopgap expenditures and go straight to the solar and renewable alternatives. For example, scientists at the University of Utah are conducting a study similar to that already underway in the Salton Sea in southern California, to match Israeli solar power generation in the Red Sea, that by 1981 will provide enough electricity, 50,000 kilowatts, for a city of 10,000 people. The potential in the ideally saline waters of the Great Salt Lake, is as great as 15,000 megawatts of power compared with Utah Power and Light's existing installed capacity of 20,000 megawatts which meets all the state's demand currently. Even more tragic is the fact that with the steady declines in consumption of electricity over the past few years, largely due to its skyrocketing price, there may be no need to build any more electric generating capacity in most parts of the country. The greatest untold scandal in the electric-utility industry is the actual excess *reserve* capacity, which averaged 33 percent in 1979, well above the 20 percent the industry considers optimal, and continued rising to almost 40 percent through 1980 as peak demand sagged. One example of inept utility-management decision making was pointed out by University of California scientist Edward Kahn in a seminar on electricity load forecasting at E. F. Hutton & Company, Wall Street investment bankers: it seemed that the utilities' very success in lobbying the Construction Work In Progress surcharges onto customers' bills to provide financing of new power plants had contributed significant additional incentives to conserve electricity, thus *increasing* the uncertainty of load forecasting. In other words, the extra charges during the construction of the plant could lead to evaporation of the "demand" it was constructed to fill by the time it is ready!

Other Catch-22 situations arose, such as that of the conflicting needs for energy and fish protein, as fishing-fleet operators clashed with oil companies over the sale of oil exploration leases in the rich fishing waters of the Georges Bank, off New England (*The Christian Science Monitor,* November 7, 1979). As this energy policy debacle shaped up, environmentalists and biologists, "soft energy" advocates, "small is beautiful" movements for ecologically and humanly appropriate technology, consumers, and an increasing number of labor unionists lobbied desperately for solar energy and renewable resources, recycling, and smaller-scale and diverse energy sources, still dismissed by traditionalists as a drop in the bucket or not feasible before the year 2000. Thus the existing capital-intensive-energy path versus the emerging, labor-intensive, skill-intensive renewable-energy path spilled into the political system, where it will likely remain as one of the continuing major debates of the 1980s.

The first salvo in the United States was fired in the 1980 Democratic presidential nomination race by candidate Edward Kennedy, who countered Jimmy Carter's energy bill with his own plan, backed up by *Energy Future,* a report from the Harvard Business School's Robert Stobaugh and Daniel Yergin that became an instant best seller (Random House, 1979). The Kennedy plan addressed the United States energy problem from the "demand" viewpoint and proposed, through increased investment in conservation and renewable resources, refitting small dams with new, more efficient turbines, and other measures, to meet the same energy supply target as the Administration's $143-billion total energy bill, at less than one half the cost: some $58 billion, most of which, instead of going to oil and other energy companies, would go to consumers, taxpayers, and homeowners, as well as businesses, for installing insulation and solar equipment and for improving design in manufacturing and architecture to achieve better thermodynamic efficiencies. Kennedy held that the plan would save 4 million barrels of imported oil a day by 1990 and showed how the United States could cut energy consumption by 30 to 40 percent by the year 2000 with little slowing of economic growth. Barry Commoner, in *The Politics of Energy* (Knopf, 1978), agrees with the viability of a rapid transition to solar but differs on strategy. A 1979 study, *Jobs and Energy* (Council on Economic Priorities), confirmed, by closely examining nuclear versus solar-energy options for Long Island, New York, that this renewable-resource and conservation approach yields over twice as many jobs per dollar invested as continuing on our current energy course. Meanwhile, President Carter sought to reassure the governors of energy-rich Western states that their states' rights would not be abridged in the desperate search for more energy supplies from ever-more-exotic schemes, such as that of retorting millions of tons of oil-bearing shale in the Rocky Mountains, which would require more water than competing agriculture could allow. Each day, new moves by states thwarted federal energy plans, whether to make the Rocky Mountain states what Governor Richard Lamm, of Colorado, called "areas of national sacrifice" or in demands such as those of California, Alaska, Maine, Rhode Island, and Massachusetts that they have a say in offshore oil leasing. As all these technological, political, and environmental woes increased, the economic impacts of continuing existing energy policies grew worse, and *Business Week,* in its November 19, 1979, issue, predicted the situation in an article entitled "The Petro-Crash of the '80s," drawing further attention to the growing trade and monetary imbalances described in Chapter 3. Just as military strategists had described the checkmate in the U.S.A.-U.S.S.R. arms race as "mutually assured destruction" (MAD), so could the energy-supply scramble of industrial nations down the path of nuclear proliferation and economic ruin be summed up as mutually assured self-destruction, equally mad.

Meanwhile, as is often the case when institutions and nations reach the dinosaur stage, individual voters and small-scale towns and states were ahead of the national and corporate leaders. Riders flocked to unprepared mass transit systems, as for example in Milwaukee, where ridership soared by some 20 percent in 1980. Environmentalists and public-interest groups had lobbied mass-transit funds into the Carter energy plan and called for spending $13 billion on mass transit to increase capacity by 50 percent. The bottleneck was Detroit, which had long since deemphasized buses. (General Motors had actually bought Los Angeles' rapid-transit rail system in the 1920s in order to tear it up and thus increase the sale of private autos.) Citizens demanded that Chrysler begin making buses as a quid pro quo for its bail-out by the taxpayers, and traffic surveys showed that Americans were, on the whole, trying car-pools, driving within the 55-mile-an-hour limit, as well as achieving significant conservation in home heating.[6] The towns of Easton, Maryland, had instituted a district-heating, cogeneration system in its municipally owned utility, where waste heat, normally vented into the air via costly cooling towers, is recycled to heat homes and factories by powering diesel generators.

Similar municipal efforts had drastically trimmed energy use in Seattle, Washington; Northglenn, Colorado; Hartford, Connecticut; Clayton, New Mexico; Ames, Iowa; Burlington, Vermont; and Greensboro, North Carolina, as documented in *Energy-Efficient Community Planning,* by James Ridgeway (J-G Press, Emmaus, Pennsylvania, 1979). Citizens also fought corporate energy waste in the form of one-way containers and bottles, which use 3.11 times the energy of returnables. New laws banning one-way containers in Maine and Michigan have saved 5.5 trillion BTUs a year, the equivalent of 40 million gallons of gasoline. Container litter was reduced 82 percent, and total solid waste was down 4.5 percent, while four thousand new jobs have been created at a savings to taxpayers of $15 million (*The Christian Science Monitor,* November 5, 1979). As if all this and more were not writing on the wall for leaders of industrial nations, public opinion polls such as that of the New York *Times*/CBS in the *Times* of April 10, 1979, showed a sharp rise since 1977 in opposition to nuclear power, with 56 percent opposed and only 38 percent still willing to have a plant close to their own town. Significantly, in the same poll, only 12 percent said they thought government would be able to share the burden of higher oil prices equitably, while 78 percent thought the oil companies would just make more money. And by 1979, an NBC/Associated Press poll showed that out of a list of energy supply choices, Americans preferred solar by a majority 52 percent; over coal, 21 percent; nuclear, 16 percent; 4 percent still favoring oil; and 7 percent unsure.[7]

The issues surrounding the use of nuclear power or the use of solar power—and the implications of the choice between the two—are symbolic of the sharpest differences between the two directions lying before us: toward greater and greater capital, energy, and materials intensity, or toward greater labor intensity. The current direction, which was historically sensible, overshot the mark. Saving labor by making a system more capital-intensive is reasonable when you have very cheap resources and not much of a problem in putting those resources at the disposal of workers for increasing individual productivity, but this system has now collided with resource scarcities.

The entire economy, the whole configuration of factories, cities, and suburbs laid out in concrete, gives the system tremendous momentum in the existing direction—and exploitation of nuclear energy is a last, baroque elaboration of that old direction no longer sustainable. Solar is the key metaphor for the way we have to go. The situation polarizes around these two types of technology.

Whether or not we manage to correct some of the major subsidy programs built into the system, which keep pushing it toward greater capital intensity, whether or not we work out an equitable way for energy prices to rise without hurting too many poor people, there are many ways in which we are being driven toward the new state that have nothing to do with human beings and our attempts at policy making. Availability of energy, of course, is the driver. The system's own pathway of accommodation is expressed as inflation. Barring any conscious policy, inflation will drift up, and quietly settle us back into a more stable sort of economy and a less centralized pattern. This course will be very difficult for some groups, and there will be a tremendous amount of unnecessary pain in simply allowing the system to do its thing. But if leadership is forthcoming, the pattern of events can be explained in ways that people can understand and adjust.

First is the matter of subsidy. We have subsidized every other form of energy technology and thereby have found ourselves in a bind: either we must subsidize solar equivalently, if we are going to be fair, or we must reduce the subsidies on other technologies to allow solar to compete. It would be more efficient to eliminate the subsidies. The March 1978 Battelle Institute *Analysis of Federal Incentives to Stimulate Energy Production* tells the story. Eight types of incentives were studied: 1) creation or prohibition of organizations; 2) taxation, exemption, or reduction of existing taxes; 3) collection of fees for delivery of a government service or goods not directly related to costs of providing; 4) federal-government disbursements without requiring anything in return; 5) government requirements backed by criminal or civil sanctions; 6) traditional government services provided through a nongovernment entity without direct charge (e.g., regulating interstate and foreign commerce, providing inland waterways, etc.); 7) nontraditional government services (e.g., exploration, research, development, and demonstration of new technologies); and 8) market activity. Such federal incentives for nuclear power were estimated to have cost taxpayers between $15 billion and $17 billion over the past thirty years (not counting the socializing of insurance-risk liability via the Price-Anderson Act). Incentives to the coal industry include depletion allowances that cost $3 billion between 1954 and 1976 and government services in exploration, research, development, and safety that cost another $3.5 billion. The oil industry has received 60 percent of the total federal incentives, costing taxpayers an estimated $77.2 billion ($40 billion of which was depletion allowances). Much of the rest was in subsidies to oil tankers, pipelines, surveys, and research and development. Natural-gas companies received some $15.1 billion from the government between 1954 and 1976 ($11 billion of which was for depletion allowances and intangible drilling expenses). No wonder we are experiencing a tax revolt!

The usual arguments of economists are that higher prices are needed to stimulate new exploration. However, we see instead oil companies diversifying, such as Mobil Oil with its acquisition of Marcor and its new venture into massive real estate development in California, Texas, and abroad. Obviously, gas producers have held gas off the market interstate waiting for today's deregulation and higher prices—normal market behavior. Meanwhile, noted petroleum geologist Earl Cook, of Texas A & M University, states flatly that if we raise the price of oil and gas by five or ten times, *some* more, but not a *lot* more will be produced: "The laws of physics and geological occurrence transcend the laws of men. . . . The price of natural gas in Texas has increased more than tenfold in the past six years—yet the finding rate continues to fall."[8] Thus, since 1918, the federal government has expended between $123 billion and $133 billion to stimulate coal, oil, gas, hydro, and nuclear energy production. Yet we are asked to believe that new forms of energy such as solar, wind, bioconversion, and others must "compete in the free market" with all the historically subsidized energy supplies, not to mention their politically powerful corporate organizations and economic interests with investments to protect.

So the nation is faced with the economically absurd, Catch-22 situation: either enact equivalent subsidies to all the needed newer solar and renewable energy sources so that they can "compete equally" in our rigged energy market or try to remove the subsidies from the old energy sources in face of the stiff opposition they have mounted to all such attempts. Dr. Ronald Doctor, of California's State Energy Commission, in a 1978 speech spelling out his program for rapid commercialization of solar energy in California summed up the situation, "Don't temporize, subsidize. Don't get tied up in meaningless conventional economic analyses that are unable to deal with the realities of energy economics. If solar energy is to compete fairly with conventional forms of energy it will have to be subsidized. These subsidies should not be viewed as handouts; but rather as equalization mechanisms."[9] He added that the Battelle study probably underestimates subsidies, and puts the total at nearer to $300 billion, at least. However, it is imperative to raise the caution that solar grants and subsidies in the Department of Energy programs are still heavily biased in the direction of large corporations already precommitted to competing conventional energy systems, whether nuclear, oil, or existing electrical utilities, which is bound to abort or distort much innovation and which raises antitrust issues as well, as exposed by Ray Reece in *The Sun Betrayed* (1978). Solar technologies are still in the "let a thousand flowers bloom" phase, and preference should be given to truly innovative approaches by independent companies and entrepreneurs without interlocking ties to existing energy interests.

We can pursue the problem on either or both of two fronts and do the best we can to illuminate the situation. We need to work as hard as we can politically to phase out those old subsidies gradually, while trying to reduce the pain to innocent individuals—and at the same time try to subsidize consumers to install the newer technologies rather quickly so they can get a foothold. The new and the old energy systems are still fighting it out in Washington—but the leadership is coming from the states. Big states like California and Florida with a large chunk of the population of the entire country, can really speed the innovation. The steps taken with California's 55 percent consumer tax credit for solar, and other steps the state is taking unilaterally, are going to help shift the entire country into the new pattern. If you really make that enormous market fair to solar and renewable energy, it becomes a test-bed in which companies can commercialize renewable energy sources and develop economic strength to lobby in Washington. Because California has so much incident-solar-energy income every day, and because its leading industry is agriculture—which means that the people are very close to the real biological efficiency of the system—I foresee California becoming one of the states together with Florida and others with similar conditions, that will lead us into a renewable-resource economy, as pilot projects for how that transition is going to work.

This current tug-of-war between older and newer energy systems is typified by contrasting the widely differing characteristics of nuclear and solar technologies and the divergent social, economic, political, and environmental impacts that each displays. As we have seen, the choice has little to do with free market forces or even consumer or voter preferences as to architectural style, greater personal control, lowered risks, and environmental impacts or more decentralized political and economic institutions. It is much more a problem of which existing oxen are to be gored; how entrenched energy systems and technologies are to be amortized or written off; how capital investments can best be channeled into developing new systems that will constitute the renewable-resource economies of the future. A clear example of the conflict was reported in *Canadian Renewable Energy News* (July 1978), which noted that the Central Mortgage Housing Corporation, which had recently been directed by the Parliament to provide mortgage incentives to passive and solar energy systems for consumers, had released a "study" showing that solar potential in Canada was too small to make such incentives "worthwhile." A similar study by the Canadian Atomic Energy Control Board assessed the risks of coal, oil, and nuclear versus solar, wind, ocean thermal, and methanol. Heavy-handedly, it found that nuclear power and natural gas had the lowest overall risks and that risks for solar, wind, methanol, ocean thermal and other "unconventional energy sources" were much higher. However, they calculated that the higher risks incurred in solar thermal and photovoltaic systems were due to their "energy back-up" systems! Furthermore, the study whisked away the inherently lower risk of conservation and passive design by simply not including these energy systems in the study,[10] now discredited in Canada and the United States. An even more shocking example of official government cover-up was revealed by Soviet dissident scientist Zhores Medvedev in his *Nuclear Disaster in the Urals* (Norton, 1979), describing the disastrous explosion of nuclear wastes at Kyshtym, in the U.S.S.R., in 1957, where thousands were killed and injured and an entire region was destroyed.

Economists, in their adherence to their market-equilibrium models (still supposedly guided by an invisible hand), overlook the extent to which the complex, interdependent economies of most mature industrial societies are actually composed of legislated markets. For example, entire economic treatises have been written expounding the idea that prices and "free markets" are the opposite of regulation. They are not. The price system is simply one very useful form of regulation, which indeed is rather a rare aberration in the history of human societies.

There are, of course, many other forms of regulation and resource allocation, set by customs, laws, and taboos in all human societies, including caste, discrimination by race and sex, or other forms of "pecking order." We need to recognize that markets are one ingenious method of regulation, which can be equitable when producers and consumers meet each other in marketplaces with equal power and equal information and if no spillover nuisance effects are visited on innocent bystanders (the conditions Adam Smith described as necessary for markets to allocate resources "efficiently"). We can also see how rarely these conditions are met in today's complex economies. It is more accurate to refer to the market system as "rationing by price," i.e., just one way of rationing scarce supply among users that have cash, or what economists call "effective demand," leaving aside the question of needs or wants.

It may be *necessary* to increase the price of energy, but it will not be *sufficient*. Other major forces in the society are shifting us in the direction of greater capital intensity; without considering those forces, we won't achieve our purpose. The biggest economic force to consider is the investment tax credit. Alone, it can continue skewing the economy and smothering the effect of rising energy prices. A study by the Joint Economic Committee, mentioned earlier, showed that the one thousand largest companies (according to Fortune-500 standards) used 80 percent of the total tax credit and 50 percent of the industrial-process energy, and created only seventy-five thousand new jobs over seven years. In the same period, the country's 6 million small businesses, using far less process energy, created 9 million new jobs. The tax credit was originally justified as a means to create jobs. We find that with the largest and most capital-intensive companies, investment just as often disemploys people, through automation or through moves abroad.

As we see in our still deadlocked energy debate, most, if not all, of the energy proposals of analysts of all political persuasions, and the recent U.S. energy legislation, are simply one form or another of rationing by price, as were proposals for deregulating gas and oil, the Crude Oil Equalization Tax (COET), or wellhead tax, the oil import fees, gasoline taxes, the various tax credits, rebates to consumers, and incentives. The exhaustion of the price system as the chief regulator of our energy system is obvious in inflationary impacts and the grave inequities it imposes on lower-income citizens. Monetary, fiscal, and price mechanisms can no longer bear such a weight as policy instruments.

Although we should try to remove subsidies from old energy systems—which will, of course, raise the prices of oil, gas, and nuclear-derived energy—price increases *alone* cannot reduce energy demand. This is due to the "set-in-concrete" configuration of our towns and cities and sprawled, suburbanized, automobilized patterns of development, and to the fact that some 19 percent of all U.S. families do not own automobiles and already use mass transit. These families may need energy, but they are already priced out of many of the excessively energy-consumptive modes of life. Raising energy prices, therefore, does not affect their already frugal energy consumption, and the middle- and upper-income groups simply go on paying increasing energy prices. As we have seen in Europe, where gasoline generally costs around $3 per gallon, consumers will deny themselves a whole range of other kinds of consumption in order to continue using their cars and appliances. For example, in Germany, gasoline is $3 a gallon, and as of July 1980, Germans were consuming more, not less, of it. (In the United Kingdom, energy-demand reduction was achieved only after a massive public education campaign was launched to encourage conservation.) Meanwhile, the advertising of high-energy life-styles encouraging the consumption of energy-guzzling cars and appliances continues unabated in the United States and many industrial countries. Thus, more equitable rationing of energy itself is not only morally sound but is also more effective at achieving reduced-consumption goals without furthering inflationary price increases.

Yet another reason that we cannot expect higher prices alone to achieve energy-demand reduction and a shift to more labor-intensive, rather than capital- and energy-intensive production in our economy, is that there are too many forces working in our tax code and federal legislation that are pushing us in the opposite direction: toward greater energy and capital intensity, including our system of tax credits for capital investments, allowances used by real estate speculators, the Social Security rate hike, as well as the accelerated depreciation and the various "tax holidays" that competing states use to lure businesses to relocate, etc. President Reagan's policies exacerbate these trends (see Fig. 5).

Finally, oil-producing countries in and out of OPEC can now checkmate our domestic efforts to regulate our energy system and reduce demand by price and monetary mechanisms, as well as by simply keeping their oil in the ground, as they have now learned. If the Administration or Congress tries to reduce domestic consumption of energy by the imposition of wellhead taxes, oil import fees, or any other price hikes, this is taken as prima facie evidence that OPEC prices are too low.

Thus we see further signs of the exhaustion of the whole range of monetary and price mechanisms as a means of regulation of energy supply and demand, or to achieve the shift of our energy system from the depletable sources of the petroleum age to the renewable resources of the solar age. At last, attention has begun to be focused on how other industrial countries, Switzerland, Germany, and Sweden manage opulent living standards with approximately one half to one third the energy consumption of the U.S.A. with combinations of regulation, tax policy, thermodynamically efficient technologies, and community planning.

Entire economies must be shifted toward a system that combines more people with less capital, energy, and material. How can this be done? Howard Odum, of the Energy Center at the University of Florida, states the problem well: the energy flowing through any system maintains its structure. The moment you begin to withdraw energy, there is a spontaneous devolution of the structure to a level appropriate with the new, lesser energy flow. Today we look at the decentralization already going on in cities, and in the economy, as a demonstration of this. Neighborhood economic development becomes more efficient. So does any smaller, flatter capital structure that has to service fewer stockholders, smaller office buildings, and can do without company jets. It makes no more sense to bake cookies on one side of the country and sell them on the other, as it did when energy was cheap. Whole industries, operating under the old paradigm, are making counterproductive decisions, because the tax system still drives them in the wrong direction. In addition, Department of Energy statistics, like so many others in Washington, are in a shambles.[11]

The loss of domestic control of these highly interactive, globally interlinked economies is now becoming evident, as in the lack of coordination of official government forecasting models revealed in the *Global 2000 Report*.[12] This loss of control has been evident in the decline of the U.S. dollar and the moves within the European Economic Community (EEC) to end its status as an international reserve currency, as well as in the daily roller coaster of the international monetary nonsystem and the chronic destabilization caused by global flows of capital between multinational corporations and their bankers (as described in "Stateless Money," *Business Week*, August 21, 1978). Stock markets no longer reflect intrinsic values but can fluctuate wildly, as they did in the late 1970s when nervous foreign investors and bankers decided to shift their glut of dollar holdings from U.S. treasury bills into stocks. Similarly, the second-largest item in our balance-of-payments deficit in 1979 was interest payments to foreign holders of U.S. debt.

Thus, the emerging economies of sustained-yield productivity based on renewable resources are an inevitable form of evolutionary succession. There is simply nowhere else for the madness of the

DIAGRAM OF THE U.S. ECONOMY ILLUSTRATING VARIOUS TRANSFERS AND TAXING FACTORS THAT SUBSIDIZE ENERGY/CAPITAL INTENSITY AND DISCOURAGE LABOR INTENSITY

total monetized economy (i.e., transactions in prices and measured in GNP)

An overall subsidy to capital/energy-intensive production occurs whenever market prices do not include social and environmental costs

Major additional subsidy programs increasing energy/capital intensity include:

UNDERVALUED, SUBSIDIZED NONRENEWABLE ENERGY SOURCES	INVESTMENT TAX CREDITS FOR BUSINESS	SOCIAL SECURITY SYSTEM	OTHER TRANSFER SYSTEMS
nuclear, petroleum, coal, gas	subsidizes capital investments and energy intensity while increasing relative costs of employing workers; does not differentiate between non-renewable and renewable energy-resource investments nor between job-intensive versus job-destroying investments	taxes employment, thus pushing the economy in the direction of relatively greater capital/energy intensity vis-à-vis labor	*Highway Trust Fund subsidizing highways and auto transportation vis-à-vis less energy-intensive transit modes
favors excessively capital/energy-intensive industrial sector, services sector, and public sector		Rate hikes and legislation to help make the Social Security System solvent, achieved this goal at the expense of pushing the total economy into further capital/energy intensity	*tax advantages to real estate which encourage speculators
(Many energy analysts believe the economy can be shifted toward greater labor intensity	A new variation of investment subsidy is the Construction Work In Progress (CWIP) charges some utilities have been allowed to add to consumers' electricity bills. CWIP forces consumers to make risk capi-		*local tax incentives and holidays to encourage relocation, which result in much tax-induced movement

and less energy use simply by focusing on energy subsidies, taxes, and prices alone, without any reference to investment tax credits or other transfer systems still *increasing* energy intensity overall.)

tal available to utilities for capital investment in expanding their facilities even if it is denied in normal capital markets, i.e. an investment tax levied on consumers which further subsidizes capital and energy intensity.

and increasing general inflation rates.

of business, abandonment of older facilities in favor of new ones, overbuilding, etc.
*tax allowances for accelerated depreciation of capital equipment and facilities
*Subsidizing and socializing of insurance risks and costs of high-risk, capital-intensive technologies

Fig. 5

printing-press, Monopoly-money game to end, except in tax revolts, recessions, the burgeoning "underground" economy mentioned earlier, heightened social conflict, and the inevitable shake-out and hangover. Neither is there any *ideological* way to save energy: only thermodynamics and better science can help. For example, Britain debated proposals to utilize her power plants' waste heat (enough to provide all the country's space-heating needs twice over). Ironically, "free market," Conservative members of Parliament were those who favored a new government agency, the National Heat Board, to develop the district-heating system, even while allowing that the original wasteful design of power plants was due to the government monopoly, the Central Energy Board![13]

A final irony was the report to the congressional Committee on Government Operations that noted in late 1979 that the federal government itself was the largest user of energy in the United States and that its conservation record lagged behind the nation's, with the Department of Energy one of the worst offenders.[14] Meanwhile, citizens are accepting the responsibility of reeducating the "flat-earth" economists of the receding industrial age in the new concepts of long-term, sustained-yield productivity, of total ecosystem resource efficiency where renewable biomass productivity will be valued as highly as the conventional economists' capital, and are helping expand economists' horizons by promoting life-cycle costing, net-energy modeling, and the thermodynamic view of the importance of second-law efficiency (i.e., *net* energy, or end-use efficiency) in energy systems, so as to counter the economists' idealized "frictionless"-equilibrium models of supply and demand. For example, profits can be made in excessively entropic production processes or by disrupting human families and communities, because prices do not include many of these social and environmental costs, which later come back to haunt us. Today, our best physicists are trying to show economists that we can leapfrog the nuclear technology—a wasteful detour producing little net energy—and go straight to solar and renewable technologies. Dr. J. Benecke, physicist at the famed Max Planck Institute für Physik und Astrophysik in Munich, recently made such a cogent case in his paper at the Second International Symposium on Hadron Structure, in Poland, May 1979, entitled, modestly, "Some Reflections on the Nuclear Energy Issue."

Whether prices continue to rise or whether more direct forms of

Is this what America's 200th birthday is all about?

I hope we shall crush in its birth the aristocracy of our moneyed corporations, which dare already to challenge our government to a trial of strength and bid defiance to the laws of our country.

Thomas Jefferson

Plate 9

EXPORTING MALNUTRITION
THE BABY BOTTLE GOES ABROAD

Among the poor, who often lack
the financial resources to buy it in adequate amounts,
the fuel, refrigeration and pure water to prepare it safely,
and the education to use it properly,
COMMERCIAL INFANT FORMULA AND FEEDING BOTTLES CAN BE A THIRD WORLD DEATH WARRANT

Aggressive marketing practices of infant formula companies entice a growing number of women to abandon breast feeding for a substitute far more costly, less convenient and even hazardous when improperly used. In poor areas of the world, the promotion of infant formula represents commercial exploitation resulting in infant malnutrition and disease, a waste of natural resources, increasing dependence and negative development. Reversing the trend toward bottle feeding requires a halt to irresponsible corporate practices, a new direction in governmental aid programs and support of Third World initiatives to improve the health and welfare of their peoples.

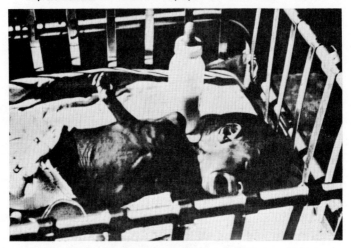

Soothe the cries of Third World babies with your voice and your pocketbook. To be successful this effort needs your support
In the stock market
At the supermarket
In your church and community

Join us through your contribution and participation in supporting shareholder resolutions, consumer actions and national legislation to end the bottle baby tragedy.

Comprehensive study/action packet
available for $3.50 from
INFANT FORMULA EDUCATION/ACTION PROJECT
INTERFAITH CENTER ON CORPORATE RESPONSIBILITY
475 Riverside Drive, New York, New York 10027   (212) 870-2750

Plate 10

Are America's businessmen cutting their own throats?

Are businessmen their own worst enemies? Surprisingly enough, it often seems that way. Consider, for example, the following news items:

The chairman of a major sugar company demands "relief from world market price levels." A trucking company president and an airline executive issue public statements opposing the deregulation of their industries. A food producer supports a government ban on saccharin. An oil company executive calls for decisions on energy production to be made by the president of the United States. And dozens of corporate spokesmen join the chorus advocating a government bailout of a multibillion dollar automobile manufacturer.

The problem.

What's going on here? Have America's business leaders gone mad? Why are they cutting their own throats—by voluntarily and systematically delivering themselves and their companies into the hands of government regulators?

The answer, of course, is simple. No, business executives do not share a collective death wish. They *think* they're gaining special advantages for their firms by approving and encouraging government intervention in the economy.

But they're deluding themselves. They are selling out their futures for a few short-term benefits. In the long run, by helping to make government powerful enough to destroy them, they will suffer the consequences of their blindness. And they'll deserve everything they get.

Today, America's free, competitive economy is in jeopardy. And a major cause of the problem is that *many businessmen actually prefer not to operate in a free market.* They may pay lip service to free enterprise. But at the same time, they demand tariffs, subsidies, licensing or some other form of political protection from the rigors of competition.

In this way, *business itself has become more of a threat to its own survival* than any of its enemies.

Is there any hope? Can anything be done to turn America's course away from socialism and dictatorship, and back toward freedom and prosperity?

We believe the answer is yes.

A new strategy.

Clearly, the time has come for a dramatically new strategy in defending the free market. Thus the need for a new organization: The Council for a Competitive Economy.

What makes the Council different from other business organizations? The answer can be summed up in one word: *principle*.

The Council speaks out aggressively in defense of the free market. Unlike other groups, we oppose not some but *all* forms of government intervention in the marketplace: tariffs, subsidies, entry restrictions, regulatory cartels and all the other special privileges designed to help some businesses at the expense of competitive businesses and consumers.

In the halls of Congress, in the news media, before the public—the Council is there, demanding an end to the strangling regulations and crippling taxes that stifle enterprise and innovation.

In its support of economic freedom, the Council is neither compromising nor apologetic. We do not ask forgiveness for profits; we defend the entrepreneur's right to them. We do not plead for exceptions and "special cases"; we advocate competitive freedom for every industry and every company.

In short, the Council for a Competitive Economy takes a principled stand for the liberty of the producer.

Isn't it about time?

The program.

The Council for a Competitive Economy has its national headquarters in Washington, with a full-time President and staff. A broad range of activities has been launched, including the following:

Lobbying in Congress to fight proposed controls on business and to work for the repeal of existing ones. *Cooperative action* with other like-minded groups to deregulate specific industries. *Conferences* to bring businesspeople together with economists, journalists and congressmen who share our position and concerns. *Publications*, including research reports, "Congressional Watchdog" bulletins, and "Competition," our monthly newsletter. And *advertising and public relations* to bring our free-market message to businessman and consumer alike.

A time to take a stand.

Never before in America's history has there been an organization of businessmen which consistently defended the market economy against all government interference. The Council for a Competitive Economy is bringing a dynamic new element into the struggle to make America what it can and should be.

We'd like to tell you more about the Council—and what's in it for you. For information, just write us at the address below. There's no charge and no obligation.

Together we *can* make a difference. We *can* turn the tide and restore America to a healthy, productive competitive economy.

After all, if businessmen don't rise to defend the free market, who will?

> 66 **[The struggle must begin] with that courageous group of businessmen who have decided to fight openly for the free-enterprise system. . . . A group of genuinely principled businessmen must be organized who will refrain from asking for one cent of the taxpayers' money, who will honorably accept the risks and penalties of freedom along with its great rewards.** 99
>
> —William E. Simon,
> *A Time for Truth*

Write for more information or use Reader Service card.

Council for a Competitive Economy

410 First Street S.E.
Washington, D.C. 20003

Plate 11

Just what the country needs:

another group of radicals.

We send out newsletters and volunteer speakers (usually in two's: one old and one young).

You can help us now, by sending a contribution to the Gray Panthers Project Fund, either in your name or in the name of a relative or a friend.

In effect, you'll be helping yourself. All of us won't be around to reap the rewards of our work; but chances are, you will.

Wrinkled and smooth; old and young;
Radical because we're proud;
We're the Gray Panthers.

We refuse to be cast aside at age sixty-five or at sixteen because society deems it convenient.
> Because for us it's not convenient.
> And for society, it's a tragic waste of talent, experience and energy.

We're a coast-to-coast network of people of every age who are determined to conquer the causes of discrimination against anyone on the basis of age.
> And, in our own deliberate way, we're accomplishing a lot.

We go straight to the places where policies are made. We get after medical societies, educational institutions, Congress, banks, nursing homes, employers--and any hearing aid dealer who dupes someone into an over-priced unit that self-destructs a week later.
> We get the word across on the Today Show, the Tomorrow Show, the Mike Douglas Show, the Phil Donahue Show, Bill Baker's Morning Exchange.

GRAY PANTHERS

Gray Panthers
National Headquarters
3700 Chestnut Street
Philadelphia, Pa. 19104

Name_____

Street_____

City_____

State_____Zip_____

(Note: if you are giving on behalf of a friend or relative please enclose that person's name and address, and tell us how you'd like your gift card signed.)

Plate 12

allocation become necessary, or a combination of both is needed in order to stem the decline of the dollar, the shifts and substitutions moving us toward a renewable and solar-based energy economy will be inevitable. A key question will be the extent to which the shift will be managed with the minimum inequity, inflation, and dislocation. This, in turn, will be determined by how well our leaders can clarify the debate and the options in spite of the confusion of economists, most of whom simply rely on inflating prices or a recession to bring about lower energy consumption as in the engineered recession of 1980–81, which did drive down energy demand, but at a staggering cost in employment and production. This lesson may show that managing the shift to a postindustrial phase will require many additional measures that are not economic. Economists now must admit that they overlook the extent to which we have "externalized" structural unemployment, social displacement, and the destruction of biological productivity. Indeed, we now find to our alarm that there may be an even more serious shift in the planet's atmosphere; scientists speculated at the July 1978 World Conference on Future Sources of Organic Materials, in Toronto, that the planet's biomass, in toto, may have now shifted from being a net producer of oxygen to a net producer of carbon dioxide (as we continue to destroy climax ecosystems and replace them with monoculture systems).[15] The possibility of such a situation is vastly more significant than most of the news headlined by the world's media, and yet it received no attention at all.

Clearly, an economic epoch is drawing to a close. We are witnessing not just the end of the age of petroleum but the end of a global economic order based on maximizing world trade and global "efficiency" measured by the single coefficient of prices and GNP-measured growth. Such a system reaches its logical conclusion at some unreal, hypothetical "global equilibrium" only when it has disordered every local social system and disrupted and depleted every local ecosystem; i.e., in game theory terms, the winner of the current world trade game is that country which can achieve maximum GNP-measured growth in its monetized sector by using the *most entropic* forms of production and consumption!

For the past fifteen years I have been asserting that the evolution of industrial societies would involve a shift from the simple, brute-force "meat-ax" technologies, based on cheap, accessible resources and energy, to a second generation of more subtle, refined technologies grounded in a much deeper understanding of biological and ecological realities. Whether these involve more-efficient use of the planet's daily solar income will require much more sophisticated scientific knowledge of organic and ecological systems, rather than concentration on inorganic, mechanistic, physical, and engineering systems, important though they will continue to be.

This shift of focus from the inert and inorganic to a deeper knowledge of the organic complexity and dynamism of bioecological systems constitutes my definition of the postindustrial revolution. The new scientific enterprise will also involve a shift from our focus on "hardware" to "software"; for example, the concept of production will no longer automatically conjure up instant visions of a factory, a machine, or any hardware at all. We will more carefully model the problem of production in its larger social and ecological dimensions: define and redefine it, review diverse options, scan ecological systems for signs of productive potential we might tap or augment before any investments are committed. As we are learning from the biological sciences, there are substitutable organic ways of meeting human needs of which we have hardly dreamed.

Aware citizens, as innovators of the renewable-resource-based solar age, will be called upon to play a part, as its political leaders and educators. For example, while scientists argue about carbon dioxide levels and climate, citizens are already acting on a remedy scientists agree on, planting trees and regreening the planet, while opposing further destruction of forests and farmlands. People can lead the world's economies back to sanity and to the basic realities in which human wealth and well-being have always been rooted: the *least* entropic forms of production and consumption that come with good design and engineering, the more sophisticated technologies of doing more with less, the integrated management and conservation of nonrenewable resources, responsible enhancement of agriculture and renewable resources, the recognition that the daily solar flux is our real income, and the careful maintenance of planetary biomass productivity. These are also the chief tasks ahead for humanity.

NOTES — CHAPTER 6

1 The full story on the costs of nuclear power began to be apparent in 1980. Although the nuclear industry kept relating its own continued expansion to our dependency on foreign petroleum, thus justifying ever increased costs, this argument failed to address the real issue. In most industrial societies, as pointed out by physicist Amory Lovins in *Soft Energy Paths* (1977), the energy shortfall is in overdependence on *liquid* fuels, for which electricity is not a substitute (unless an enormous long-term investment in replacing the world's automobiles with electric cars is attempted, an unthinkably costly scheme). Even today, the nuclear industry is still portraying itself as the savior of the U.S. energy situation in the face of situations such as the Iran shutoff of oil and the Mid-East conflicts.

Meanwhile additional costs of Three Mile Island kept drifting in, threatening to bankrupt its owner, Metropolitan Edison, which then sued the NRC hoping for a quasi-bailout by tax-funded damages. A survey of effects on residents in the Harrisburg area showed that 144,000 people fled their homes at a cost in evacuation expenses and lost wages of $18.2 million (Washington *Post,* September 24, 1979). Charles Komanoff, of Komanoff Energy Associates, of New York, author of the comparative cost studies of nuclear vis-à-vis coal-power plants for the Council on Economic Priorities, *Power Plant Performance* (1976), predicted a ballooning of nuclear costs in the 1980s. Komanoff noted in a 1979 paper that reactors operate at an average of only 60 percent of their rated capacity (well below the 80 percent predicted by boosters) and that the larger plants of recent years run at only 50–55 percent. Thus although the nuclear plants built in the early 1970s produce power at roughly the same cost as coal-fired plants, the new reactors are *more expensive* than coal plants, even those fitted with high-efficiency pollution controls cutting emissions to one fifth of former levels. By the late 1980s, Komanoff predicts, nuclear electricity will cost *twice* as much as coal power. Also, the truth of the horrendous unaccounted costs of decommissioning reactors as they end thirty-year life-spans came home as the Richland, Washington, reactor was scheduled for dismantling, the first of some fifty reactors that must be retired over the next twenty years or so. The job of dismantling Richland will begin in 1982 and not be completed until 1984, at a cost (not counting inflation) of about $32 million. At present, some cost estimates of decommissioning a full-size reactor like one of those at Three Mile Island, or the similar-design, troublesome reactor at Crystal River, near Tampa, owned by Florida Power and Light Co., are approximately 20 percent of the cost of building the reactor in the first place (*The Christian Science Monitor,* January 28, 1980).

2 Although the three dumps were reopened after bitter debates, the issue is still highly politicized, and activists are increasingly demanding that their states pass local ordinances to prohibit nuclear wastes being transported through their areas. This has triggered a new federal-state battle as federal Department of Transportation officials tried to overrule states' rights to control local access to highways and railroads for dangerous cargoes. By spring 1980, at least eighty-two

communities in the United States had passed waste-transportation restrictions, and the problem of where to store the growing backlog of nuclear wastes was becoming a hot potato. After the unconscionable delays of decades, the Carter administration faced up to the nuclear waste-storage crisis by announcing a plan to locate sites and repositories around the country for long-term storage of highly radioactive wastes; licensing of such storage sites would be overseen by the Nuclear Regulatory Commission. His administration also dropped the long-held myth that the technology for safe storage is available, and admitted that problems persist (*The Christian Science Monitor,* February 14, 1980).

[3] *Power Line,* Vol. 5, #12, July 12, 1980, Washington, D.C.

[4] New York *Times,* January 24, 1980.

[5] Thomas Stauffer and Peter Navaroo, "What Carter Juggles Under His Oil-Import Ceiling," editorial in *The Christian Science Monitor,* October 11, 1979. On January 3, 1980, the New York *Times* called on President Carter to resume filling the petroleum reserve stocks on the Gulf Coast, noting that its stocks were down to 100 million barrels—two weeks of import replacement—and proposing that the reserve be filled by conserving just 1 percent of the 18 million barrels a day we currently burn, by reducing just a small amount of summer driving on 1980 vacations. However, stockpiling annoys OPEC and tempts price increases.

[6] A survey by the New York business group, the Conference Board in February 1980 reported that 45 percent of U.S. families said that they could reduce their driving by more than 10 percent, with many suggesting cutbacks of 20 percent. Based on surveying five thousand households, the report said that an average 11 percent reduction was not even perceived as a hardship and would produce an annual saving of 200 million barrels of oil. Further, if the government were forced to take action to reduce driving, 53 percent of U.S. families favored gasoline rationing over other measures such as rationing by price or Sunday gas-station closings (*The Christian Science Monitor,* February 11, 1980).

[7] The nuclear industry counterattacked with a multimillion-dollar propaganda barrage, hiring public-relations strategists Mark Harroff and Jay Smith to launch Campus America, sending industry engineers and scientists to promote nuclear power and combat antinuclear activists (*Fortune,* January 28, 1980, pp. 108–10). Meanwhile the evidence that nuclear power was simply not needed grew. In a paper for the Bellerive Colloquium, held in Geneva in February 1979, entitled "Is Nuclear Power Necessary?" Amory Lovins cited a growing number of energy studies by governments and universities throughout the world, which showed that other, more cost-effective technical measures exist to use energy far more efficiently and that we can reduce the need for both total energy and electricity, despite increasing affluence. Furthermore, Lovins points out that nuclear power is physically unable to provide timely and significant substitution for oil, and the maximum potential role for nuclear power is in providing base-load electricity for electricity-specific applications, which form only 4 percent of all end-use energy and only 10 percent of primary energy. (Papers by Lovins and other international scientists advocating alternative renewable energy are published in *Soft Energy Notes,* subscription $25 yearly from 124 Spear Street, San Francisco, CA 94105.)

Economist Vince Taylor reported similar conclusions in a study for the U. S. Arms Control and Disarmament Agency, stating that nuclear-plant construction starts could be halted "with no major adverse effects on the U.S. or other non-communist countries' energy security or economies through 2025" (*Nucleonics Week,* January 25, 1979).

[8] Earl Cook, Distinguished Lecture, Louisiana State University, March 6, 1978, Baton Rouge, Louisiana.

[9] Commissioner Ronald Doctor, California State Energy Commission, *Moving Toward an Effective Federal Program for Commercializing Solar Energy*, Department of Energy Forum, Los Angeles, California, June 15, 1978.

[10] Herbert Inhaber, *Risk of Energy Production*, March 1978, Atomic Energy Control Board of Canada, Ottawa, Ontario K1P 559, Canada.

[11] The General Accounting Office report to Congress *Iranian Oil Cut-off: Reduced Petroleum Supplies and Inadequate Government Response*, September 13, 1979, followed earlier reviews of the Energy Department's performance and forecasting shortcomings by the Office of Technology Assessment of Congress. It pointed to the Department's "ineffectiveness in providing timely, accurate, complete energy data and analyses" and concluded that DOE "did not provide the Congress and the public with credible, convincing explanations of the status of gasoline, diesel fuel, and home heating supplies." Factors leading to these inaccuracies included the use of unverified industry figures, failure to define and measure demand, revision of base-line data, and other statistical errors.

[12] *Global 2000*, Presidents Council on Environmental Quality, July 1980.

[13] *The Christian Science Monitor,* November 6, 1979.

[14] Washington *Post,* "Study Rips Rate of Federal Energy Use," November 7, 1979.

[15] Chemical Resources Applied to World Needs Conference, Toronto, Ontario, Canada, July 10–13, 1978. International Union of Pure and Applied Chemistry, abstracts from Multiscience Publications, P.O. Box 1464, Station B, Montreal, P.Q., Canada H3B 3L2.

PART TWO

A Look Back:
Economics as Politics
in Disguise

CHAPTER 7

Economists as Apologists

The notion that is developing today, in the healing arts and sciences, that a person is responsible for his/her sickness can also be applied usefully to a culture and its responsibility for its social diseases. This responsibility lies, of course, in the value system it has chosen. A society's value system will both engender certain types of social/economic/political/technological arrangements and determine the nature of both individual and institutional social stresses and pathologies, as well as adaptation patterns and possibilities.

Thus the values a society lives by are a key determinant underlying not only its sociotechnical structure but also its world view, knowledge, scientific enterprise, and economic system. Values are integrally related with epistemology, and since both individuals and societies are faced with the basic cognitive dilemma that there is always an infinite number of phenomena and sets of data to examine, the choice of what realities to pay attention to is always value-based. Once cultures have expressed and codified these collective, subjective sets of values and goals, these then represent at any period, the parameters of their options, choices, perceptions, insights, and innovative potential for social adaptation, regeneration, and cultural evolution. As a cultural value system changes, new sets of options and potential new patterns for cultural evolution emerge. Human cultural value systems have always changed, often when presented with natural environmental challenges or environmental changes caused by human activity. Sometimes these human value shifts are survival-oriented and occasionally, they are extremely rapid, for example, the value shift in Maori culture, in New Zealand, after guns were introduced by the British (see Andrew Vayda, "Maori Conquest in Relation to New Zealand Environment," *Journal of the Polynesian Society,* Volume 65, Number 3, pp. 204–11). Sometimes there are maladaptations such as that of the tribe known as the Ik, in Tanzania, who were displaced from their habitat, and disintegrated behaviorally (see "Plight of the Ik and Kaiadilt Is Seen as Chilling Possible End for Man," *Smithsonian Magazine,* November 1972).

One of the most herculean works of our time that has attempted to trace the key role of values in structuring human societies in all their aspects is the four-volume lifework of the late Pitirim A. Sorokin, *Social and Cultural Dynamics* (1937–1944). Sorokin, a Russian who served briefly in the ill-fated government of Alexander Kerensky, in 1917, was expelled by the Soviets in 1922 from his post as professor of sociology at the University of St. Petersburg, then emigrated and founded the Department of Sociology at Harvard in 1931, which he chaired until retiring, in 1955. Sorokin's unifying principle for the synthesis of Western history uses the basic concept of cyclical waxing and waning of three basic value systems undergirding its cultures, empires, belief and knowledge systems, law, arts, and technologies. These three types of value systems are the Ideational, the Sensate, and the Idealistic. Briefly, the Ideational value-principles hold that true reality lies beyond the material world, in the omniscient, eternal, ever-present universal creation and in an omnipotent creator and in the existence of absolute, superhuman standards of justice, truth, goodness, beauty, and rationality. Sorokin points out that its Western expression included Christian and Judaic monotheism but that similar ideas are expressed differently in Brahman India, Buddhist and Taoist cultures, and Greek culture from the eighth through the sixth centuries B.C. The Sensate value system is profoundly different, i.e., that true reality and values are sensory. "Only what we see, hear, smell, touch, and otherwise perceive through our sense organs is real and has value. Beyond such a sensory reality either there is nothing, or if there is something, we cannot sense it; therefore it is equivalent to the non-real and the non-existent" (Sorokin, Lowell Lectures, February 1941, in *The Crisis of Our Age,* Dutton, 1941).

Sorokin contended that the rise, maturation, overripening, and decay of these two basic expressions of human culture and their cyclical rhythms also produced an intermediate, synthesizing stage, the Ideal-istic, which represented their blending, i.e., that true reality is partly supersensory and partly sensory and that there is an infinite, mani-fold unity in which these aspects coexist. Idealistic cultural periods thus tended to attain the highest and noblest expressions of both Ideational and Sensate styles, producing balance, integration, and aesthetic achievement in art, music, science, technology, law, and philosophy, such as the Greek flowering of the fifth and fourth centuries B.C. and the culture of the thirteenth and fourteenth centuries in Europe. These three basic patterns of human cultural expression, according to Sorokin, produced identifiable cycles of ap-proximately three to four hundred years, with alternating periodici-ties that he plotted on dozens of charts for subsystems including wars and internal conflicts, systems of truth and jurisprudence, socioeco-nomic institutions, and technological development. Other fascinating charts include fluctuations in styles of architecture, sculpture, and lit-erature, as well as in concepts of time (e.g., linear versus cyclical), space, liberty, and universalism versus singularism.

Sorokin's grand scheme was widely admired, but inasmuch as it collided with the extreme, reductionist, empirical style of U.S. sociol-ogy in the 1950s (e.g., the Talcott Parsons school), it was considered too broad or "not rigorous." However, I think his method was well chosen, for the range and time scale he studied did not strive for minute, unattainable exactitude, and well illustrated the shift now needed from classical, Newtonian concepts of empirical proof. I feel it will enjoy a new vogue in the next decade, precisely because of its disciplined holistic sweep. German sociologist L. von Wiese may have been predictive when he said that "in comparison with Sorokin's great work, the works of Comte, Spencer, Pareto, and Spengler ap-pear to be arbitrary and fanciful" (Sorokin, *Social and Cultural Dy-namics* [Biographical Notes], one-volume edition, Porter Sargent, 1957, pp. 719–80). Perhaps Sorokin is the best augur of a shift from the reductionist empiricism of the 1950s to a more sophisticated, holistic sociology with much larger, indeterminate probability pat-terns.

Sorokin's work supplies a key aspect to my discussion of eco-nomics, as his model fits well with my own concerning the decline of the industrial age, except that he saw the current crises of Western culture more broadly, in a longer historical context, as another period of maturation and decline of Sensate culture. The rise of our current Sensate era followed the ascendancy of the Ideational period of the rise of Christianity and medieval theocracy and their subsequent flowering into an Idealistic stage in the thirteenth and fourteenth centuries (i.e., the European Renaissance). It was the slow decline of these Ideational and Idealistic epochs in the fifteenth and sixteenth centuries that produced the rise of a new Sensate period in the seventeenth, eighteenth, nineteenth, and twentieth centuries (roughly the Enlightenment period up to the present). Thus Sorokin predicted, in 1937, with great prescience the upheavals we are witnessing today in Western technocratic societies and industrially maturing societies as "The Twilight of Sensate Culture" (*Social and Cultural Dynamics,* pp. 699–704).

Sorokin saw that "Western culture is entering the transitional period from its Sensate supersystem into either a new Ideational or an Idealistic phase; and since such epoch-making transitions have hitherto been periods of the tragic, the greatest task of our time evidently consists, if not in averting tragedy, which is hardly possible, then, at least, in making the transition as painless as possible. What means and ways can help in this task? The most important . . . consists in the correction of the fatal mistake of the Sensate phase, and in preparation for the inevitable mental and moral and sociocultural revolution of Western society." Sorokin outlines five essential steps:

1) Realization that we face no ordinary crisis, but one of the *great* transition phases, which he has charted in previous cycles of human history. This diagnosis can help us devise remedies of adequate scope, so that our "sociocultural physicians" will not treat dangerous cultural pneumonia with "surface-rubbing medicines," as if it were a cold.

2) Recognition that the Sensate form of culture is not the only great form of culture, nor free from many defects and inadequacies, and that Ideational and Idealistic cultures are also great, but *different.*

3) The need to shift from one basic form to another when one of these great forms of culture begins to show signs of exhaustion (as they all do after some period of dominance)—in today's case, from the disintegrating Sensate to the Ideational or Idealistic form. This shift should not be opposed, but *welcomed* as the only escape from agony, collapse, or mummification.

4) The concerted preparation for the shift implies the deepest reex-

amination of the main premises and values of Sensate culture, rejection of its now exhausted "pseudo values," and a re-enthronement of some of the real values it discarded. Among these are the balanced, integrated conceptions of sensory and supersensory reality. "From the integralist standpoint, the present *antagonism* [italics added] between science, religion, philosophy, ethics, and art is unnecessary, not to mention disastrous" (ibid. p. 317).

5) Such transformation of the mentality of Western culture must naturally be followed by a corresponding transformation of *social relationships* and forms of *social organization.* "Neither capitalism nor socialism nor communism nor totalitarianism; neither mechanical individualism nor mechanistic collectivism; aristocracy nor democracy; etc., is an absolute value. Even the values of the nation-state and private property have outlived their period of greatest service to mankind. Superficial measures of economic or political readjustment will not suffice: the remedy demands a change of the contemporary mentality and a fundamental transformation of our system of values and the profoundest modification of our conduct" (Lowell Lectures, pp. 315–21).

Thus Sorokin's work provides a superb context for the current cultural reevaluation in all its major aspects, whether the shifting paradigms in physics, psychology, and medicine, or those in technology, economics, and social organization, with which we are dealing; as well as a context for the metaphysical reconstruction increasingly discussed today.

Because economics has monopolized the debate over resource allocation and claims to define, in various epochs, what is valuable, it is the most crucially placed of contemporary disciplines. Since economics is defined as the discipline dealing with the production, distribution, and consumption of wealth, it is also the quintessential expression of Sensate values. In fact, the emergence and separation of the discipline of economics from philosophy and politics coincides with Sorokin's mapping of the emergence of Sensate culture in Western Europe and its increasing domination over the medieval, otherworldly values that were rigidifying during the fourteenth and fifteenth centuries. Until the sixteenth century, there was no isolation of purely economic phenomena from the fabric of life. As Karl Polanyi documented in his *Primitive, Archaic and Modern Economics*

(Doubleday Anchor, 1968), throughout most of history basic resources, food, clothing, shelter, medicine, etc., were produced for use-value and/or redistributed within tribes and groups according to the two basic value systems of *reciprocity* and *redistribution.*

In *Dahomey and the Slave Trade* (1966), Polanyi examined an economic value system of ceremonial redistribution by the king. In *Trade and Market in Early Empires* (1957), Polanyi illuminated the normative basis of economics by outlining the framework for studying economies that were not industrialized or organized by market institutions. In *The Great Transformation* (Beacon Press, 1944), Polanyi studied the rise in seventeenth-century England of the novel economic organizing principle of institutionalizing a *national system of markets* and the gradual spread of this unusual economic form as we see it today in our interlinked, global "marketplace." Polanyi's systemic view enabled him to predict with accuracy that this attempt to optimize production and exchange values would exact an inevitable price in suboptimizing other social and ecological systems: in the catastrophic dislocation of agrarian culture and resulting poverty and misery of millions of landless former peasants, and in the gradual destruction and depletion of the environment.

Markets had been fairly common since the Stone Age, but always isolated, local, and almost incidental to economic life. Human economic activities had always been embedded and submerged in the general social relationships. Even early trading had little economic motivation, but was more often sacred, ceremonial, territorial, or related to kinship and familial customs. For example, Trobriand Islanders, of the Pacific Melanesian chain, had sea trading routes stretching for thousands of miles. This so-called Kula Ring involved circular voyages with no profit, barter, or exchange motives, but, rather, the etiquette and magic symbolism of carrying white seashell jewelry in one direction and red seashell ornaments in the opposite direction, so as to encircle the entire archipelago every ten years (*The Great Transformation,* p. 50). The very idea of profit, let alone interest (called usury) was either inconceivable or banned, while haggling was decried and gift-giving considered a virtue as a cohesive, survival-oriented behavior or high-status-conferring activity, as in the famous potlatches of the Indian tribes of the western coast of North America.

Many archaic societies used all kinds of money, including metal currencies, as for example in ancient Egypt, but they were used for payment of taxes and salaries, not for general circulation. Economic organizations of vast complexity and elaborate divisions of labor were operated entirely by the mechanism of redistribution, as, indeed, were all systems of feudalism through various forms of collection, storage, and redistribution focused on the chief, the lord, the despot, or the temple. Thus we see that social organization, rank, and caste (often divinely ordained) operated as resource-allocation systems (a point that Karl Marx would drive home concerning nineteenth-century capitalism). Another important principle was that of "householding," i.e., production for one's own use, which the Greeks called *oeconomia,* the root word, of course, of "economics." The motive of individual gain for economic activities was unknown in early societies; the focus was generally on production and storage for the self-sufficiency of the group, the household, the village, the tribe, and later the seigneurial manor of feudalism (pp. 52–53). Of course, this did not preclude age-old motives of power, domination, etc., which lead to territorial wars and conquest. Aristotle, too, insisted on production for *use,* as against production for *gain.* Even the highly developed trade of the Graeco-Roman period was characterized by the redistribution of grain by Roman administrators in an otherwise household economy. Aristotle believed that trade was "natural" as long as it was a requirement of group self-sufficiency and prices were "just," i.e., if they conformed to the values of the community and thereby strengthened its goodwill and cohesiveness. In contrast to the economics that emerged during the rise of the Sensate culture of the Enlightenment, Aristotle rejected the idea that human needs are boundless and that there is a scarcity of subsistence in nature. If there was a *perception* of scarcity, it must be attributed to a misconception equating the "good life" with a desire for greater abundance of physical goods and enjoyments. Aristotle believed that the elixirs of the good life—the elation of the theater, jury service, holding public office, festivals, even the thrill of battle—could not be hoarded or physically possessed. However, we see that these components did rest on *leisure* and, therefore, were partially underwritten by slave labor, not to mention the domination of women.

Market trade for money arose in Greece during Aristotle's time but was restricted to low-class persons or aliens, termed hucksters. The first record of markets was the Agora, of the sixth and fifth centuries B.C. Aristotle offered a formula by which a rate for such exchanges should be set. It is given by the point at which two diagonals cross, each of them representing the *status* of one of the two parties to the exchange (Aristotle, *Politics,* 1133a, 10, in Polanyi, *Primitive, Archaic and Modern Economies,* p. 107). Polanyi points out in "Aristotle Discovers the Economy" (pp. 113–14) that later economists used incorrect translations of Aristotle's *Politics* and *Ethics,* which altered the meaning and his use of the word "metadosis," which in Greek usage meant "giving a share." Fatally for latter-day theorizing, "metadosis" was translated as "exchange," leading to the belief held by Adam Smith and his contemporaries that exchange and a "propensity to barter" must be a trait of human nature. Aristotle's view on what today we would call "economics" was reasserted in the teachings of St. Thomas Aquinas (1225–74). Such matters as "just prices" fell under moral, not secular, law (see Erich Fromm, *To Have or to Be,* pp. 7 and 59, Harper & Row, 1976) and that private property was justified only inasmuch as it served the welfare of all. In fact "private" comes from the Latin *privare,* which means to *deprive* others, showing the widespread ancient view that property was first and foremost *communal.* As cultures moved from this communal, participatory, systemic viewpoint to the more individualistic, reductionist viewpoint, they no longer thought of private property as those goods that individuals deprive the group from using, but actually *inverted* this logical position and held that property should be private and that the society should not deprive the individual without due process of law.

Such communally oriented economic concepts were also perpetuated in Europe by the German Dominican theologian Meister Eckhart (1260–1327), who also stressed ideas similar to the Buddhist concepts of nonattachment and overcoming craving for things as well as for ego gratification. Like Aquinas and Aristotle, Eckhart taught that the "good life" was one directed at virtuous activities, self-control, inner peace, contemplation, and spiritual knowledge. Such Western ideas are very close to the Buddhists' "right livelihood" and form a continuous thread in European history through the Middle Ages, the Renaissance, and the Reformation, until the seventeenth century. As the return of the Sensate blossomed in the age of Enlightenment, concern for *this* world gradually began to reassert itself for the first time since the height of the Roman Empire. Thus, the value base for all human endeavors began to shift again toward the sense-dominated empiricism marked best by the philosophies of René Descartes (1596–1650), whose *Discourse on Method* propounded the idea of proceeding from *doubt* (rather than faith) and verifying facts empirically. Descartes hoped thereby to contribute to a "universal science of quantity" (*Descartes: The Project of Pure Enquiry*, edited by Bernard Williams, Penguin Books, England, 1978, p. 16).

This philosophical approach, which was a rejection of religious dogma and developed from the investigations of Descartes's older contemporaries Francis Bacon, Galileo, and Johannes Kepler, led to the greatest flowering of scientific and technological achievement since the earliest civilizations of China, as well as toward the materialistic goals of production of worldly goods and luxuries, the increasing domination of nature, and the rise of the manipulative rationality of the industrial age. This, in turn, gave rise to specific rationales for such goals and value shifts and the new institutions they created in law, custom, and political life. The new academic pursuits it engendered gave rise to a proliferation of theorizing about a new set of specific phenomena that suddenly stood out in sharp relief: economic activities, production, exchange, distribution, moneylending, trade, and merchant "venturing"—all of which not only required description and explanation but also *rationalization*.

Social values are supported by a coherent world view and belief system. As Weisskopf emphasizes in *Alienation and Economics* (Dutton, 1971, p. 33), any set of social values is maintained by psychosocial repression of those modes of experience, expression, and behaviors not *conforming* to the dominant value system (e.g., hedonism and sexual freedom are repressed in Puritanical societies; market and profit-oriented behavior is repressed in the U.S.S.R. and other Eastern-bloc countries). Thus social systems are buttressed by a *conforming reason*, i.e., rationalization, legitimation, and justification of their particular profile of value *ex*pressions/*re*pressions (note the complementarity!). Conversely, social systems are *undermined* when *critical reason* emerges, since it turns against the *ex*pressed value system and sets the stage for the return of the *re*pressed values.

This distinction between *conforming reason* and *critical reason* is a key to today's processes of social transition, and to the earlier transition we shall now examine in Europe, as sixteenth- and seventeenth-century mercantilism, feudalism, and the divine right of monarchs gave way to the liberalism and social revolution of the Enlightenment, which ushered in the period of classical and "laissez-faire" economic individualism, representative government, and property rights. This Enlightenment period was midwived by a crescendo of *critical reasoning* against feudalism and aristocracy. "Natural law," always invoked as a higher court in periods of social criticism, was invoked again as a more comprehensive frame of reference from which to argue and engulf the old values, and as a way of displaying their limited relativity to the newly proclaimed absolute.

As the new values of liberalism, free markets, individualism, representative government, and bourgeois institutions became entrenched, a new body of *conforming reason* developed to buttress them. Again "natural law" was invoked, this time to serve the conforming reason, for example, Adam Smith's claim that there is a propensity to barter in human nature. Thus the free-market ideology became during the eighteenth and nineteenth centuries an example of a confirming ideology, together with the reversal of the concept of communal property and the assertion of the individual's primary rights in property at the expense of the group.

Today, as liberal, "laissez-faire," and free-market values and institutions lose their organizing power, a new wave of critical reasoning has emerged. This reasoning also invokes new concepts of "natural law" as a higher court for indictment of the old values: ecology; planetary awareness; and a new political world order based on biogeological regions, ocean ecosystems, climate and meteorological processes, concepts of population-carrying capacity, conservation of nonrenewable resources (as the earth's capital), and sustained-yield utilization rates for renewable resources (earth's income), as well as new views of the nature and potential of an emergent, fully actualized human person. These new invocations of natural law and the powerful imagery they conjure up are now arrayed in accusation of the limited theories of classical economics (which was fairly consistent with the population/resource ratios of its time), as was the empirical reductionism of Descartes, the atomistic individualism of John Locke, the classical mechanics of Isaac Newton's and Gottfried Leibniz's differential calculus, and the equilibriating, "laissez-faire" theories of François Quesnay and Adam Smith.

Economics, as we now see, has always been the most clearly value-based and normative of the social sciences, since it was founded on such drastic reevaluations of new human power to manipulate the natural world and human behavior and the change of goals such powers engendered. The earlier moral strictures were gradually buried or refuted and an entirely new body of theorizing began to develop around the new values, pursuits, behavior patterns, and their societal institutions and consequences.

It is worthwhile to note another sweeping synthesizer of value systems that gave rise to specifically materialist forms, institutions, and economic systems: Max Weber (1864–1920). Weber, who used many of the ideas and methods of Karl Marx, traced the development of capitalism from a value-system and ideational viewpoint, rather than the more materialistically determined Marxian view. Weber is also a distinguished representative of one of the interesting divergences within the discipline of economics: the tradition of branching into the study of economic history. Since it became progressively more difficult to study and profess economic "science" in anything but an ever more compartmentalized manner, those critical economists who wished to study economic phenomena as they actually existed, embedded within society and cultures, and who dissented from the narrow economic viewpoint, were almost forced into designating themselves economic historians. Weber, a critic of capitalism, took this stance, and the practice continues to this day. In fact, a time-honored way by which economists counter these more systemic studies is by reading these scholars out of the economics fraternity; J. K. Galbraith and Robert Heilbroner are often designated as "sociologists"; Kenneth Boulding is referred to as a philosopher. These designations are a substitute for addressing the issues they have raised. Similarly, Marx was faulted for his mathematics in order to avoid dealing with his evidence, and academic rewards went to the reductionists who could exclude most of the larger social and ecological variables by "externality" theories.

The renowned Swedish economist Gunnar Myrdal puts it succinctly: "Among the social scientists, the economists, by their so-called 'welfare theory,' provided themselves with a vast and elaborate cover for their escape from the responsibility to state, simply and straightforwardly, their value premises in concrete terms" (Myrdal, *Against the Stream,* Vintage, 1975, p. 149). Another rebel, Kenneth

Boulding, speaking as president of the American Economic Association, called this concerted attempt to avoid the issue of unstated values "a monumentally unsuccessful exercise . . . which has preoccupied a whole generation (indeed several generations) of economists, with a dead end, to the almost total neglect of the major problems of our age" (p. 149). Myrdal adds that economists have completely disregarded modern psychological research on people's behavior as income earners, consumers, and investors, because the results of such research cannot possibly be integrated into their conceptual framework (p. 150).

No wonder the Weberian tradition persists! By contrast, Karl Marx refused to call himself an economist, baldly asserting that economists were no more than apologists for the existing capitalist order. Max Weber was more cautious and thus provoked a passionate mainstream dialogue with economists after his book *The Protestant Ethic and the Spirit of Capitalism* (first published in German in 1904–5) was translated into English in 1930. Max Weber also studied archaic "economic" arrangements in Babylon, ancient Egypt, China, India, and medieval Europe, looking for contrasts with the capitalist system. He singled out two requirements of a rationalized capitalistic enterprise: 1) a disciplined labor force and 2) regularized investment of capital as its most profound distinctions from traditional modes. The implications of both accumulation and reinvestment and a disciplined labor force were crucial: "Man is dominated by the making of money, by acquisition as the ultimate purpose in life. Economic acquisition is no longer *subordinated* to man as the *means* for the satisfaction of his material needs" (Max Weber, *The Protestant Ethic and the Spirit of Capitalism,* Scribner's, 1958, p. 53). Weber then presented the ingenious thesis that the religious idea of a "calling," which emerged with Martin Luther and the Reformation, and its moral obligation to fulfill one's duty in *worldly* endeavors, provided the essential emotional drive and energy of capitalism. This idea of a worldly calling projected religious behavior into the secular world and became even more demanding in the Puritan sects, particularly Calvinism, Methodism, Pietism, and Baptism. This performance of "good works" and worldly activity and the material rewards that not surprisingly accrued to such industrious behavior were gradually seen as a sign of predestination, of being one of those chosen by God. Thus, hard, self-denying work, as-

cetic self-control, and worldly success began to be equated with virtue. However, since all but frugal consumption was abhorred, accumulation needed to become sanctioned as long as it was combined
with an industrious career. Weber's dialectical thesis that such Puritanical religious motives formed part of the mainspring of capitalism
was criticized from all sides by Catholic and Protestant theologians
(although it was never conceived as a book of comparative religion,
but merely forged a new tool of analysis). Marxist scholars also ob-
jected (probably more than Marx would have) to its emphasis on
values and ideas as determinants of economic phenomena.

Weber qualified his thesis as tracing "only one side of the causal
chain connecting Puritanism to capitalism" (p. 27). In fact, he listed
six fundamental socioeconomic factors that distinguished European
history from that of India and China:

1. The separation of the productive enterprise from the household in
   Europe vis-à-vis China, where kinship units remained the major
   form of economic production;
2. The development of the Western city, which set off bourgeois society from agrarian feudalism, whereas in Eastern cultures cities
   remained more embedded in the local, agrarian economy;
3. The existence in Europe of an inherited tradition of Roman law
   that allowed rationalization of social organization;
4. The potentiality—as a result—of the nation-state, of a wider scope
   of administration and bureaucracy than was possible in Eastern
   civilizations;
5. The development of double-entry bookkeeping in Europe, a
   major requirement for regularizing capitalistic enterprise;
6. That series of changes, also emphasized by Marx and later by
   Karl Polanyi, which produced an unattached labor force that
   could be commoditized. Whereas feudalism faded in Europe and
   set the stage for economic dynamism, its analogue in Eastern
   societies, the caste system, persisted and helped foreclose on such
   economic change.

The Marxian and Weberian tradition of critiquing economics from
sweeping historical and sociological analyses provided avenues for
later critics, notably the Frankfurt school at the Institute for Social
Research, whose influence through such thinkers as Theodor
Adorno, Max Horkheimer (its founder), Albrecht Wellmer, Ernst

Bloch, Erich Fromm, Herbert Marcuse, and Hannah Arendt, as well as Jürgen Habermas's criticism of the current scientism of today's system theorists and in our technocratic approaches to knowledge (see, for example, Habermas, *Theorie und Praxis*, 1963). Similar currents in sociology can be seen in the work of Weber's student George Lukács, *History and Class Consciousness* (1923); C. Wright Mills, *The Power Elite* (1956); and David Riesman, *The Lonely Crowd* (1961). In psychology, Erich Fromm's *The Sane Society* (1955) and *To Have or to Be* (1976) and Philip Slater's *Earthwalk* (1974) and *Wealth Addiction* (1980) are in the same tradition, as well as the work of economists who have broadened their concerns beyond the narrow mathematical virtuosity that is academically and financially rewarded, notably Adolph Lowe, Robert Heilbroner, Joan Robinson, J. Kenneth Galbraith, Barbara Ward, Kenneth Boulding, Gunnar Myrdal, and many others. The Marxians, who are academically acceptable in Europe and Japan, still find much discrimination in the United States; among them are Paul Baran and Paul Sweezy, authors of *Monopoly Capital* (1966); Sam Bowles, founder of the Union for Radical Political Economy; and Marc Linder, who while a student at Princeton University wrote a two-volume critique of Paul Samuelson's textbook *Economics* entitled *Anti-Samuelson* (1977). A sample of the incisive values-clarification derived from the Marxian method appears in Volume 1 as Linder dissects Samuelson's assumptions and biases in discussing the role of labor unions vis-à-vis business in today's mixed industrial market economies. "Business is treated in a business-like manner, that is to say, it is viewed primarily in its economic functions . . . business as a synonym for production becomes a higher category to which labor can then be subordinated (as in fact it is, to capital). . . . Not so labor . . . it is seen as a political foreign body in the economy." Linder adds, "In the first edition of Samuelson's text, the chapter on labor was entitled 'Labor Organizations and Problems'—business poses no problems." By thus stressing labor as a political force, Linder accuses Samuelson of mystifying the workers' economic role as *producer* and the history of labor unions as a movement of *producers* to win control over social production. Linder shows that by drawing a dividing line between the political and the economic, labor's role in capitalist production can be made into a relatively superficial phenomenon (Marc Linder, *Anti-Samuelson* Volume 1, Urizen Books, 1977, p. 124). We will return to the political role of reductionist economics as mystification.

Thus we see the powerful legacy of Marx's contention that the "facts" on the surface of most societies are rationales or concepts, not facts (Michael Harrington, *The Twilight of Capitalism*, p. 370).

Continuing this tradition of holistic studies of economic phenomena as embedded in culture and value systems are the even more fundamental critiques of the feminists, who see industrial societies and their underlying dominance-and-submission patterns as going even deeper than class conflicts between capitalists and workers, in their *patriarchal* structures whether capitalist, socialist, or communist, whether developed or developing, European, North or South American, Asian, or African. It is no accident that all the critiques and studies I have so far mentioned refer to males or "man" and "his" problems of alienation and of controlling "his" technology. Industrial societies and their pathologies are also patriarchal and have designated their most highly valued traits as masculine and accordingly repressed those they have designated feminine—cooperation, holism, intuition, humility, and peacefulness—assigning these roles to women and other low-status populations. At the same time the International Labor Organization's study for the July 1980 United Nations Conference on Women, held in Copenhagen, estimated that women provide two thirds of the world's work hours and produce 44 percent of the world's food supply while receiving only 10 percent of all wages and owning a mere 1 percent of all property (*The Christian Science Monitor*, July 30, 1980). Such cultural dichotomy can be stated in less sexually polarizing terms as that of the Chinese yin and yang symbols. Industrial societies have overemphasized the yang qualities and are now suffering the inevitable pathological result of such imbalance. A re-orchestrating of these yin-yang modes is necessary, which will require a return of the yin.

The rise of not only feminism but many other movements for the liberation of subordinated groups is now visible in most maturing industrial countries, including the rise of ethnic and indigenous peoples and tribes, gay liberation, and—even more significant—the new empathy for other living species, such as movements to save whales, seals, and redwoods. The status roles attached to work have also been highly stratified, and much hidden symbolism is involved. Work with the lowest status tends to be that work which is most "entropic," i.e., where the tangible evidence of the effort is most quickly destroyed, such as cooking a meal, which is immediately eaten; making beds, which get unmade; sweeping factory floors; etc. When there is no visible, durable product to bear witness to the effort and justify the work's reward, the work, although it is the most necessary for daily existence, is not valued. In the Eastern cultures, and in Buddhism, it is precisely this work: sweeping the temple garden, meticulous housekeeping and cooking, which forms the daily significant ritual. Mohandas Gandhi, India's greatest leader of modern times, when visiting a village of his followers, would ask to be shown the latrines, and clean them. In industrial cultures, the highest-status work involves building skyscrapers, supersonic planes, nuclear warheads, and all the other high technologies. While this can involve extremely *meta*physical daily activities, such as pushing papers around or serving a computer, it is associated symbolically with high-tech, high-status endeavors. Another aspect involves the fact that "entropic work" or nurturing, facilitating, human services work can be used to "trivialize" the time of subordinated groups, since these jobs are never done, yet are extremely exhausting and demanding. The charge is then raised as to why more women didn't write more books, paint more paintings, build more buildings, etc. These points are, of course, over and above the more common forms of job discrimination, and the wealth of evidence concerning the number of technical innovations of women that were appropriated by their husbands on the grounds that women were not permitted to take out patents or join professional societies (as documented by Elise Boulding in *The Underside of History*, 1977).

Patriarchal domination was discussed by Friedrich Engels, who pointed out the Latin meaning of "familia," i.e., the number of slaves, women and children one man owned, in *On the Origin of the Family, Private Property and the State* (1884); August Bebel, founder of the German Social Democratic Party, in *Women and Socialism* (1885), as well as a continuous stream of non-Marxian criticism, including Charlotte Perkins Gilman, *Women and Economics* (1898), on up to Betty Friedan, *The Feminine Mystique* (1964), Kate Millett, *Sexual Politics* (1970), and Adrienne Rich, *Of Woman Born* (1976). Curiously, Rosa Luxemburg's Marxian theories in the *Accumulation of Capital* (1917) made original contributions to the debate over the imperialistic stage of capitalism, rather than examining patriarchal aspects of the system.

Today economics, narrowly focused as the "science" of produc-

Something's Coming Between Us and the Sun.

A funny thing happened on the way to the Solar Age.

Big Oil got interested.

Remember when they said solar was too costly and impractical to bother about? That's what they're still saying. In public.

In private, however, the same corporations making a fortune off fossil fuels and uranium are moving in on solar power.

Arco, Mobil, Exxon and Shell have bought up their own photovoltaic companies. In fact, only one pioneering solar electricity firm remains independent.

Domestic copper production is about sewn up, too. Oil companies now control 65 percent of the copper essential to the manufacture of solar collectors.

For Big Oil, the sun is just another token on the energy monopoly board. For the rest of us, the sun represents our last chance for energy independence.

America desperately needs a coherent solar policy to guide development of low-cost, decentralized solar technologies.

We call our plan "Blueprint for a Solar America." And it's free to new members of Solar Lobby.

We're the people who brought you the first Sun Day on May Third, 1978. And now we're working in Washington to keep solar competitive, pushing enlightened solutions to our energy problems.

The Solar Lobby needs your help. Your check for $15 will make you a member—and get you a free copy of "Blueprint for a Solar America."

As a member of the Solar Lobby, you will also receive the *Sun Times*, a monthly newsletter filled with the latest information from Washington, and practical advice on how you can benefit from solar energy now.

Join the Solar Lobby. Before Big Oil comes between you and the sun.

Forever.

☀ Solar Lobby

1001 Connecticut Avenue, NW, Fifth Floor
Washington, D.C. 20036
(202) 466-6350

☐ Please send me more information on Solar Lobby.
☐ Here is my $15 contribution. Please enroll me as a member of Solar Lobby and send me your "Blueprint for a Solar America."

Name _____
Address _____
City_____ State_____ Zip_____

Public Media Center

Plate 13

We're trying something new: *democracy*

America needs help. Government exists in spite of people rather than for them. The problems of the poor grow daily yet remain unsolved. We're trying to change that. We believe that democracy can only be taught through the direct experience of democracy itself. That's just what we're doing in 23 areas in Virginia and North Carolina.

Since 1968, we have been working with the poor to build county-wide, democratically structured organizations called Assemblies. Each Assembly is broken down into neighborhood groups of 50 people. Each neighborhood sends a representative to voice its concerns in the county-wide Assembly.

Through such grass-roots organizing, a community is brought together to fill the gap created by inefficient and insensitive government. In each Assembly, the people deal directly with the severe problems that confront them. It is their effort, their decision, their Assembly. They succeed because democracy works.

One Assembly moved to obtain adequate health care; they now have a mobile clinic. Another Assembly decided their community needed more low-income housing; their efforts succeeded in opening 80 new housing units. A third Assembly tackled recreation for their youth; they raised $10,000 to fund a year-round program of fun and learning. We could go on. The number of personal and community problems solved runs into the thousands.

You can participate in this democratic experiment. You'll have the satisfaction of helping make the Declaration of Independence a reality to a people just emerging from slavery.

Send us your tax-deductible contributions. We also need staff. Address gifts and inquiries to:

NASP
201 Massachusetts Ave., N.E.
Washington, D.C. 20002

THE NATIONAL ASSOCIATION FOR THE SOUTHERN POOR

Isn't it time the people won?

Plate 14

The People's Platform

Following are some of the key points in the ACORN's People's Platform. For a complete copy, contact ACORN, 628 Baronne, New Orleans, LA 70113.

• Halt the construction of nuclear power plants until all safety, waste disposal and liability issues are resolved.

• Establish new public energy companies to develop resources on federally-owned land. Prevent any single corporation from owning a major interest in more than one energy source or step of the energy development process.

• Create a new national health care system which makes 24-hour, 7 day-a-week health care available through a network of neighborhood clinics, is progressively financed, puts doctors on set salaries and is managed by democratically elected community-based committees.

• Require subsidized development projects to find replacement housing for every person they displace. Enact a confiscatory tax on income from real estate speculation.

• Create one million new units of federally subsidized housing for low and moderate income people per year.

• Guarantee the right to a job which pays a living wage and offers opportunity for advancement to every person who wants to work. To those who cannot work, guarantee enough income to afford them the basic necessities and allow them to live in dignity.

• Prohibit investor-owned, non-farm corporations from owning agricultural land.

• Require large companies desiring to leave a community to show cause and obtain an exit visa signifying its employees and the community-at-large have been compensated for all losses due to the relocation.

• Charter banks for just five years with renewals based on the bank's performance in meeting its obligations to the community.

• Establish a lower "lifeline" rate for housing credit for low and moderate income people.

• Remove taxation from the basic necessities and from the basic income necessary to live.

• Tax windfall profits by corporations providing basic necessities: housing, health, food and energy.

• Require proportional representation of low and moderate income people in all major political institutions including the Cabinet, the judiciary, the regulatory boards and the national Party conventions.

Reprinted with permission from *Just Economics*, Washington, D.C.

Plate 15

URANIUM: PUNK ROCK

It's dangerous stuff.

Uranium ore is dug out of the ground, processed and made into the fuel for nuclear reactors and atomic bombs.

Either as an innocent looking rock or as the substance out of which a reactor fuel rod is made, uranium emits cancer-causing radiation. Which kills.

The radiation potential in a nuclear reactor is a thousand times greater than in a Hiroshima-type bomb.

The worst possible accident at a small reactor would kill 47,000 people, injure 100,000 and contaminate an area the size of Pennsylvania, an Atomic Energy Commission study reports.

Some punk rock.

And plutonium, a waste produced by nuclear reactors, is even more dangerous. Exposure to just one tiny particle can cause cancer. It remains radioactive for 250,000 years and there's no known way to dispose of it.

Some punk rock.

Friends of the Earth, an active environmental organization, doesn't believe that building reactors is a good idea. (Recently we helped stop the Sundesert reactor in California.) We believe there are alternative clean fuel resources that this country should be using today. Like solar energy. And planned conservation.

Friends of the Earth has six full-time lobbyists in Washington, experts in nuclear power, energy, wildlife, wilderness, air and water pollution, parks and public lands. We lobby in state capitols and maintain active local branches across the country.

Friends of the Earth spreads the word, publishes, and promotes the solar alternative. We keep you informed and give you a voice that's heard.

Write us. We'll tell you how to become a member.

You can also sustain our separate foundation with tax-deductible donations which go to publishing, research and getting the word out where it counts.

Support us either way. You make a difference. Check and send in the coupon below and Friends of the Earth will provide you with the resources you need to help build a clean energy age.

The solar future begins today.

PHOTO: KEN HAAK

Plate 16

tion, consumption, and distribution of wealth, faces daily crises and even public ridicule as it attempts to devise tools to manage the multidimensional nonlinearity of complex industrial societies. Two recent examples include those documented in the *Global 2000 Report* (President's Council on Environmental Quality, July 1980), showing that major agencies of the federal government use quite different, uncoordinated forecasting and management models, with different assumptions and growth projections—all yielding conflicting policy recommendations. The Library of Congress made a similar comparative study in 1976, which was ignored. To provide some coordination, a new high-level group of thirty economic leaders from the United States, Europe, and Third World countries was formed by H. Johannes Witteveen in Paris on January 10, 1979. One of the group, Geoffrey Bell, of the United Kingdom, stated, "The world economic system is going wrong in so many different ways at once that no single government, banker or economist can find the answer alone. We need an interdisciplinary approach." What is so extraordinary is that this is simply a statement of system nonlinearity and the limits of reductionist, uncoordinated policies, and the group is not interdisciplinary at all, but extremely homogenous: i.e., all dominant-culture, male (one token woman), and from the world of economics, business, and finance. As late-stage, Sensate, industrial cultures attempt to restructure their knowledge, they will now need to turn to the repressed, alternative ways of being and thinking locked in the perceptions of their subordinated groups. It will be almost impossible to find innovative ways of handling today's crises from within the dominant culture. Today's crises, whether designated as "economic," "social," or "ecological," are all crises of *perception*. Only a return to holism and broader mapping can rediscover the place of economic activity as it is embedded in society and culture, as well as incorporate the new ecological view and the boundaries and parameters economics has been able to ignore for most of the rise of industrialism as it used up the earth's "capital" of stored fossil fuels and materials.

We now begin to see economic theory and values as relative, and operative within a certain range as to space/time/system, bounded by and valid under certain conditions (as Newtonian physics is so bounded). But this awareness, not surprisingly, begins outside economics' professional orthodoxy (Thomas Kuhn, *The Structure of Scientific Revolutions,* University of Chicago, 1962), as more embarrassing "exceptions" are forced upon its simple, linear, equilibriating models of locomotion and its concepts of maximizing unmeasurable quantities such as "utility" and "consumer preferences," and analyzing cost/benefit trade-offs under conditions of ever greater uncertainty, all revealed by prices. As economists are forced to abandon their absolutist claims and their policy preeminence, the search for new, more inclusive values begins anew.

The yearning for new absolute values is evident today in the breakdown of traditional Western religion and morals and the rise of new cults and community and personal loyalties. Walter Weisskopf notes, "As soon as the value system begins to disintegrate for whatever reasons, the social hierarchy and class stratification begins to be questioned. The groups remote from the old value system begin to clamor for higher status. This is what the proletariat did, and what the disadvantaged groups are doing today" (*Alienation and Economics,* Dutton, 1971, p. 33). This search goes deep into our Western European past and widens into explorations of oriental wisdom and insights, as in *The Tao of Physics* and in E. F. Schumacher's and my concepts of an excessively yang-oriented value system emphasizing instrumental rationality, empirical knowledge, competition, expansion, and aggression. So the search for new values includes reviving the yin qualities and the nurturing, cooperative, cohesive patterns. The idea of yin-yang periodicities and cyclic time reemerges as a healing element, whether in Freudian terms as a "return of the repressed" or in Sorokin's terms of predicting the return of the Ideational or Idealistic forms of culture. "The shift will be led first by the best minds of Western society. Its best minds will become again new Saint Pauls and Saint Augustines, and great religious and ethical leaders. When this new stage of catharsis is reached, new creative forces will emerge in the ordeal of Western society, and usher in a constructive period of integralist culture" (Sorokin, *Social and Cultural Dynamics,* p. 702). Sorokin shows that a similar historical pattern existed when other great cultural crises were overcome: at the end of the Old Kingdom in Egypt, in the Graeco-Roman periods, in sixth-century China, and in the emergence of Western Christian cultures in the Middle Ages.

Finally, this cyclical, organic view of human affairs permeated the dialectics of Hegel and Marx, giving resolution to opposite tendencies and recognizing their simultaneous coexistence. I believe that it is this holistic aspect of Marx and even its multivariant *in*determinacy that has provided its fascination for generations of thinkers, as well as the interminable, tedious arguments about "what Marx meant" that are now tantamount to idolatry. In Marx one can find materialistic determinism, historical laws, dynamic evolution, organic change (e.g., socialism's emergence from the womb of capitalism), simultaneous coexistence of innumerable opposites, contradictions, and cyclical complementarities. Reality is indeed very like that: indeterminate, depending on where the observers stand and what they are looking for. In the sense that Marx critiqued the narrow instrumental reasoning of his day and tried to illustrate its historical, cultural, and value relativity, he was a prophetic moralist!

The theologian Paul Tillich notes that in the classical philosophical tradition, "ontological reason" was the broad ability to grasp and transform reality by means of the cognitive, aesthetic, practical, and technical functions of the human mind. This broad definition of reason, which *fused* the cognitive, technical, and intuitive and was fired by emotion and intellectual love, drives the mind toward the true and the good (Tillich, *Systematic Theology,* Volume 1, p. 72). Naturally, such values as the "true" and the "good" cannot be precisely defined or formulated. But their *survival value* in human evolution is *precisely their indeterminacy, as heuristic devices to allow humans to continually correct their models to fit new environmental conditions and the continual changes that are the only certainty we know.*

However, the response to such heuristic moral absolutes and to worldly uncertainty in Western culture since the time of Descartes, has been to split the moral value debate off into the realm of spiritual and purely subjective concern and reduce uncertainty in the worldly realm by means of empirical science, which could be utilized for prediction of and control over the forces of nature. These cultural responses have now led to a moral dilemma: humans themselves have become the objects of their own manipulation (e.g., genetic engineering) and now are approaching as a species an evolutionary cul-de-sac as they progressively despoil and destroy their own ecological niche. Walter Weisskopf, in discussing this historical splitting of rationality, points out that if one defines science together with knowledge in the broadest sense, such knowledge should be applicable to values, for example, as demonstrated in the Greek polis and the high Middle Ages. "Cognitive rational systems can be alloyed with values, and reason can be used in pursuit of the good. But during the last two hundred years in the West, the normative was eliminated from cognitive knowledge, and the latter demoted to a value-free, or rather, value-empty science. However, the normative is an essential aspect of human existence and if repressed, it will return and manifest itself in one form or another. . . . The idea that values are merely the result of nonrational factors is an intellectual aberration of modern civilization and a phenomenon of decay" (Weisskopf, op. cit., pp. 44–45).

Let us now turn to a contemporary economist who revived the philosophical tradition and showed us how economic systems are always embedded in value systems and specific cultures: E. F. Schumacher, author of the worldwide best sellers *Small Is Beautiful* (1973), *A Guide for the Perplexed* (1977), and the posthumous collection of his lectures in the United States, *Good Work* (1979).

Fritz Schumacher used to say that he most enjoyed those responses to his book *Small Is Beautiful* that did not praise its originality but, rather, congratulated him for articulating what his readers had always known to be true. Above all, Schumacher believed in the common sense of ordinary people and their ability to expand their awareness of comprehensive, eternal truths. This was the essence of his work and the element of it that was most meaningful to me as a citizen activist. He gave me, and millions like me, the courage of our convictions, even when we were facing the mystifications of legions of brilliant, quantitative specialists and narrow economic rationalizers.

As the citizen movements arose over the past decade in the United States, Canada, and all mature industrial societies, they were driven by physical awareness of the social costs: all those diseconomies, disservices, and disamenities that economists had dismissed, in their Freudian slip, as "externalities." We began to smell the dirty air, hear the rising noise levels, taste the adulterated food and water, see the growing piles of garbage, experience the dislocation of our families and communities, feel the pain of unemployment and meaninglessness, and sense the ungovernability (now confirmed by many studies[1]) of our anonymous cities, giant bureaucracies, corporations, and institutions.

Theodore Roszak had struck the same chord in *Where the Wasteland Ends,* in 1972, and in his Introduction to *Small Is Beautiful,* Roszak notes how far the enthroning of economics had progressed when, in 1969, the Nobel Prize Committee instituted a prize in "economic science." Roszak quotes the statement of Professor Erik Lundberg, of the Nobel Committee, in justifying the new award: "Economic science is developed increasingly in the direction of a mathematical specification and a statistical quantification of economic contexts. . . . These techniques have proved successful and have left far behind the vague, more literary type of economics."

While Roszak's point about the absurdity of raising the pseudo rigor of mathematical economics to the status of a science is well taken, it is also important to realize that the prize set up for economics is, in fact, not a Nobel Prize at all. In reality, this prize was set up in 1968 by the Central Bank of Sweden in the amount of $145,000, in the memory of Alfred Nobel, and is the only one of the prizes that was not set up by Nobel himself.[2] The confusion perpetrated on the public by erroneously portraying economics as a science is now compounded by an added confusion—that this discipline has been sanctified by the awarding of Nobel Prizes to its practitioners—rather than the truth, that the Swedish bank persuaded the Nobel Committee to lend its prestige to their own award in economics by allowing it to be called the "Nobel Memorial Prize."[3]

Fritz Schumacher precisely exposed and debunked this type of narrow, quantitative empiricism, which ignores all incommensurables and qualitative differences and reduces them to a single coefficient: that of money. Karl Polanyi had illuminated this particular form of madness in *The Great Transformation,* in 1944. Polanyi noted that, far from being derived from God (or some human propensity to barter, as Adam Smith had thought), the "free market" was actually a package of social legislation enacted in Britain after almost a century of bitter conflict. The keystone of this social legislation installing "the free market" was, of course, the enclosure of land so that it could be bought and sold as a commodity. Its inevitable corollary was to make commodities of human beings who were driven off the land and forced to wander to the towns and factories to sell their "labor."[4]

Polanyi noted that, while markets have always existed in human societies, this was the first attempt in human history to institutionalize a nationwide system of "free markets" as the chief means of allocating resources. Polanyi warned that this monstrous oversimplification of maximizing market-measured cash transactions and production would simply lead to even greater social dislocation and environmental depletion. Soon eighteenth-century social reformers began to fret that the marvelous increase in production of goods seemed to have led to increasing social misery, with ragged, starving bands of "paupers" wandering all over the land. However, by 1776, when Adam Smith published his *Inquiry into the Nature and Causes of the Wealth of Nations,* he saw only the new landscape of small buyers and sellers all orchestrated by the "invisible hand" of competition, with even human workers simply part of the overall scheme, competing with each other to sell their labor at the lowest price to the factory and property owners.

But even in those early days of the industrial revolution there was another problem: that of "nuisances"—smoke, smells, etc.—visited on innocent bystanders by these private-market transactions, that augured, of course, the avalanche of social costs of industrialism we see today. Schumacher zeroed in on these flaws in industrial logic; he said, "In a sense, the market is the institutionalization of individualism and nonresponsibility. We need not be surprised that it is highly popular among businessmen. What causes us surprise is that it is considered virtuous to make the maximum use of this freedom from responsibility!" Schumacher suggested that one of the most fateful errors in this system is its inability to recognize that the modern industrial system, with all its intellectual sophistication, consumes the very basis on which it has been erected: it treats as income the irreplaceable "capital" of fossil fuels, the tolerance margins of nature, and the human substance. The centrally planned socialist economies are founded on the same unsustainable basis. It is clearly not just a problem of who owns the means of production but also a problem of those means themselves. Both Marxian and market-oriented economists espouse labor theories of value (in their frequent labor-productivity maximizing), thus shortchanging the role of natural resources and photosynthesis and other solar-energy-driven processes.

Kenneth Boulding pointed to the same insanity in his essay of 1966 *The Economics of the Coming Spaceship Earth.* All of nature's systems, he wrote, are closed loops, while economic activities are linear and assume inexhaustible resources and "sinks" in which to throw away our refuse. Boulding also noted the one-dimensionality of market systems, which map only money transactions. He pointed out that there are three basic types of human transaction: 1) the threat system—"Give it to me or I'll kill you" or today's more sophisticated version: "How much will you pay me to stop harming or annoying you?" (economists call this the "compensation principle"), 2) the exchange system, that narrow waveband of market transactions with which economics concerns itself, and 3) the integrative system, i.e., the transactions based on the love, sharing, and altruism of which human beings are capable in spite of the denial of these phenomena in economic theory.

Schumacher clothed Boulding's insights in the unforgettable imagery of "Right Livelihood" in his most famous essay, "Buddhist Economics." He showed how higher levels of ethics and greater sensitivity to all living beings could transcend the narrow, market view and how that most dreadful reduction of human beings to a commodity called "labor" could be replaced by the concept of "Good Work"—which challenges individuals to grow and develop their faculties, to overcome their ego-centeredness by joining with others in common tasks, to bring forth those goods and services needed for a becoming existence, and to do all this with an ethical concern for the interdependence of all the life forms of one planetary biosphere. Schumacher's interest in Eastern thought stemmed from the years he spent advising the Government of Burma and from his admiration of Mohandas Gandhi and Gandhi's view that India needed not the capital-intensive, mechanistic, centralized, Western form of mass production but, rather, ecologically and culturally compatible forms of decentralized "production by the masses."

By concentrating on the values and goals of economic activities, Schumacher saw the possibilities of transforming unsustainable industrial modes of production into production methods that build up soil fertility and create health, beauty, and permanence. From his knowledge of the true reality of our species' situation on this planet came Schumacher's prescription of evolving "small-scale, nonviolent, intermediate technology—technology with a human face," as Schumacher called it, "so that people have a chance to enjoy themselves while they are working . . . in new forms of partnership in managing enterprises," and in such pioneering forms of common ownership as that of the Scott-Bader Commonwealth, of which he was a director.

One of the problems that our Western, dichotomizing logic produces is the polarity of "either-or"-type thinking.[5] I remember Fritz Schumacher's good-humored frustration about this. "When I say that small is beautiful," he told me, "someone is sure to jump up and say,

'Aha! so you think that big is bad.'" "No!" he would patiently explain, "there are so many different economies of scale, and it is a matter of restoring a lost balance and knowing when some things have reached obvious diseconomies of scale."

My own efforts have been focused in this area and in the illumination of these various "efficiencies of scale." Efficiency is indeed the slogan of the industrial era and of its economic rationalizers. But efficiency is either a value-laden or a meaningless term unless one inquires, "Efficiency for whom? Efficiency in what time-frame? Efficiency at what level in the social system?" For example, is it individual efficiency that ought to be maximized, or is it corporate efficiency, social efficiency, or ecosystem efficiency? Each would require a different policy.[6]

Fritz Schumacher illustrated the lunacy of conventional economists' views of efficiency and their ideas of comparative advantage, ideas that have led to a lot of frantic transporting of commodities to and fro, within and between nations. The social and ecological cost of all this unnecessary transporting of commodities ranges from depletion of petroleum supplies and pollution to the continual disruption of the domestic affairs of small, less powerful nations. It disrupts their workers and their agriculture and causes greater dependence on foreign capital as they are forced onto the roller coaster of world trade and an international monetary system dominated by the powerful nations, justified by the abstraction of a "global free market." For example, small island economies based on producing one or two cash crops for this world market, such as Jamaica or Cuba, can never survive global market and monetary turbulences—even if the International Monetary Fund were able to put Adam Smith himself in charge! Likewise, Karl Marx could do no better.

With his understated humor, Schumacher debunked the tenet of industrialism that the soundest foundation of world peace would be universal prosperity. "One may look in vain," he wrote, "for historical evidence that the rich have regularly been more peaceful than the poor." He chided John Maynard Keynes for his "trickle-down theory" of economic development. He also criticized him for his ambivalent view that economic progress could be achieved by employing the baser human drives of greed and avarice but that, once we had all become rich, then perhaps our grandchildren could return to the sure and certain principles of religion and traditional virtue:

"that avarice is a vice—the exaction of usury a misdemeanor and the love of money is detestable."

Schumacher summed up the paradoxical Keynesian message thus: "Ethical considerations are not merely irrelevant—they are an actual hindrance, for foul is useful and fair is not." In other words, "The road to heaven is paved with bad intentions." Rather, Schumacher suggested, the foundations of peace cannot be laid by universal prosperity, "because if attainable at all, it is attainable only by cultivating such drives of human nature as greed and envy. At the same time, the wealth of the rich depends on their making inordinately large demands on limited world resources, and this puts them on an unavoidable collision course (not primarily with the poor, who are weak and defenseless) but with other rich people."

Thus Schumacher revived the central economic issue of distribution that John Stuart Mill had illuminated in his great *Principles of Political Economy,* published in 1857. Mill emphasized that the distribution of wealth after it has been produced is essentially a political matter and that property ownership (even that deriving from one's own labors) is secured only by the society's willingness to employ police and other means to protect property owners in their possession. Thus we see in today's demands, of the less developed countries now hostages to northern hemisphere bankers, for a New International Economic Order an understanding of this principle that "economics" is merely politics in disguise—a game that they are exposing, for example, in the 1980 Declarations of Third World leaders in Arusha, Tanzania, and Kingston, Jamaica (*Development Dialogue,* 1980: 2, Uppsala, Sweden).

K. W. Kapp had clarified many of these absurdities in *Social Costs of Private Enterprise,* in 1950, and turned the Pollyanna assumptions of the benign workings of the "invisible hand" on their head. He developed the axiom (later adopted by general systems theorists) that maximizing behavior on the part of micro units of an economy (individuals and firms) tended to be at the expense of other micro units and to suboptimize the macro economy and larger social system. Boulding's and E. J. Mishan's critiques are similar, both pointing to the inevitable "bads" in the form of social costs that come along with the "goods."

The great mathematician Oskar Morgenstern railed against the statistical idiocies of the GNP.[7] He drew attention to the problems caused by the fascination of economists with manipulating fiscal and monetary variables on the basis of this crude indicator without confessing the inadequate formulations of its data collection and the problem of time leads and lags, which in turn make the timing of economic intervention so erratic. Indeed, it is my contention that the business cycles of today are now caused by economic tinkering, rather than by the mysterious forces of the market, which are usually blamed.[8]

Schumacher, in focusing on the essential differences between nonrenewable and renewable resources, and between reversible and irreversible economic decisions, reinforced the work of Nicholas Georgescu-Roegen, who demonstrated, even to the satisfaction of the so-called "rigorous" economists, the conceptual flaws at the heart of their discipline. Georgescu-Roegen, in *The Entropy Law and the Economic Process* (1971), clarified unforgettably the difference between "stocks" and "flows" and illuminated the fatal "flow fetishism" of economists and their GNP indicators, which allow us to use the nonrenewable "capital stock" stored as fossil fuels and treat it as "income." Similarly, as Herman Daly, Joan Robinson, J. Kenneth Galbraith, Barbara Ward, Louis Kelso, and Robert Theobald have emphasized, economists tend to conveniently overlook the fact that the stock of wealth, if tightly held by a few, will force all the rest of the society to live on its speeded-up flows, whether by inflating aggregate demand or by instituting public works projects, more stopgap transfer payments, or larger military budgets. They also realized that ever more centralized, automated production would simply shake more and more people out of the bottom of an economy, and we would need warfare, workfare, and welfare, and more consumption, force-fed by advertising, to keep the whole thing going.

Schumacher's great contribution to this debate was his focus on the role of intermediate-scale, inexpensive technology as a way out of the Keynesian, aggregate-growth, "trickle-down" model, which had reached its logical limits and had become inherently inflationary. Indeed, the labor-*saving* goals had now become counterproductive, while conventional measures of "labor productivity" had become little more than an "automation index" telling employers how well they are doing in getting rid of their employees. Schumacher focused on the crucial issue of how much capital it cost to create each workplace and believed that it should be roughly equivalent to how much each worker could earn in the job per year. This measure has been a key one for me in developing the rationale and strategies of the coalition Environmentalists for Full Employment, since the cost-and-capital intensity of each workplace is also a rough measure of its environmental effects: the more capital-intensive, by definition the more resource-and-energy-intensive and therefore the more environmentally depleting.[9] Thus we have Schumacher's prescription for technology: 1) cheap enough to be accessible to everyone, 2) suitable for small-scale application, 3) compatible with human needs for creativity, and 4) in a nonviolent relationship to nature.

Nuclear power, in Fritz Schumacher's view, is the very antithesis of these properties, and he stated well the danger that "invariably arises from the ruthless application of partial knowledge." C. S. Lewis saw the same danger in this simple-minded application of knowledge to the conquest of nature, which resulted in a few human beings having great power over all other human beings.[10] The "victory" over nature is always hollow and evanescent, as we see in the unforgiving, irreversible technology of nuclear power. As it proliferates, it brings threats of terrorism and the problems of containing radioactive wastes; it has become a curse hanging over all future generations.

Most of all, it is the clarity of Fritz Schumacher's vision that electrified millions and galvanized them into action for a saner future. He broke the spell laid on citizens by the empty expertise and mystifications of intellectual mercenaries. He punctured what Alfred North Whitehead called "the fallacy of misplaced concreteness" and called to account the generalizing of narrow specialists by showing the rest of us the limits of their "expertise and technique." He elaborates on these themes in his book *A Guide for the Perplexed*.

Fritz gave me courage in my efforts to call economists to account. I used to say merely that economics was getting in the way of citizens' talking to each other about what is valuable under drastically changed conditions. After knowing Fritz, I have had the courage to say simply that economics is a form of brain damage.

Fritz Schumacher helped us all to reconceptualize our situation in the now declining industrial era. He was a changer of cultural paradigms, and he pointed to the new path we must travel. This is the dynamism of his work. Intermediate and appropriate technology are now the rallying cries of broad political movements in the United States, Canada, Western Europe, and Japan, as well as of rapidly developing industries in conservation, recycling, solar and wind power, and bioconversion. In Australia and New Zealand, Schumacher's ideas underlie two new political parties, one of which, the Values Party of New Zealand, captured 5 percent of the vote in its first election in 1975. His message is self-evident in many developing countries as a matter of bare necessity and common sense, as is seen by the thousands of requests for assistance that still pour into his London-based Intermediate Technology Development Group, now headed by his colleague George McRobie, author of *Small Is Possible,* a fascinating and detailed account of the group's work (Harper & Row, 1980).

But the triumph of his ideas was in the overdeveloped world, where the very success of industrialism had become, to millions, clearly pathological, where we are now experiencing the scenario Pitirim Sorokin unerringly portrayed as the "Twilight of the Sensate Culture."[11]

NOTES — CHAPTER 7

[1] See, for example, Jay Forrester, *Urban Dynamics,* 1969; *Report to the Trilateral Commission on the Governability of Democracies,* 1975; Duane Elgin, *Limits to the Management of Large Complex Systems,* Stanford Research Institute, Menlo Park, CA, 1977.

[2] "Britain's Meade and Sweden's Ohlin share Nobel Prize in Economics," New York *Times,* October 15, 1977, p. 1.

[3] Bowing to public opinion, the 1979 Nobel Memorial Prize went to two economists who had emphasized issues of inequality, West Indian Sir Arthur Lewis and American Theodore W. Schultz (*The Christian Science Monitor,* October 17, 1979).

[4] Karl Polanyi, *The Great Transformation,* Beacon Press, 1944.

[5] See, for example, Hazel Henderson, "Reexamining the Goals of Knowledge," Public Administration Review, January 1975.

[6] Hazel Henderson, testimony before the Joint Economic Committee, U. S. Congress, November 18, 1976 (reprinted in *Alternatives 6,* Spring 1977).

[7] Oskar Morgenstern, *On the Accuracy of Economic Observations,* 2nd Edition, Princeton University Press, 1965.

[8] Hazel Henderson, *Creating Alternative Futures,* G. P. Putnam's Perigee Books, 1978, pp. 113–35.

[9] See, for example, Gail Daneker and Richard Grossman, *Jobs and Energy*, Environmentalists for Full Employment, Washington, D.C., March 1977.

[10] C. S. Lewis, *The Abolition of Man*, Macmillan, 1947.

[11] Pitirim Sorokin, *Social and Cultural Dynamics*, rev. ed., Porter Sargent, 1957, pp. 699–704.

CHAPTER 8

Three Hundred Years of Snake Oil: Defrocking the Economics Priesthood

Modern economics, strictly speaking, is a little over three hundred years old. It was founded by Sir William Petty (1623–87), professor of anatomy at Oxford and of music at Gresham College, London, and physician to the army of Oliver Cromwell. Among his circle of friends were Christopher Wren, the architect of many London landmarks, and Isaac Newton. His major works include *A Treatise of Taxes and Contributions* (1662), *Political Arithmetick* (1671), and *Another Essay in Political Arithmetick Concerning the Growth of the City of London* (1682). Petty invented a set of ideas that became an indispensable grab bag for Adam Smith and other later economists, including:

1. The labor theory of value (adopted by Smith, Ricardo, and Marx). Petty also expounded the notion of "just wages" related to rank and status, after the style of Aristotle.
2. Differential rent between good and marginal land (later expanded by Ricardo).
3. The theory of interest (replacing the idea of usury with that of reward for abstinence and risk).

This chapter is adapted from a background paper prepared for Dr. Fritjof Capra for his forthcoming book on social implications of modern physics to be published by Simon & Schuster. Used with permission.

4. The distinctions between price and value (various formulations of this issue have preoccupied economists ever since).
5. Monopoly (i.e., "imperfect competition" miraculously rediscovered in the twentieth century after Marx made it impossible to ignore it any longer).
6. The quantity of money and its velocity in circulation. (Petty asked the crucial question still asked by macroeconomists today: "How much money is necessary to drive the trade of a nation?" The answer is that it depends on its velocity, i.e., the number of "revolutions and circulations" it is required to perform.) This is the basic body of theorizing of the monetarist school, even today.
7. National accounting. (Petty regarded people as part of the wealth of a country, as did Adam Smith, but this view was reversed by Malthus. Petty made estimates of the total wealth and national income of Britain, Ireland, France, and Holland.)
8. Division of labor and economies of scale. (Petty described the gain that would accrue to manufacturers from large-scale manufacture and dividing the work into many simple steps, almost one hundred years before Adam Smith made it a cornerstone of his own work.)
9. Public works as a remedy for unemployment. (Anticipating Keynes by more than two centuries, Petty suggested that beggars should be maintained by the state and employed "making the highways broad, firm and even and cutting and scouring to make rivers navigable and planting trees." Petty held that it did not much matter what the nature of the jobs provided, as long as it was "without expence of Foreign commodities.") (From Guy Routh, *The Origin of Economic Ideas,* Vintage, 1977, pp. 36–45.)

Today's policies, as they are debated in Washington, Bonn, and London, would not be any surprise to Petty except that they have changed so little. Petty's *Political Arithmetick* seems to owe much to Descartes, its method consisting of replacing words and arguments by numbers, weights, and measures and "to use only arguments of sense and to consider only such causes as have visible foundations in nature" (Routh, op. cit., p. 45). However, he fell short of his intentions, and when data were not available for his studies of national wealth, he fell into estimates, assumptions, and guesswork as easily as those who followed him.

The older economic order of mercantilism was still defended by theorists who believed that a nation's path to riches was in the accumulation of money, gold, and silver in foreign trade. Mercantilist policies prescribed the fostering of domestic manufacture and commerce by raising tariffs on imports, giving bounties to shipping, improving domestic transportation, forbidding export of gold and silver bullion, and controlling the dealings of foreign traders and their amassing of capital assets overseas. Trading companies were chartered by the Crown, as in the case of the British East India Company, in 1601, which was not merely an enterprise for profit of the venturers of capital but an instrument of national policy. Other mercantilist trading nations of the time were Spain and France. Mercantilist theorists included Antonio Serra (1580–1650), of Spain, and Thomas Mun (1571–1641), of Britain, whose 1630 treatise *Discourse on England's Treasure by Foreign Trade* argued the key mercantilist tenet: ". . . the means to increase our wealth and treasure is by Foreign Trade, wherein we must ever observe this rule: to sell more to strangers yearly than we consume of theirs in value" (i.e., an export surplus should be the chief goal of national policy). This concept was quite consistent with the limited world view of insular, sparsely populated European monarchies at the time, but today, although some nations still attempt to maintain large trade balances in their favor (for example, Japan), we now see that in an interdependent world, not all nations can win at such games simultaneously. Some will lose disastrously, and such cutthroat competition can lead to trade wars, depressions, and international conflict. Mun's treatise also expounded the innovative concepts of balance of payments, identified invisible exports and imports (for example, insurance), advocated the use of international credit and debt to finance foreign trade, and first explained the operation of foreign-currency exchange.

In France, the leading exponent of mercantilism was Jean-Baptiste Colbert (1619–83), Minister of Finance for Louis XIV. He granted charters and monopolies to many trading companies, established model industries, encouraged invention, and in a decade had doubled the King's revenues—making France the most powerful nation in Europe, with the largest navy, flourishing academies, a large bureaucracy, *and* the high taxes to support it all! Antoine de Monchretien (1575–1621) also promoted mercantilism and the importance of self-interest as contributing to the public good (an almost unprecedented idea), which became the precursor of Adam Smith's famous formula of the invisible hand. His book *Traité de l'économie politique* (1615) included the first use of the term "political economy," as economics was called until the nineteenth century. The term mercantilism was applied to these types of trading regimes only after the fact (notably by critics, including Adam Smith). The theoreticians of these practices did not call themselves economists but were politicians and merchants who were trying to explain and justify their policies and actions.

Two key figures in Britain who emerged to lay economics' foundation stones along with Petty, Mun, and the mercantilists were John Locke (1632–1704), a physician, and Sir Dudley North (1641–91), who published tracts separately in 1691. Petty had held that prices of commodities should reflect justly the amount of labor they embody, thus undermining both theological and state authority with an objective standard. But if neither church nor state had the authority to interfere in business transactions, then who would see that justice was done? Locke and North came up with the same answer: prices were also determined objectively, by demand and supply. Not only did this concept liberate the merchants of the day from the moral law of "just" prices, but it became another cornerstone of economics and was elevated to the status of the laws of mechanics, and persists today as the bedrock model used in neoclassical economic analysis: supply and demand equilibrated by price. It also fitted perfectly with the new mathematics developed in 1666 respectively by Isaac Newton as "fluxions" and by Gottfried von Leibniz as differential calculus. If enough *a priori* assumptions were made, economists could now assign "objective," scientific status to the determination of wages (by designating workers as a "supply of commodities" in relatively greater or lesser demand), to subjective desires (today's revealed preference functions, marginal "utility"), etc., as I shall elaborate later in this chapter.

Henceforth, efforts to make economics a mathematical science increased on the notion that economics deals with continuous variations of very small quantities, and that its most appropriate instrument is the differential calculus. The problem was and is that such economists' notions as "utility" and "demand" are indeterminate, have to be hedged with many assumptions, and are not susceptible to this type of measurement.

John Locke, however, is a key figure in the history of economic thought because he was also the chief English philosopher of the Enlightenment, which succeeded mercantilism. Locke followed in the footsteps of Thomas Hobbes (1588–1679), who proclaimed, in *Leviathan* (1651), the radical idea that in the natural state, all individuals were free and dependent on themselves. But Locke, while agreeing on the freedom of individuals deriving from natural law, their essential rationality, and the idea that people voluntarily relinquished some of their freedom in a "social contract" in which they created the state for their own protection, then parted company with Hobbes. Hobbes's view of the world was governed by fear and the belief that the natural condition of human life was nasty, brutish, and short, governed by increasing competition for survival. Hobbes was a materialist and was exposed to the ideas of Galileo, Kepler, and Descartes through the network of the mathematician-monk Mersenne, and read Descartes's *Meditations* in manuscript. Locke also exchanged ideas with Descartes in correspondence, and went even further than Descartes's sensory empiricism, challenging Descartes's notion that humans do have innate knowledge of general principles and maxims with which they also observe the material world. According to Locke, the human mind at birth is a *tabula rasa*. Locke's *Essay Concerning Human Understanding* propounded such theories of knowledge, which were not only supportive of the work of Newton and other scientists of the day but were in extreme reaction to clerical dogma. While Hobbes favored the monarchy, Locke advanced the idea of representative government and safeguarding individuals with the rights to property and to the fruits of their labor. Thus, he also promoted the labor theory of value and believed that no product had value except by virtue of the labor expended on it. Locke held that once individuals created a government as the trustee of their rights, liberties, and property, its legitimacy depended on protecting these rights. If the government failed to protect their rights, the people possessed the supreme power and could dissolve it.

Thus economics was founded on many of these radical moral concepts of the Enlightenment and became a powerful part of the rationalization of individualism, property rights, free markets, contract law, and democracy. It is also clear how these European ideas, including the writings of Jean-Jacques Rousseau and others, contributed to the thinking of Jefferson and the other authors of the Declaration of Independence.

Before we move to the classical period of economics, inaugurated with the publication by Adam Smith (1723–90) of *Inquiry into the Nature and Causes of the Wealth of Nations,* in 1776, it is necessary to pay the French physiocrats their due. This group of thinkers were the first to call themselves economists, to regard their theories as "objectively" scientific, and to develop a complete view of the French economy, just prior to the Revolution. Physiocracy means "the rule of nature," and they bitterly criticized mercantilism and the growth of cities. They decried manufacturing as sterile and claimed that only agriculture and the land were truly productive of all real wealth, thus representing an early "ecological" view. But this agrarian revolt against the cities and the state was by no means a peasant movement. Its proponents were politicians and landed aristocrats: Anne Robert Jacques Turgot, a finance minister under King Louis XVI; Mirabeau the Elder; and Pierre Samuel Dupont de Nemours, who coined the term "physiocrats" and emigrated to America, where his son Eleuthère founded the Du Pont chemical company which still bears his name.

The physiocrats' leader, like Sir William Petty and John Locke, was a physician, François Quesnay (1694–1774), who was surgeon to the royal court. Borrowing the unpublished ideas of Richard Cantillon, an Irishman living in Paris, Quesnay expounded on the idea that if natural law were left untrammeled, it would govern economic affairs for the greatest benefit of all. Thus the doctrine of "laissez-faire" was introduced, as another cornerstone of economics. Private property was sacred to the physiocrats, and they believed in the labor theory of value, but to them prices were determined by supply and demand. Yet, paradoxically, this laissez-faire, free-enterprise, natural system was to be advanced under the regulation of an absolute monarchy! Quesnay's major work, *Le Tableau économique* (1757), represented the first effort to create a national economic model showing how money, rents, and goods circulated, with diagrams showing funds bounding and rebounding between three columns (landlords, manufacturers, and farmers) in fixed proportions. The *Tableau* was full of errors and assumptions, but it became the marvel of the times, and Quesnay and Dupont were showered with honors. Adam Smith conferred with and was influenced by the physiocrats during a visit to

Paris in 1766 and would have dedicated his *Wealth of Nations* to Quesnay if Quesnay had remained alive (Routh, op. cit., p. 70).

Adam Smith, the founder of classical economics and the most influential economist, was a professor of moral philosophy at the University of Glasgow, Scotland, and a friend of the Scottish Enlightenment philosopher David Hume. Smith wrote his first book, *The Theory of Moral Sentiments,* on ethics, in 1759. Not only was Smith influenced by the ideas of the physiocrats and the Enlightenment, he was also friendly with James Watt, the inventor of the steam engine, met Benjamin Franklin and probably Thomas Jefferson, and lived at a time when the industrial revolution had begun to change the face of Britain. Hargreaves had invented the spinning jenny, and Arkwright's loom was used in cotton factories employing up to three hundred workers. The new regime of private enterprise, factories, and power-driven machinery shaped Smith's ideas, and so, just as the physiocrats had critiqued mercantilism and rationalized landed aristocrats and the importance of agriculture, Adam Smith championed the industrialists and their new order and critiqued the remnants of the land-based, feudal system. Like most of the great classical economists, Smith was not technically trained or a specialist, but a broad thinker with fresh insights. Edmund Burke, the great politician, reflected widespread reaction to Smith's *Wealth of Nations:* "In its ultimate results, probably the most important book ever written" (George Soule, *Ideas of the Great Economists,* Mentor, 1952, p. 40). Smith set out to show how the wealth of a nation is increased and distributed—the basic themes of modern economics. In countering the mercantilists' view that wealth is increased by foreign trade and hoarding gold and silver bullion, Smith held that the only basic is production resulting from human labor and natural resources. Wealth would be increased according to the skill and efficiency with which labor is applied and the percentage of people engaged in such production (that is, real income per capita). The basic means of increasing production was by the division of labor, Smith contended, as had Sir William Petty earlier.

Smith deduced from the prevailing idea of "natural law" that it was "human nature" to barter and exchange, and also held it "natural" that workers would have gradually to facilitate and quicken their work. A darker view of the invention of labor-saving machinery in that period is found in David Dickson's *Alternative Technology*

(Fontana, U.K., 1974, pp. 79–81), in which he quotes letters from early manufacturers that made clear that they understood that machines could replace workers and so keep them afraid and docile. Marx's theory of capital accumulation held that competition between capitalists also forced them to develop more labor-saving technology or risk being bought out by bigger firms, which is echoed by business leaders today.

Smith elaborated and codified the physiocrats' laissez-faire theme, immortalizing it as the invisible hand that guides the individual *self-interest* of all entrepreneurs, producers, and consumers for the harmonious betterment of all ("betterment" being equated with material production of wealth). Even though supply and demand would determine prices in "free" markets, Smith nevertheless also followed the labor theory of value and believed that the real price of every thing was the toil and trouble of acquiring it (although he avoided the issue of unearned or inherited wealth). Thus he adopted the idea of the self-equilibrating economy from the physiocrats, rather than taking his mechanistic notion from Descartes (unless via John Locke) or from Isaac Newton. One of the fallacies of these metaphors of mechanical equilibration when analogized to the social system was the lack of appreciation for the problem of *friction,* as well as wear and tear, heat loss, local effects of the Entropy Law (which will become ever more important for economic processes with today's and future boundary conditions vis-à-vis resources). The first *economist* to notice this was Nicholas Georgescu-Roegen, in *The Entropy Law and the Economic Process* (1972), although, not surprisingly, chemists, engineers, and physical scientists were aware of the problem; for example, "net" energy modelers, such as Howard Odum of the University of Florida, a pioneer in accounting for all of the upstream energy used in the processes of extracting, converting and distributing energy to the end user. Adam Smith, however, imagined that the equilibrating mechanisms will be almost instantaneous, without the delays and complications of real-world processes, continually describing their adjustments as "prompt," "occurring soon," and "continual," while prices were seen as "gravitating." Today's economists refer to the assumption of "mobility," for example, social mobility of displaced workers, of capital, etc. Smith assumed social mobility of all factors of production as well as technical mobility (i.e., that production units and processes would remain small and nonspecific; that is to say that small producers would meet small consumers in the marketplace, that they would have equivalent power and information, and that no nuisance effects —externalities—would spill over onto innocent bystanders). The basic principles governing Smith's model were atomism, microautonomy and rationality additively producing macrorationality, frictionless mobility, and the system goal (of both macro and micro units) being that of maximizing the increase of materially defined production of wealth. No principle of conservation was present let alone that of increasing entropy (the first and second laws of thermodynamics), and natural resources were taken as constant within the given time span. The dynamic variables were productivity and technical progress. The same mental traps enmesh economists today, in spite of Adolph Lowe's convincing demonstration of the rigidifying structure of industrial societies in *On Economic Knowledge* (Harper & Row, 1965, pp. 169–72), as well as Lowe's brilliant concept of their "viscosity."

The idea of the growth of *structure* in the form of monopolies was alluded to by Smith as he excoriated "people of the same trade conspiring to artificially raise prices" (Routh, op. cit., p. 9), but he did not see the systemic implications. For Marx, this growth of structure was integral and a central tenet: i.e., the class structure itself, continually hardening into two great groups, the proletariat becoming larger and the ranks of capitalists becoming smaller as wealth was accumulated into larger and larger companies, leading to the final monopoly stage. In fact, Marx saw the structure of capitalism as so intractable that nothing short of social revolution of the proletariat could change it. Within neoclassical economics, the study of monopoly proceeded with extreme caution as the "imperfect competition" of Joan Robinson's *The Economics of Imperfect Competition* (1933) and Edward Chamberlin's *The Theory of Monopolistic Competition* (1933) and, of course, the re-orientation of the entire body of economic theory toward the examination of its macro-dynamics and sectoral structure of John Maynard Keynes, as we shall discuss in this chapter.

Smith justified capitalists' profits by stating that if all were to enjoy the fruits of more productive machinery, some people would have to save and invest in more machines and factories. Thus the *worker* could not receive the full, natural value of his product, since some of the price had to be apportioned for profit. Marx, of course, challenged the assumption. Smith also decried the habit of workers and "other inferior ranks of people" (Smith, Volume 1, p. 89) of producing so many children, which only added to the labor supply and caused wages to fall to the levels of mere subsistence. He also noted that employers and workers did struggle over the division of the product, and that workers attempted to combine to increase wages. He did not allude to the unequal power that existed between workers and capitalists to affect the market—a key point in Marx.

Smith thought that if demand for labor was continually increasing, wages would also increase, leading to growth of population and thus greater demand for goods, and that, indeed, higher wages might motivate workers to higher output. Thus his equilibrating system was also in slow, steady, dynamic growth, and the idea of this sort of continually increasing progress took hold in economics. But despite Smith's general optimism, he did foresee this progress finally saturating into a "stationary state," where it has pushed its wealth to the limit of what the nature of its soil and climate and its situation with respect to other countries would permit. When this stage was reached (this is the catch), at *some irrelevantly far-off time in the future,* Smith held, wages would again fall to subsistence levels of the existing workforce, whose numbers could then no longer increase (i.e., they would starve). Smith also acknowledged the importance of the material base of a society in determining its civil institutions—the view later emphasized by Marx (see Robert Heilbroner, "Decline and Decay in the Wealth of Nations," *Journal of the History of Ideas,* April–June 1973, pp. 243–62). Smith also articulated the doctrine of comparative advantage, whereby each nation should excel in some types of production. Instead of raising tariffs, as the mercantilists did, countries should lower them, thus allowing an international division of labor and free trade. This model of international free trade is now producing its own set of social costs and still underlies today's thinking on the global economy. In systemic terms, this type of world-trade "game" reaches some hypothetical global equilibrium (using the coefficient of prices) when the winners have disordered every local social system and despoiled every local ecosystem; that is, when the economic behavioral sink has become global.

Finally, Smith thought that the role of government was to provide for defense, dispense justice, bear the expense of public works, and raise taxes proportional to incomes and not so as to discourage industry and trade. His body of ideas still provides the comprehensive framework of economic theory, and it is evident from even this brief review how little the debate has changed in two hundred years (except for the fundamental challenge of Marxism). Smith was often self-contradictory; for example, his paradigm was of a world where only government interference hampered the invisible hand from guiding humanity along the road to plenty; yet he noted that employers were unsure of their rate of profit, and if so, how could capital flow unerringly to the activities yielding the highest rate of profit? He also discovered empirically that competition failed to yield correct prices or to establish equal pay for equal work, and noted the inherent defects in commercial society that rendered its working population ever more stupid and uneducated, because of the division of tasks into segments of idiot simplicity. All this has been overlooked in modern mathematical, positive economics, while providing grist for the followers of Marx. The darker side of commercial, market-oriented societies has most often been treated since by academic disciplines other than economics: from theological perspectives (e.g., Paul Tillich), psychological bases (e.g., Fromm, Freud, and Maslow) and holistic critiques such as those by Lewis Mumford, Jacques Ellul, Norman O. Brown, Herbert Marcuse, Ivan Illich, E. F. Schumacher, Leopold Kohr (in *The Breakdown of Nations,* 1957, a book that influenced Schumacher), Theodore Roszak, Gregory Bateson, et al., as well as the newer ecological critiques of Rachel Carson, Garrett Hardin, Barry Commoner, and Lester Brown.

How was it possible for generations of economists from Smith onward to *use* equilibrium assumptions in their models of economic processes while *at the same time* being perfectly aware of dynamism, growth, and change—to the point where growth and progress are *postulates,* even fetishes? The explanation seems to be that they used these dynamic elements as *coordinates,* which would allow this cognitive dichotomizing—i.e., mainstream Keynesians using Keynes's *tools* (which implied disequilibrium) on a system they still modeled with equilibrium assumptions). Another part of the answer must lie in the habits of *conforming* rationality, but, deeper, it may be another manifestation of the Cartesian split of mind/body and thought/action, which could perpetuate such lack of integration, together with the "objective observer" who cannot use or integrate

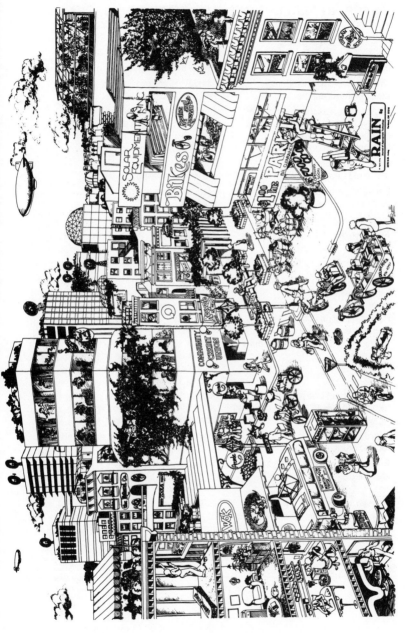

Plate 17

DIFFERING PERCEPTIONS, ASSUMPTIONS, FORECASTING STYLES BETWEEN ECONOMISTS AND FUTURISTS

ECONOMISTS	FUTURISTS
Forecast from past data, extrapolating trends	Construct "What If?" scenarios; trends are not destiny
Now also use optimistic, pessimistic forecasts	Identify "Preferred Futures"— plot trends for cross-impacts
Change seen as **dis**-equilibrium (i.e. equilibrium assumed)— all other things equal "normal" conditions will return	Fundamental change assumed (transformation assumed)— no such thing as "normal" conditions in complex systems
Reactive (invisible hand assumed to control)	Pro-active (focus on human choices and responsibilities)
Linear reasoning; reversible models	Non-linear reasoning; irreversible models, evolutionary
Inorganic system models	Living system, organic models
Focus on "hard" sciences and data	Focus on life sciences, social sciences, "soft," fuzzy data, indeterminacy
Deterministic, reductionist, analytical	Holistic, synthesis, seeks synergy
Short-term focus (e.g. discount rates in cost/benefit analysis)	Long-term focus, inter-generational costs, benefits and trade offs
Data on non-economic, non-monetarized sectors seen as "externalities" (e.g. voluntary, community sectors, unpaid production, environmental resources)	Includes data on social, voluntary unpaid productivity, changing values, lifestyles, environmental conditions; maps contexts, external variables (use post economic models: technology assessment, environmental impact, social impact studies)
Methods tend to amplify existing trends (e.g. Wall Street psychology)	**Methods** "contrarian" (e.g. look for anomalies, check biases in perceptions, cultural norms)
"Herd instinct" in investing, technologies, economic development	Identify potentialities that are latent
Entrepreneurial when "market" is identified	Socially entrepreneurial (Schwartz) (e.g. envision future needs, create markets)
Precise, quantitative forecasts (e.g. gross national product for next quarter of year; annual focus)	Qualitative focus (e.g. year 2000 studies, anticipatory democracy), data from multiple sources, plot interacting variables, trends in long-term global contexts

Plate 18

**The Vicious Circle Economy
of Fast Feedback Loops**

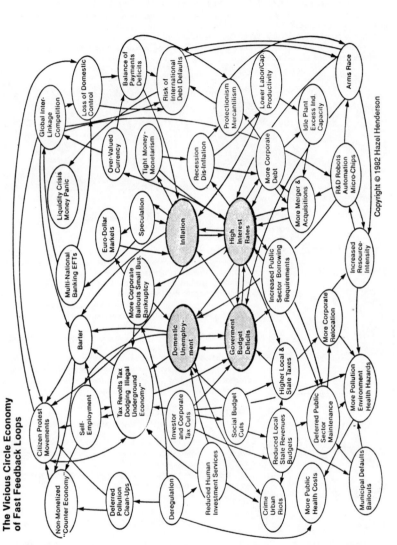

Plate 19

APPALACHIA

Council of the Southern Mountains

An Appalachian People's organization since 1913...working people, people who cannot work; people for a democratic and economically secure future in the mountains.

Delegates from local groups in 5 states on our "Board of Representatives" plan regional strategy for solving our problems. Your gift can help us help ourselves; keep our office open and staffed, hold our Board meetings, publish our magazine; work in local groups.

Council of the Southern Mountains Board of Representatives at a meeting.

Black lung victims lobbying for a new bill.

Mountain Life & Work — Since 1925
The best resource for keeping abreast of the thinking and work of Appalachian people.

Special Issue:
URBAN APPALACHIANS

Coal Mining —monthly coverage of Appalachian coalfield developments, efforts at unionization, mine safety and health improvements, damages caused from stripmining.

Community Unions — regular reports from community and labor groups.

Community Economic Development — descriptions of co-ops and self-help projects in seven states.

News and Events — discussions of major events and issues affecting the region and the nation.

General Interest — fiction, poetry, biography, mountain culture, and history.

"Special Issues" — focus one or two times a year on such concerns as mountain women, textile mills, music, land use, coal.

Published 11 times a year. Send check ($5.00) to Mountain Life & Work Drawer N, Clintwood, Virginia 24228. ($6.00 foreign).

Appalachian Bookstore
The finest collection anywhere in the U. S.

In historical Berea, Ky., just off I-75, south of Lexington
More than a Bookstore — a good place to drop in and talk about the mountains
Over 900 books, pamphlets, films, and records on the Appalachian mountains

For great gift ideas:

Send $1.50 for our "BIBLIOGRAPHY ON THE APPALACHIAN SOUTH": CSM Bookstore, CPO 2106, Berea, Ky. 40404.

ALL CONTRIBUTIONS ARE TAX-DEDUCTIBLE!

the 51st state?

Plate 20

his/her own action and participation (what I called the Heisenberg Uncertainty Principle operating on the macro level). Here, the *simultaneous healing of both society and self* becomes today's imperative, and where interdisciplinary models of social change use complementarity (e.g., Roszak's Person/Planet synthesis as a paradigm). This is where too objectively oriented socialism and social activism *fail,* as does merely inner, personal growth. Smith also shared the Cartesian flaw of not taking account of the vantage point of his observations. As with all the other Enlightenment philosophers, their middle-class, educated status allowed them to conceive of radical ideas of equality, justice, liberty, etc., but could not allow them to extend these concepts to include the "inferior classes," "illiterate rabble," and the "brutish poor" (see, for example, Norman Hampson, *The Enlightenment,* 1969). I might add, let alone *women!*

The Reverend Thomas Robert Malthus (1766–1834) focused on the darker side of industrial development; his *Essay on the Principles of Population* (1798) was a distinctly *dis*equilibrium view, even a catastrophe model in the sense used by mathematician René Thom in *Structural Stability and Morphogenesis* (1972) and discussed in Chapter 11. Malthus' theory that the means of subsistence, i.e., food supply, grew only arithmetically while human numbers grew geometrically caused a furor and shaped the evolutionary theories of both Charles Darwin, in *The Origin of Species* (1859), and Alfred Russel Wallace. Malthus held that real wages could not rise above subsistence levels, because each increase in well-being would lead to an increasing supply of workers. When their subsistence wages fell below that level, these surplus workers would be eliminated by death, bringing supply and demand for workers back into equilibrium (the "iron law of wages"). Smith and Malthus noted the high wages in the United States, but Malthus took the gloomy view that this was not so much "progress" and productivity as merely the high ratio of land to the numbers of people. Malthus noted the high rate of population growth in America and concluded that, in time, the laborers would be much less liberally rewarded. Malthus had an ecological perspective and developed the idea as it related to the productivity of land, of the law of diminishing returns (i.e., a given piece of land will yield more with the application of fertilizer and more labor, but there comes a point beyond which added increments of fertilizer or labor do not proportionally increase the yield, and therefore it does not pay to add them, and in fact, if carried too far, they will reduce yields).

The law of diminishing returns remains a key concept in economics, but it has been applied with sometimes ideological, sometimes absurd selectivity (e.g., overlooked since the 1960s vis-à-vis fossil-fuel inputs to agriculture, while exhaustively studied in trivial textbook examples of irrelevant, interpersonal comparisons of marginal utility, between preferences for tea or coffee, steak or pork, etc.; see also Benjamin N. Ward, *What's Wrong with Economics,* Basic Books, 1972, p. 199). Meanwhile, the theory is rarely applied to general satiation of consumption, except by Staffan Burenstam Linder in *The Harried Leisure Class* (1970), who saw time as the constraint on consumption; and by noneconomists (Robert Theobald, Duane Elgin in *Voluntary Simplicity,* 1977, myself, et al.) although it fits well!

Malthus' prescriptions were sexual restraint by the workers and poor for their moral betterment, and the refusal of charity to families who could not support themselves. This harsh medicine was justified as humanitarian in the long run, so that population growth might be checked (quite similar to some of the extreme population/ecology "hawks" of today, for example, The Population Crisis Committee and the Environment Fund; Jay Forrester, Garrett Hardin, William Paddock, et al.; as well as the revival of the doctrine of "triage," in which, in wars, doctors are faced with limited medical supplies and time and must make terrible decisions about whom to save, i.e., the less mortally wounded). Malthus reinforced the conforming rationality of his day by arguing "scientifically" that "laws of nature" were operating and the poor were responsible for their own misfortune. Later economists, including Nassau William Senior (1790–1860), as well as factory owners, used Malthus' theories to defeat the ten-hours bill of 1837 to shorten hours and similar social legislation to improve workers' and beggars' conditions. This type of rationalizing away of the issues of social justice may haunt industrial societies in the belt-tightening 1980s.

David Ricardo (1772–1823) was a stockbroker who became a multimillionaire at age thirty-five and then devoted himself to studying mathematics, science, and (after reading Smith's *Wealth of Nations*) political economy. He became a large landowner and a member of Parliament, yet developed a theory of rent that saw it as a monopoly price. Ricardo declared that if land were as abundant as air it would likewise be a "free good." This, he believed, was the original situation, but as population had grown, the first farmers had naturally appropriated the best land, and, subsequently, as population increased, poorer, marginal land was cultivated. Since this land produced less, the relative value of the better land increased and the rent charged for it became an unearned incremental payment for merely owning it, over and above that paid for labor in cultivation (i.e., a monopoly price). Ricardo's concept of "marginal" land fitted well with Malthus' idea of diminishing returns and became the basis for the later economics of marginal analysis used today.

Ricardo agreed with Malthus' "iron law of wages" but went further and analyzed the cost of subsistence and its dependence on food prices that were inflated by the landowners' rent. Rent, and therefore food prices, would continually rise due to population growth and the exhaustion of the best farmland. Ricardo propounded the labor theory of value, differing with Smith's view that rent was a "cost" that also should enter into price. But, significantly, Ricardo *did* include in the natural price the cost of the labor required to build the machines and factories, so that in receiving profit, the owner was taking something that labor had produced, a point on which Marx built his theory of surplus value. Rent, for Ricardo, set the wage earner and the employer into conflict over the division of income from industry and pitted industrial employers against landlords over the division of profit. Ricardo predicted that profits would have a tendency to fall, eventually, to zero and landlords would end up as winners, with the unearned surplus. He provided the rationale for the repeal of the Corn Laws (1833), tariffs to protect English agriculture. This favored further industrialization and foreign trade and finance (Soule, op. cit., p. 52). Paradoxically, the two most devout proponents of laissez-faire, Ricardo and Quesnay, also provided Marx with the bases of many of his theories. In fact, Joseph Schumpeter asserts that Ricardo was Marx's master (*Capitalism, Socialism and Democracy,* 1942, p. 22). Harper & Row, 1950.

In France during this period, Jean Baptiste Say (1767–1832) advanced political economy in general agreement with Adam Smith, adding two ideas: 1) broadening the category of "goods" as material things to "utilities," i.e., anything, goods or services, that people want and will pay for, and 2) proposing Say's Law, which stated that total supply must always equal total demand, since the production of any article creates the equivalent demand for some other article, so that there could be no such thing as "overproduction." Such are the pitfalls of simple, equilibrium models too highly aggregated! Say overlooked relative velocity of money circulation and cyclical shifts in spending, saving, and investing, technological change, and structural factors, etc., some of which Keynes would show later.

By the end of the eighteenth century, all these rationalizers of industrialism and Enlightenment liberalism had triumphed in Western Europe and the United States. Economics was now consolidated into a set of dogmas that a later welfare economist, A. C. Pigou, in his Stamp Memorial Lecture, in London, said "furnished the ungodly, blunt instruments with which to bludgeon at birth, useful projects of social betterment" (Routh, op. cit., p. 105). Workers' uprisings were becoming frequent (e.g., the Luddite movement's destruction of factories and machinery). This new economic body of conforming rationality engendered its own horrified critics, long before Karl Marx. They included instrumental pragmatists such as Jeremy Bentham (1748–1832) and his extraordinary schemes for turning workhouses into factories where the unemployed would be reduced to grateful cogs in a social machine so as to earn their dole payments. Bentham developed the maxim of "the greatest good to the greatest number": what we now call "utilitarianism." This "good" to be maximized, he precisely defined as anything that increased the pleasure or decreased the pain of any person. Money or the lack of it was to be the measure. Social good was the algebraic sum of all individual "good," and every institution was to be judged on its usefulness to individuals. Such well-meaning but crude, unworkable formulations were thought to be fit subjects to be measured by Isaac Newton's and Leibniz's differential calculus and led to a long series of inappropriate formalizations known later as welfare economics. Welfare economics, like the assumption of "perfect competition," is a long-standing scandal in the view of dissidents Boulding, and Myrdal, who asserts "It grows like a malignant tumor. Hundreds of books and articles are produced every year on 'welfare economics' even though the whole approach was proved to be misdirected over four decades ago." He laments that in the natural sciences, theories are refuted and hypotheses become obsolete, but that in economics, "all doctrines *persist*" (Myrdal, *Against the Stream*, p. 151). Swedish economist Knut Wicksell (1851-1926) made a similar point in a 1902 lecture: "The Copernican idea of the universe, the Newtonian system, the theory of blood circulation, and the phlogiston theory in chemistry once found adherents and opponents. Nowadays, these theories are either universally believed or disbelieved," and he contrasted this with the unfortunate situation in economics. Wicksell anticipated Keynes by developing a theory of business cycles, which Keynes acknowledged.

The theory of welfare economics is worth a closer look. From the earlier objective view of "welfare" as material production, and the *labor input* theory of value, the welfare-economics school shifted to subjective criteria, i.e., of individual welfare as defined by Jeremy Bentham's "utility": whatever maximized pleasure and minimized pain. Elaborate charts and curves were even constructed based on "units of pleasure" and "units of pain." It was assumed that in a perfect marketplace, all would maximize their units of pleasure and minimize their units of pain, and that this would be reflected in the prices of commodities, land, labor, etc. This subjective approach promoted a value-free approach to public policy of trying to determine action at the macro level by aggregating the sum of all these individual preferences into some guide for social order. Kenneth Arrow, the contemporary Nobel-Memorial economist, refuted this idea in his General Impossibility Theorem, which states that individual preferences cannot be logically ordered into social choices. Yet welfare economics persists; it is really a thinly disguised recipe for anarchic, selfish individual behavior, because it undermines any cohesive set of goals for the "common good" and has led to many "tragedies of the commons" as expounded by Garrett Hardin, where individual self-interest behavior is disastrous for the group as for example, today's deadlock in the United States on energy policy (Garrett Hardin, "The Tragedy of the Commons," *Science,* December 13, 1968, p. 1243). Welfare economics, as Walter Weisskopf points out, "eliminates the social and moral content from the concept of economic welfare" (Weisskopf, *Alienation and Economics,* p. 94), since subjective "utility" for any individual may be altruism, greed, frugality, or mere neurosis! The theory states formally that social welfare will be increased if the satisfaction of some individuals can be increased without *de*creasing the satisfaction of other individuals. Thus any economic change that makes someone "better off" without making anyone "worse off" is a desirable change for social welfare. Italian economist Vilfredo Pareto (1848–1923), in his *Manual of Political Economy* (1906), codified these concepts; their yardstick is called Pareto Optimality, which underlies cost benefit analyses. Pareto added the concept of "compensation": any economic change in which the beneficiaries of the change can compensate those who lose from the change and still be better off themselves is a desirable change for society. This kind of welfare economics, with "better off" and "worse off" defined in terms of material or money gain, is now playing havoc with environmental policy, as I predicted it would together with others, notably K. W. Kapp in *The Social Costs of Private Enterprise* (1950). This inevitable, one hundred fifty-year derailment of Jeremy Bentham's unworkable "utilitarianism" will be taken up again in Chapter 9.

Meanwhile, a school of more realistic critics in France and Britain —the utopians—addressed capitalism's deficiencies in frankly idealistic experiments. Most famous and successful was Robert Owen (1771–1858), who ran an experiment in industrial humanism, a hugely profitable, notably humanitarian factory: the New Lanark Mills, in Scotland. He reduced workers' hours, raised their wages, educated their children, cared for their families' health, and provided recreation and insurance. Distinguished visitors came from all over the world to marvel and learn. But Owen, in spite of the well-established success of New Lanark Mills, realized in time that his flourishing experiment in a hostile economic environment depended on paternalism, which would make such experiments unlikely to succeed for long, so he turned to the idea of workers' cooperatives. Only when capitalists were replaced by cooperatively owned and managed enterprises would industrial enterprises be humane. Both Ricardo and Jeremy Bentham supported Owen as he then set up experimental communities, one in Scotland and one in the United States, in Indiana, called New Harmony. Neither worked out well and he ran out of money. Undismayed, in 1832 he founded the National Equitable Labor Exchange, in which anyone could deposit the products of his labor and receive promissory notes valued at the hours of labor they represented and exchange them for the work of others. A later version of this, the Time Store, was started in Cincinnati, Ohio, around 1900 and many new versions flourish today, such as the Free Trade Exchange storefront operated by Ellery Foster, author of *The*

Coming Age of Conscience, 1977, Box 841, Winona, MN 55987. Owen then started a consumer cooperative movement in Rochdale, England, that is still thriving, and helped found the Grand National Consolidated Trades Union, in 1833, which officially launched the labor movement in Britain.

The French Utopians included François Noël Babeuf (1760–1797), who envisioned a communitarian society and the nationalization of businesses and private property, with production and distribution managed by an elected government. Food and clothing were to be the same for all, except for differences according to age and sex, while political rights were to be given only to those who worked. Babeuf was guillotined in the Revolution (Soule, op. cit., p. 54). Étienne Cabet wrote a utopian novel in 1788, *Voyage to Icaria,* envisioning a technocratic, standardized society. Instead of conspiring with revolutionaries, Cabet emigrated to the United States, first to Texas, then to Nauvoo, Illinois, where he tried to set up a colony of fifteen hundred people. It broke down in internal quarreling.

Claude Henri de Saint-Simon (1760–1825) had a vision of equality of opportunity. He came to America and fought in its revolution, and supported the French Revolution. He denied himself, sacrificing his money and health to write his ideas, chiefly in *The New Christianity.* The existing order must be destroyed, he wrote, but men also needed a new spirituality to take the place of the church, and something better than the anarchic individualism of the Enlightenment thinkers must be devised. The New Christianity would be founded on the principle that men are brothers. War must be eliminated and Europe united under a single parliament. Saint-Simon thought that industry should be publicly owned and that income should be apportioned by "merit"; no idlers, rich or poor, would be tolerated. His followers included Auguste Comte, the founder of positivist philosophy. Saint-Simon's other disciples founded a church with branches in Germany and England.

Charles Fourier (1772–1837) was an even wilder visionary, who minutely described experimental communities called "phalanxes" to be laid out around a central building, where labor would be divided according to taste, with children doing all the dirty work, because everyone knew that children like to get dirty. A few "phalansteries" were tried in France and in the United States. Louis Blanc

(1811–82) also rejected laissez-faire and was the first of the utopian socialist reformers to appeal to workers themselves to initiate reforms. In 1848, during the revolutionary unrest in France, he became a member of the provisional government; however, in 1871, he did not support the insurrection of the Paris Commune. He proposed a national federation of worker-controlled social workshops, in which all would have guaranteed jobs. He founded the influential *Revue de Progrès,* in which in 1840 he published his chief work, *Organisation du travail,* and first articulated the famous socialist formula "From each according to his ability, to each according to his needs." Pierre Joseph Proudhon (1809–65) not only opposed private property and business but the state as well. The basic principle of society he favored was that everyone was entitled to the product of one's own work, and this would happen naturally if the state did not interfere by protecting the capitalist exploiters and appropriators and otherwise "rigging the game." This doctrine of anarchism has been a continual theme for reformers and socialists within the European labor movement, especially in the syndicalist unions of France and Spain, as well as in the American section of the Industrial Workers of the World (IWW). Today there is a new resurgence in mature industrial countries of the anarchists' fear of the state and its power, since the growth of huge enterprises and multinational corporations make bureaucratized government almost inevitable. Thus the anarchist view focuses broadly on industrial organization and the social regimentation that it engenders, rather than on the capitalist mode of production and class conflict exclusively. Marx wrote a critique of Proudhon's book *The Philosophy of Poverty,* entitled *The Poverty of Philosophy* (1847).

Karl Marx disapproved of all the utopian socialists, since his agenda was that of organizing the proletariat against the clearly defined enemy: the capitalists. The issue of similar forms of oppression of workers by the state (first as agent of the capitalists, as Proudhon saw government, and later as owner of the means of production *in the name of the workers*) would wait, and of course did not arise as an issue for socialists until the twentieth century, after the Bolshevik takeover of Russia, and the lesson of Stalinism. Another objection Marx had to the utopians (although he owed much to their ideas, imagination, and experimentation) was that they were middle-class, scholarly people whose efforts sprang from theory, not from their own experience of the evils of capitalism. Therefore Marx believed that their experimental communities and cooperatives could not last, since they did not emerge "organically" from the existing stage of material economic development. Although I think Marx was unnecessarily scornful of the utopians, he did have a point, and perhaps we had to wait until today's "postindustrial" phase of excesses and weariness with mass consumption and the mounting social and environmental costs for the actual conditions to exist for the utopians' cooperative-based, ecologically compatible social order to emerge.

Other contributors to the reformist and socialist stream of thought of the period were Adam Müller (1779–1829), a German romantic who abhorred economic individualism and materialism and urged a return to and acceptance of mutual interdependence and the wholeness of life, rather than division of labor. Nations, he felt, should build up their "spiritual capital," rather than concentrate on material capital, land, and labor as the only means of production. William Morris (1834–96) extolled the virtues of crafts, and his ideas informed the British Guild Socialists during World War I. These were decentralists who believed that humans were degraded by working in machine industries and that the division of labor and repetition should be replaced by artisans who derived pleasure and satisfaction from their creations. These ideas are newly discovered thanks to E. F. Schumacher and the Buddhists now taking root in Western societies. Friedrich List (1789–1846) taught at the University of Tübingen, Germany, and advocated economic nationalism and moderate tariffs, but he also developed, in his *National System of Political Economy* (1841), the concepts of total *social* (not just economic) productivity and of the interdependence of economics with law, education, technology, and philosophy. A similar view of social interdependence was expounded by William Thompson (1775–1833), who warned of the dangers of interpreting economic phenomena in terms of economics alone, rather than embedded within their social and natural-world context. He also focused not only on the production of wealth but on its use and distribution, and his book *An Inquiry into the Principle of the Distribution of Wealth Most Conducive to Human Happiness* (1842) propounded ways to attain a system alternative to capitalism. The great American reformer of the period was Henry George (1839–97), who focused on the unjust distribution of the fruits of industrial productivity, which kept workers poor while capitalists justified their larger slice by Adam Smith's "wages-fund" theory of wages (which stated that the maximum possible payment to wage earners was determined by the fund of capital devoted to productive enterprises). Henry George, like Marx, thought this put the cart before the horse (since workers also had created that capital with their prior labor). His book *Progress and Poverty* (1879) was rejected by commercial publishers, published by a friend, and became a best seller. Although Henry George's diagnosis was very insightful (he zeroed in on Ricardo's version of the unearned increment of rent as being a key to impoverishment), his prescription, the single tax on land so as to abolish the unearned increment, would not have been the panacea he hoped, since he missed the problem of business cycles and was writing his analysis during a long downturn. After the Civil War, the wages of workers increased with the increase in industrial productivity, while the relative income to property has not increased relative to labor. There is still today a flourishing Henry George Society in the United States, and many decentralists favor trying his "single tax." It might help, and George's work is still very much worth studying, but our complex, nonlinear society will not respond to "single bullet" remedies, whether the "single tax on land" or the monetarists' panacea of regulating only one variable, the money supply.

But the greatest of the classical economic reformers was John Stuart Mill (1806–73), who joined in the socialists' criticisms. In 1848, Mill published his *Principles of Political Economy,* a Herculean reassessment that came to a radical conclusion: economics had only one province: production and the scarcity of natural means. This narrowed the focus of political economy to a "pure economics," later called "neoclassical," which allowed a more detailed focus on the economic core process while excluding social (not to mention environmental) variables in an analogue of the controlled experiments of the physical sciences. After Mill, economics became split between the neoclassical, mathematical, "scientific" approach and the more policy-oriented "art" of broader social speculation. This led to today's disastrous derailment, in which the two are confused, producing policy "tools" forged in "in vitro," unreal "laboratory experiments," or from econometric models still using the "market" assumptions of the classicists!

Mill meant well with his conclusion about production and means as the correct province for economics; i.e., that distribution was a political, not an economic, process. Once goods are produced, we humans can do what we like with them: share them, throw them away, etc. "The things once there, mankind, individually or collectively, can place them at the disposal of whomsoever they please and on whatever terms. Even what a person has produced by his individual toil, unaided by anyone, he cannot keep unless by permission of society" (*Principles of Political Economy*, 1848). Thus the distribution of society's wealth depends on the laws and customs of society, which are very different in differing cultures and ages. Mill also held that labor was mental as well as physical and that society would reach the "stationary state" that Adam Smith mentioned, further accumulation would be impossible, and distribution would become all-important (i.e., when the material pie can't grow any more, we must learn to share it better).

Mill thus forced the explicit issues of values back onto the agenda of political economy, which was already a covertly value-laden discipline pretending to be a science (in spite of the fact that his attempt failed to clarify that the physical process of production is the only possible subject for a science). Mill, having revealed the *ethical choices* at the heart of political economy, did not fall for the tenets of socialism or the communistic plans of the utopians or those of Marx. He mused, "It is not by comparison with the present bad state of society that claims of communism can be estimated. . . . The question is, whether there would be any asylum left for individuality of character; whether public opinion would not be a tyrannical yoke; whether the absolute dependence of each or all, and the surveillance of each by all, would not grind down to a tame uniformity of thoughts, feelings, and actions. No society in which eccentricity is a matter of reproach can be in a wholesome state" (in Heilbroner, *The Worldly Philosophers*, Simon & Schuster, p. 128). Mill's question has reemerged today as the central concern of the decentralists' "Right Livelihood" and appropriate technology, holistic health advocates, and the humanistic psychology and consciousness movements.

At the same period, another group of critics emerged, who focused on political economy's growing pretensions as a "science." A key figure was Simonde de Sismondi (1773–1842), a Swiss critic of economic theory such as laissez-faire, who raised the issue of distribution by focusing on human health and well-being, asking, "For whom?" Sismondi deplored the economists' propensity for wild generalizations and deductions and their neglect of reality while enveloping their hypotheses in abstract calculations to the point that it was becoming "an occult science." Sismondi observed, "There is perhaps no manner of reasoning that exposes itself to more errors than that which consists of constructing a hypothetical world for the purpose of applying one's calculations" (*Nouveaux Principes d'économie politique*, 1819, quoted in Routh, op. cit., pp. 3–4). We shall follow this "arithmomania," as it was reinvigorated after Bentham's utilitarian formula and developed further by Victorian economists. It is indicative of the widespread acceptance of deductive, *a priori* economics, that such rebukes from Sismondi, Richard Jones (1790–1855), and others, were necessary. However, they remained unheeded as the absurdities were pyramided by the increasing application of differential calculus by the Victorian, neoclassical "welfare economists" and their obsessions with specious mathematics. Even then, it had reached a point where economics produced its own satirist, Frédéric Bastiat (1801–50), who spoofed it in his Paris newspaper articles.

By the mid-1800s, classical political economy had branched into two broad streams:

1) the reformers, the utopians, anarchists, socialists and communists, and the minority of classical economists who followed John Stuart Mill, concerning themselves with holistically examining social structure, value systems, political assessments of the relative power of social groups and classes, and the "art" of economic/political policy studies, as well as those continuing to critique reductionist and apologetic economics, who often preferred to call themselves economic historians, socioeconomists, etc. (indeed, the original meaning of "socialist" was one who did not subscribe to the "economic" view), and, of course, most important, Karl Marx and Friedrich Engels, his faithful friend and interpreter, whose coauthored *Communist Manifesto* appeared in 1848, and their millions of followers; and

2) the school of neoclassical economists who decided to *narrow* their field of inquiry to the "economic core process" and to press on with its "scientific" elaboration. Surely many of them were well intentioned in trying to establish objective formulae in *prices* for utility maximization and welfare, thus arriving at social policies, it was

hoped, without distasteful political haggling. Others retreated into ever more abstruse mathematics as a defense against the devastating critiques of the utopians, the socialists, and Marx. The first concept they jettisoned was the labor theory of value, because it lent itself so well to the Marxian view (that wealth was produced by labor and that even capital was produced by prior labor, and this "surplus value" in labor was expropriated by capitalists). I believe that Marx went too far and underemphasized the use of resources, coal, etc., but for understandable reasons.

So the labor theory of value, with its ethical justification for wages, went the way of "just" prices and wages but reemerged when circumstances required—in twentieth-century socialist and market-oriented industrial societies as "labor productivity," as the growing volume of production of goods had to be soaked up with purchasing power and the *flow of income* became the key driver. *Costs of production* (the price system's last touchstone in physical reality) were no longer to be the key determinant of prices or wages, but supply and demand, which could be abstracted, thus further sanitizing the debate over "just" distribution and obscuring the role of finance and industrialists. Only John Elliott Cairnes (1823–75) devoted himself to defending the *a priori* school, with its newly scientized methods *on their merits,* by defending their still deductive methods as "experiments carried on mentally" (Soule, op. cit., p. 96). Thus he was precursor of today's modelers. The quantification-oriented school was built on postulates about *aggregate* supply, demand prices, wages, etc., which could be represented as equations and graphs representing assumptions as to their relationships and which did not require any knowledge of *actual quantities* involved; e.g., the basic Supply-Demand graph of all elementary economics textbooks.

The founder of the mathematical school of economics was Antoine Augustin Cournot (1801–77), appropriately a professor of mathematics at the University of Lyons, France. His book *Recherches sur les principes mathématiques de la théorie du richesse* appeared in 1838, but, for years, not a single copy was sold. The same fate befell Herman Heinrich Gossen (1810–58), a German public official, and his treatise *Die Entwicklung der Gesetze des menschlichen Verkehrs,* published in 1854. Disgusted, Gossen destroyed all but one copy, which found its way to the British Museum. There it was rescued at last, along with Cournot's book, by William

Stanley Jevons (1835–82), who breathed life into the budding mathematical school. Gossen contributed to the formalizing of Ricardo's marginal concept in his rent theory by expanding it to apply to demand, thus allowing Bentham's "pleasure and pain" calculations to be stated "rigorously" as the *theory of marginal utility*. Operationalized marginal-utility theory has been the delight of mathematical economics ever since.

Jevons combined utilitarianism and calculus into a new theory of value depending entirely on "utility." Even though Jevons allowed that units of pleasure and pain were difficult to quantify, neither could gravity itself be measured, except by its effects on the motion of a pendulum. Likewise the "oscillations" of the human will were minutely registered in the price lists of the markets. Thus marginal-utility mathematics did not try to measure total pleasure or comparative pleasure, but the pleasure of having a little more of this relative to a little more of that—at the margin. Utility will be maximized when any commodity is distributed among all its alternative uses in such a way that the final degree of utility derived from each use is equal to each of the others. However, this was only possible under the assumption of perfect competition: i.e., Adam Smith's conditions for free markets, requiring that buyer and seller meet each other with equal power and information. But Jevons stated that ". . . the theoretical conception of a perfect market is more or less carried out in practice," and from this postulate he derived the Law of Indifference: that, in the same market, at any one moment, there cannot be two prices for the same kind of article" (Routh, op. cit., p. 221). And so it went! Newton might well have turned in his grave, as Jevons spun theories of marginal labor and marginal productivity, eliminating not only the labor theory of value but the conflict between labor and capital to which it gave rise, by also picking up the idea from Nassau Senior that *abstinence* gave rise to capital. In the same vein, Francis Edgeworth (1845–1926) followed, in 1881, with *Mathematical Psychics*. Alfred Marshall (1842–1924), who had hoped to become a physicist but became one of the most respected mathematical economists of the period, reintroduced social concern and tried to integrate it into the theories.

Marshall intuited the importance of biology for economics, the idea of irreversibility (which is now crucial), and the idea of "externalities," but in the positive sense, i.e., social infrastructure, public works, and an educated labor force that the entrepreneur could take advantage of but had not paid for. It remained for his student at Cambridge A. C. Pigou to use the term in a negative sense, as pollution. He also introduced the concept of "elasticity of demand" (to take into account the growing viscosity of the old equilibrium "mobility" as industrial societies became more rigidly structured). John Bates Clark (1847–1938), an American economist, then applied marginal theory to distribution of income.

In France, Léon Walras (1834–1910) was more careful in pointing out that his mathematics and use of statistical theory (Law of Large Numbers) were merely tools of inquiry, not statements of fact. Like Marshall, his inquiry went deeper than empty mathematical virtuosity. He was an agrarian socialist and wanted to nationalize land, but he talked of humans as "economic molecules" and gave concepts like scarcity scientific definitions analogous to heat in physics. He developed the first complete mathematical model for a whole economy and thus founded what is now called econometrics and input-output modes of total economies, such as those of Wassily Leontief. However, Walras displayed the familiar, almost schizophrenic assumptions of private property, perfect competition, and even the idea that labor is a form of capital! In Austria, the leading mathematical economists were Karl Menger (1840–1921) and Eugen von Böhm-Bawerk (1851–1914), who was chiefly concerned with refuting Marx but who also developed a theory of interest based on the marginal utility of capital. This introduced the idea of comparing present and future values, which led to today's problem, with many public and private investments, of excessively discounting the future (i.e., the present value of goods and services is deemed greater than their future value).

The Austrian school is famous as a bastion of laissez-faire, with later luminaries including Ludwig von Mises (1881–1973), and Friedrich von Hayek (1899–), author of The Road to Serfdom (1944), a critique of collectivist and Marxist economics. Italy's leading mathematical economist was Vilfredo Pareto (1848–1923), discussed earlier, whose Optimality introduced the theory of marginal rates of substitution. Swedish economist Knut Wicksell (1851–1926) prefaced Keynes's theory of underconsumption with the explanation of business cycles as due to overinvestment (as did Russian economist Nicholas Kondratieff, with his "long-wave, fifty-year" theory of business cycles—whose work, as mentioned, is enjoying a new vogue).

At this time, yet another group of challengers to the mathematical school appeared, with small success, however, since their names have remained almost unknown. They zeroed in on the misuses of mathematics now firmly entrenched. The unsung efforts of Thomas Edward Cliffe Leslie (1827–82), John Kells Ingram (1823–1907), Walter Bagehot (1826–77), and Henry Sidgwick (1838–1900) deserve mention. Luckily, the tradition did not quite die out, being kept alive in the United States by Thorstein Veblen (1857–1929), as described in my "The Decline of Jonesism" (*The Futurist,* 1974). In Britain, John A. Hobson (1858–1940) critiqued capitalist expansion, in *Imperialism* (1902), more drastically than Marx. Marx merely said that capitalism would destroy itself, but Hobson said it would destroy the world, thus aligning himself with Rosa Luxemburg's view. Lenin approved, and imperialism became the capstone of Marxian theory. As we have seen in Chapter 5, imperialism theory failed to take account of transnational and multinational corporations.

Now we turn to Karl Marx (1818–83), who refused to be identified as an economist, yet critiqued classical and mathematical economics more expertly and effectively than any of its practitioners (who had more to lose!). Today's economists are similarly inhibited. Marx's body of work is, of course, so comprehensive, and encompasses so many fields, that I can do little more than sum up his economic ideas and give a brief account of the debate he engendered, which still rages. But while Marx the social revolutionary is canonized by millions all over the world, economists have had to deal with his embarrassingly correct identification of boom-and-bust business cycles and the tendency for market-oriented economics to develop "reserve armies" of last-hired, first-fired, hard-core unemployed (usually of low status, for example, blacks, other minorities, and women) to bear the brunt of recessions. Marx's prediction of the downfall of capitalism (as socialism was to slowly emerge from its womb) has not *yet* been proved wrong. In the West, the revival of capitalism after the Great Depression (when Keynes's theories helped give it a new lease on life) is cited as all the proof needed that Marx was wrong. Similar arguments are used to disprove Malthusian theories of population outrunning food supply (for example, global starvation affects "only" millions and not yet billions of humans).

The critics of Marx who claim his theories were *exclusively* deterministic and materialistic have either not read him at all or misread
the meaning of his theory of dialectical change. Michael Harrington's
reassessment of Marx in *The Twilight of Capitalism* (1976) is per-
suasive. He cites Marx's views in the 1857 "Introduction" and Volume I of *Das Kapital* that the economic means of a society are *both*
determinant *and* part of an organic whole; i.e., the specific production creates the pervasive lighting and special atmosphere of societies.
"Under capitalism the fields are organized like a factory. A plow
under feudalism and a plow under capitalism might be physically
identical, but they exist in different and special atmospheres" (pp.
66–67). But Marx can also be a technological determinist, pointing
out in Volume I of *Das Kapital* that "machinery is utilized as the
most powerful weapon in the capitalist's arsenal as the best means
for overcoming the revolts against capital." And this: "All progress
in capitalist agriculture is progress in the art not only of robbing the
laborer but of robbing the soil; all progress in increasing the fertility
of the soil for a given time is a progress toward ruining the lasting
source of that fertility. The more a country starts its development on
the foundations of modern industry, like the United States, for example, the more rapid is this process of destruction" (*Das Kapital,* Volume I, p. 254, Gateway Edition). Thus what Marx emphasized is
that it is only under capitalism that economics, as such, plays the
leading social part in its own name. Marx's main body of work was a
critique of capitalism, and it is misinterpretation to assume that he
generalized his economic interpretation to *every* social structure.

At the same time, Marx saw that capitalist forms of social organization would speed the process of technological innovation and increase material productivity (his Law of Accumulation) and that
dialectically this would change social relationships again. In the
Grundrisse, he sees deeply into the capitalist mode: ". . . modern industry began, not with the factory, but with the measurement of work.
When the worth of the product was defined in production units, the
worth of the worker was similarly gauged. . . . But under automation, with continuous flow, a worker's worth can no longer be evaluated in production units" (Harrington, op. cit., p. 129). Marx saw
that under capitalism "the reciprocity and universal dependence of
individuals indifferent to one another forms the basis of their social
connection," but he also saw that a time would come when one could

no longer derive any neat scheme of social entitlement from measuring the contributions, i.e., inputs to production of land, labor, capital, technological innovation, etc.; it would become an inextricably complex social process. He not only predicted the time when labor time could no longer serve as a measure of value but even stated that labor would become more "mental" as science and knowledge are applied.

Thus, Marx seemed to allow always for change, even in his labor theory of value, although in his time, when resources were plentiful and population small, it was indeed human labor that was the most important input. More important, Marx used the labor theory of value and the idea of surplus value as a way to raise issues of justice, and as powerful holistic concepts with which to surround the reductionist logic of the neoclassical economists of the time. Inasmuch as he fell into trying to present "scientific" arithmetical formulas of the labor value of commodities in order to deal with the reductionist economists, he undermined his larger, more systemic, sociopolitical model. At least he did not try to derive *price formulas* from the theory, since he knew well enough that wages (which he defined as an equal exchange among unequals) and prices were much more politically determined. Oscar Wilde said it best: It's possible to know "the price of everything and the value of nothing."

Marx viewed society and capitalism from explicitly stated vantage points: from the position of the workers, from that of the capitalist, and in historical and cultural perspectives (very much in the paradigm of Einstein, Heisenberg's Uncertainty Principle, and the new physics of Fritjof Capra, which acknowledges that observation always affects the outcome of research). Marx knew that one cannot have a clear dialogue until all can establish where they see themselves in space/time/system. Marx's broad sociopolitical model, with its historical dynamism, allowed him to see economic processes in large patterns: monopoly, depressions, and the process of expanded reproduction (i.e., innovation), as well as the fact that capitalism would foster socialism (as it has) and eventually disappear (as it may). Another non-Marxist, well-respected economist, Joseph Schumpeter, gave Marx full credit for the former prediction in his *Capitalism, Socialism and Democracy* (Harper & Row, 1942); Schumpeter's prognosis was similar, with important differences and hindsights: e.g., the workers had not been continually "immiserated" but had ridden the escalator of material productivity, although at relatively lower levels and with much struggle. But Schumpeter believed that capitalism would destroy itself because it was antithetical to its own expressions, for example, entrepreneurial spirit, and was disintegrating and bogging down in an emerging accidental socialism of bureaucracy, regulation and administrators, politicians, lawyers, and intellectuals who had "vested interests in social unrest." He preferred a more orderly, explicit emergence of planned socialization. One might say that he made historically specific some Marxian ideas, e.g., that capitalism's industrial technological base was at odds with its individualistic, private-enterprise superstructure. Schumpeter showed that there were flaws in Marx's futurism, particularly that the workers' revolutionary ardor often was co-opted into reformism. In fact he dismisses socialism as a cultural or class movement and limits himself, in good reductionist-economics style, to see it merely as a reorganization of economic affairs, which belongs to the public, not the private, zone, "where business people were no longer anointed as the custodians of the general welfare." It is generally pointed out by Marx's critics that the U.S. labor force, which should have been, according to Marx, the first to organize politically and rise up to create a socialist society, *failed* because they received high enough wages to begin identifying with the upward mobility of the middle class. However, the newly translated *Why Is There No Socialism in the United States?* by Werner Sombart, written in 1905, gives several other explanations, including that American workers were extremely transient, moving for the jobs in a dynamically growing frontier; were divided by their language and other ethnic differences, which were exploited by factory owners; and that enormous numbers of them went back to the old country as soon as they became rich enough to provide a better life for their waiting families (e.g., between 1907 and 1911, for every one hundred Italians arriving in the United States, seventy-three returned home). Thus, opportunities for organizing a socialist political party, European style, were very limited.

Of course, Marx's model could not have *explicitly* foreseen that capitalism would produce new forms of exploitation and social oppression, e.g., ecological devastation and exploitation of Third World people, via advertising of baby formulas, cigarettes, Coca Cola, etc., and might create new protest groups and revolutionary consciousness, e.g., women, blacks, environmentalists, consumers, etc. Nor could Marx the prophetic moralist (in his early manuscripts, *The Economic and Philosophical Manuscripts of 1844*, which came to light in 1932; see, for example, Robert Tucker, *The Marx-Engels Reader*, Part I, The Early Marx, W. W. Norton, 1972, pp. 3–110), do more than vaguely anticipate the human-potential, holistic-health, and consciousness movements of today, even though these manuscripts reveal many humanistic insights. Marx also developed the notion of "false consciousness," a brilliant conceptual tool to dissect today's advertising industry and its manipulations of the consciousness of consumers, as by Stuart Ewen, in *Captains of Consciousness* (1976), and David Potter in *People of Plenty* (1954). Similarly, generations have found the concept of "alienation" useful, elaborated from Marx's theory of how workers suffered alienation from their own time and existential life while producing for capitalists (Tucker, *Marx-Engels Reader,* ibid. p. 94–96). Human creatures are subjective and objective beings, and their consciousness and activity create objects that help them mirror and understand themselves. It is much more than mere capitalist appropriation of the surplus value in these objects that Marx means by "alienation"; rather, that the essential nature of humans, which they express in work and fashioning objects, is *distorted,* and the learning, self-definition, and self-appraisal that naturally produced objects provide is preempted when a person's creative powers are programmed by someone else. Marx accomplished more than his share of both theorizing and social activism, but choices must be made, since our lives are short. Although he concentrated on the social dimensions of the human dilemmas of his day and founded movements of social, not individual, change (such as the consciousness movement), he was obviously concerned about these dichotomies between theory and practice, and constantly exhorted that they should be interdependent, reciprocal. Theodore Roszak, in *The Making of the Counter Culture,* sees Marx's concentration on social change as his basic flaw and that of his followers. I forgive Marx for it, but not the Marxists (who have the benefit of Freud). The delicious paradox is that Marx was himself largely a scholar and theoretician, and while he did not belong to the proletariat, participate in strikes, or fight at barricades, his theories were the stuff of instantaneously combustible activism.

Neither could Marx be expected to have emphasized ecology; it

was not the burning problem of his time. However, his model, with
its huge scope and all its interacting variables and dynamism (e.g.,
positive feedback loops, accumulation), consolidation of the proletariat, even the longer-term depletion of the soil mentioned previously *could* have been used to predict the ecological exploitation
that capitalism produced and socialism perpetuated. The technological determinism of both systems could also have been predicted
as well as applying insights from his description of "alienation" that
afflicts the worker under socialist systems as well as market-oriented
societies, i.e., as a function of industrialism *per se*. Thus, one can
certainly fault his followers for not grasping the ecological issue earlier, since it provided yet another devastating critique of capitalism
and confirmed the vigor of the Marxian method. Unfortunately, if
they had faced the ecological evidence honestly, they would have
been forced to the conclusion that socialist societies had not done
much better, saved only by their lower per-capita consumption,
which in any case they were trying to increase! However, ecological
knowledge is subtle, obscure, and first requires scientific understanding, since other species, whether squirrels or redwoods, cannot
provide revolutionary energies to change human institutions, because
they are mute and don't vote. (This problem has been addressed by
Peter Berg in *Reinhabiting a Separate Country* [1978] and lawyer
Christopher Stone in *Should Trees Have Standing?* [i.e., in courts of
law; 1972].)

Yet Marx was not quite blinded by his zeal for the proletariat's
cause, even though his espousal of the labor theory of value almost
obliterated his concern for nature and her inputs. In the *Economic
and Philosophical Manuscripts of 1844* he states, "The worker can
create nothing without nature, without the sensuous, external world.
It is the material on which his labor is manifested and from which
and by means of which it produces" (Tucker, *Marx-Engels Reader*,
p. 58). Thus we have to see the labor theory of value as serving
the human cause of the moment, although it made less sense
as the twentieth century unfolded, and no sense at all today, or in the
future, to ignore the resource inputs that undergird the entire economic process. Marx valued natural resources in a quite specific
way: with the concept of *use-value*. In *Das Kapital*, Volume I, Chapter 1, on commodities, he says, "A thing can *be* a *use*-value, without
having value. This is the case whenever its utility to man is not due

to labor. Such are air, virgin soil, natural meadows, etc." He also included as use-values "all the things produced, but not as commodities for sale, but for people's own use or for the community as social use-values." He completes his point thus: "Man can work only as nature does; that is by changing the form of matter, and in this, he is constantly helped by natural forces. . . . We see then that labor is not the only source of material wealth, of use-value produced by labor. As William Petty puts it, labor is its father and the earth its mother" (Tucker, *Marx-Engels Reader*, p. 205).

All this is very clear regarding the most inclusive meaning of the value of human production. What Marx then distinguished was the *capitalistic* mode of production and of accumulation of *"productive capital,"* which he regarded as congealed, past, dead labor. It was in relation to this type of production in which he insisted on the labor theory of value, because in this form it involved appropriation of the workers' natural use-value production and his existential lifetime. This was accomplished because the worker had to sell his labor at the market *price* (i.e., wages), while for the capitalist, labor is *use*-value. While there are many references and restatements of the role of nature, it was not the central issue for an activist of the day. Yet Marx chides the German socialist organizers in his *Critique of the Gotha Program,* refuting their first paragraph by saying, "Labor is *not* the source of all wealth. *Nature* is just as much the source of use-values (and it is surely of such that material wealth consists!) as labor, which itself is only the manifestation of a face of nature, human labor power" (Tucker, *Marx-Engels Reader,* p. 382). This was written in 1875, while the earlier quotes, from Volume I of *Das Kapital,* were written in 1848; Marx's view on the role of nature remained quite steadfast. Dozens of similar references (however incidental) appear in Marx, even an intuitive grasp of what later would be called "ecological-niche theory," in his Preface to a *Contribution to the Critique of Political Economy* (1859): "No social order ever perishes before all the productive forces for which there is room in it have developed" (Tucker, *Marx-Engels Reader,* p. 5). Or this prescient comment from the *Early Manuscripts* of 1844: "Industry is the actual historical relation of nature, and therefore, natural science to man. Industry is conceived as the exoteric revelation of man's essential powers" (Tucker, *Marx-Engels Reader,* p. 76). This is the view of biologist A. J. Lotka, who explains why the economic process is a continuation of the biological one (in which humans are like other species in using only *endosomatic* instruments, e.g., claws, beaks, paws, hands for digging, etc.). In the economic process, humans begin to use *exosomatic* instruments (knives, boats, fire, etc.). In this exosomatic production, we are brought into irreducible conflict with each other. This point is applied in *The Entropy Law and the Economic Process* (1971), by Nicholas Georgescu-Roegen, the only *economist,* so far, who has made any original, comprehensive reformulation of economics since Marx and Keynes.

Why, then, have these aspects of Marx been ignored for so long by Marxists? Largely, I suspect, because they were peripheral to their social-organizing, reforming, and revolutionary intentions. Marxian-oriented scholars, e.g., Michael Harrington and Erich Fromm, have reread Marx and brought to light some of these subtleties, but they are inconvenient for social activists, because certainty is simpler to communicate and organize around. Perhaps this was why Marx finally stated at the end of his life, "I am not a Marxist." Even Harrington overlooks the power of the ecological critique of capitalism. The best blending of the social and ecological criticism is by an ecological scientist, Barry Commoner, in *The Closing Circle* (1971) and the *Poverty of Power* (1977). I consider my own work a social/ecological/spiritual critique of *industrialism,* whether capitalist, socialist, or mixed. In sum, my view of Marx is that he was an intuitive with great intellectual power; possessed well-integrated functioning of both brain hemispheres; and had great passion and ethical concern and a highly developed value system that must be called "heuristic" by systems and information theorists and "prophetic" by religious scholars and philosophers (see, for example, "Beyond Marx and Niebuhr: Toward a More Prophetic Politics," by Neal Riemer, Department of Political Philosophy, Drew University, Madison, New Jersey).

As the continuous assaults on capitalism and market economics built during the late-nineteenth and early-twentieth centuries, Marxian predictions of socialist reforms seemed to be emerging victorious. But there was one more turn of the wheel of capitalism's fortunes, in spite of the social disintegration that threatened during business cycles, culminating in the Great Depression, of the 1929–33 period: the social interventions of governments, justified (sometimes after the fact) by the theories of Keynes.

John Maynard Keynes (1883–1946) was the son of a Cambridge University economist and studied under Alfred Marshall, imbibing neoclassical theory easily within his much more comprehensive world view. Keynes was keenly interested in the entire social and political scene and viewed economic theory as an instrument of policy. Therefore, the Keynesian revolution (or "restoration," as the Marxists would call it) bent the so-called "value-free" methods of neoclassical economics to serve instrumental purposes and goals, and in so doing, made economics once again *political* (but in a new way). It also involved giving up the classical Newtonian stance of the "objective observer," perpetuated by the neoclassicists, and made economists "participants." This was a contradictory and uneasy synthesis whereby Keynes tried to calm the fears of the neoclassicists regarding 1) "intervention" in the equilibriating operations of the market system, 2) loss of "objectivity," and 3) the scientific claims. By showing them that he could *derive* his policy interventions from their neoclassical model by proposing it as a "special case" (in just the ways physical scientists proceed), Keynes demonstrated that economic equilibrium states and equilibration in the traditional sense were *exceptions,* rather than the rule in the real world. Yet he managed to keep his General Theory, as an updated version of the model of the economic core process, isolated from any changes in extrasystemic variables.

Keynes *holds as constants* in his model, over the period of analysis, technologies of production (i.e., ignoring Marx's accumulation theory and innovation), as well as supply of labor, consumers' tastes, degree of competition, and the general motivational assumption of economists—the maximization of utility. This enables him to investigate "what determines at any time, the national income of a given economic system (which is almost the same thing) and the amount of its employment" (Keynes, *General Theory of Employment, Interest and Money,* Harcourt Brace, 1934, p. 247). But, as Adolph Lowe points out (*On Economic Knowledge,* p. 219), this very formulation is in opposition to orthodoxy, since it implies that aggregate income and employment are liable to short-term changes (i.e., booms and busts), whereas neoclassical theory postulates full employment. So Keynes defends his heresy by appealing to experience: the embarrassing fact that "an outstanding characteristic of the economic system in which we live, that it is subject to severe fluctuations

The right to work

Everybody who wants to work should be able to. Every adult person, regardless of age, sex, race or religion should be able to have a satisfying job that returns a living wage. It should happen that way. But it doesn't.

New Zealand has many laws which protect private property. There is not one law which protects a person's right to work!

Certainly the dole makes sure that a jobless person does not starve, but it can do nothing, absolutely nothing, to restore that person's dignity and feeling of self worth.

The desire for short term profit has meant that much economic growth has happened in industries that cannot last. The automobile industry, for example, will soon grind to a halt when oil becomes too expensive to import. The pulp and paper, aluminium and steel industries take 60% of New Zealand's industrial energy, yet employ less than 3% of the workforce.

Is growth the answer?

To overcome downturns in the past, the system has depended on growth. The mineral and energy resources that have fuelled past growth are quickly becoming scarce and more expensive. Growth can no longer be used to solve what is basically an allocation problem. Until we resolve the question of ownership and control of New Zealand's financial and productive resources, the human tragedy of jobless people will remain with us.

the
disposable
worker

Unemployment is part of the system

Unemployment is a natural result of our system, where **capital employs people.** In a co-operative economy, where **people employed capital,** unemployment would be unknown.

The official figures of those unemployed or on Government Relief work are near record levels. The true rate of unemployment is staggering. It would be well over 100,000 people if young people unable to register and married people whose spouse is working were involved in the statistics. Most of these people are looking for work and cannot find it.

Many people mistakenly believe that an economic upturn is just around the corner. They are prepared to tolerate some unemployment in the meantime. There will be **no** permanent upturn!

High unemployment will stay with us there is a basic change in our economic system.

Plate 21

25% of Farm Workers are Children

exposed to pesticides, excessively long hours of hard labor and denied a normal education.
The United Farm Workers are fighting to change the lives of these children by:

- bargaining for just wages and decent working conditions, enabling migrant families to settle and children to attend school
- demanding the abolishment of child labor and indiscriminate use of pesticides
- establishing community clinics, daycare centers and many other services.

El Taller Grafico is the Graphics Workshop of the United Farm Workers. Buy from us and your money goes directly toward the support of the UFW.

Send for our free brochure

JEWELRY ★ POSTERS ★ BOOKS ★ BUTTONS
FLAGS ★ CHRISTMAS CARDS ★ CALENDAR

United Farm Workers, La Paz, Keene, Ca. 93531

Plate 22

	Proposed Project	Alternative Processes	Alternative Investment

Jobs Created

1. How many new local jobs will be *directly* associated with the proposed new facility (program, project, etc.)?

2. How many new local jobs will be created indirectly by local purchases of supplies and services by the new facility?
(This can be roughly calculated by: (a) obtaining estimates of the new facility's annual local purchases of supplies; (b) figuring the proportion that will go to pay the wages of new workers that local suppliers will have to hire to handle the increased business; (c) and then dividing by local average wage, to convert the dollar figure to a jobs equivalent. Consistent and reasonable "guestimates" are OK.)

3. How many *induced* local jobs will the new facility generate?
(This can be roughly calculated by: (a) estimating the percentage of new direct and indirect payrolls that will ultimately become local personal expenditures by workers; (b) figuring the proportion that will go for wages of new workers that local retailers will have to hire to handle the increased business; (c) and dividing by the average local wage to convert the dollars to job estimates.)

4. Total number of jobs created:
(direct + indirect + induced = total)

5. How many jobs will the new facility eliminate—directly or indirectly—in other local businesses?

6. NET number of jobs created:
(number of jobs created minus number of jobs eliminated)

Alternatives

1. What are the employment impacts of alternative means of providing the same services?
(These alternatives can be different production processes—such as employing bank tellers instead of computers to process checks—or different institutional arrangements. Small, locally owned shops keep business profits within a community, where they provide indirect and induced jobs, while large, outside-owned franchises remove profits—and induced jobs—from the community and frequently purchase supplies from outside suppliers they own, reducing indirect jobs in a community.)

2. What are the employment impacts of alternative uses of the same investment resources (particularly if public funds)?
Almost any expenditure of money creates jobs, but using that money for different purposes may have very different results while providing the same amount of jobs. Building new power plants provides jobs, and insulating homes provides jobs, but only the latter eliminates need for unnecessary future expenditures of work, dollars, and energy. Expenditure of tax money provides jobs but raises our taxes, giving us each less to spend, which would have provided jobs anyhow. And expenditures for different purposes provides very different numbers of jobs and use of resources. Hospital services provide three times as many jobs per dollar spent as highway construction. Expenditure of funds for waste-treatment construction, social security benefits, or national health insurance instead of present Army Corps of Engineers projects would provide 30 to almost 60% increases in employment.

Community Impacts

1. What special conditions will be required to sustain the activity being planned? (Do the product produced and the proposed rate of resource use indicate sustainable operation? Are special markets, government subsidies, local resources, abnormally low local wages necessary?)

2. Are the activity and the jobs it creates seasonal or cyclical over a longer period of time?

3. How many of the new jobs created will be permanent?

4. How many of the new jobs will be temporary (e.g., associated only with construction or initial operation)?

5. What will happen to workers in temporary jobs when their work ends?

6. What will be the distribution of new jobs among types and wage levels?

Type of job	Wage	Number	Percent of Total

7. Will the income distribution of the jobs provided increase inequality of wealth in the community?

8. How many of the new jobs are likely to be filled by local unemployed people?

9. How many of the new jobs are likely to be filled by workers whose employment has been terminated, directly or indirectly, because of the new facility?

10. How many of the new jobs are likely to be filled by local residents? How many by newcomers?

11. How many of the new jobs are likely to be filled by men? women? minorities?

12. Will the new facility make it harder for people without special education or training to get jobs in the community?

13. Will the new facility make local employment more dependent on outside decisions that don't incorporate the needs of the community?

14. Will the financial base of the new project make it comparatively harder for small local industry to compete fairly for loans?

Prepared by RAIN Magazine from an earlier version by Avrom Bendavid-Val

Plate 23

WHO RUNS AMERICAN BUSINESS?

If you guessed the ones on the right, you're wrong. They both do.

The ones on the right manage a traditional capitalist business—and have all the resources, money, and educational expertise behind them. The ones on the left are committed to a new approach to business: one based on employee ownership, participatory management, appropriate technology, and community accountability. They have limited money, resources, and training. But now there's a business school designed specifically for them.

THE NEW SCHOOL FOR DEMOCRATIC MANAGEMENT
practical business training based on principles of democracy in the workplace

for:

appropriate technology producers
community development corporations
worker owned businesses
non-profit organizations
women's enterprises
food and housing coops

performing arts groups
alternative media groups
minority enterprises
senior citizens groups
neighborhood clinics
credit unions

For information about our regional
training sessions in 1980 write or call us:

NEW SCHOOL FOR DEMOCRATIC MANAGEMENT
589 Howard St.   San Francisco, CA 94105   (415) 543-7973

Plate 24

in respect of output and employment" (Keynes, p. 249). But the reason that the cat was now out of the bag was because, up to that point, business-cycle theorizing had always invoked *extra*-systemic variables and *ad hoc* occurrences: e.g., sun spots (Jevons), credit inflation, technical progress (which Marx made *part* of his model as a long-term feedback loop), wars, etc. Today's economists perpetuate these errors, viewing OPEC, resource depletion, etc., as "exogenous shocks," rather than incorporating them systemically. Even technological change, the mainspring of economic growth in the past two hundred years, is most often treated as a *coordinate,* as discussed in Chapter 10.

Keynes's model broke with the neoclassical model of general aggregation of *all* micro units by aggregating micro units into several major components, thus "sectoralizing" the earlier model. Aggregate consumer demand plus capital investment, mediated by interest rates, is related to aggregate output and employment. Investment is related not only to interest rates and levels of expected profit but also to the consumers' *propensity to save* as well as consume, and to their attitude toward keeping cash on hand (i.e., preferences regarding liquidity); in addition, investment is related to the quantity of money supplied by the banking system. The quantity theory of money, from Sir William Petty to Milton Friedman and today's monetarists, has been subject to many debates over its definition; e.g., in the early-nineteenth century, the big debate was over whether money was just coins and banknotes, or whether it included bank deposits and other financial instruments. Today monetarists are having an identity crisis, because the variable they watched, called Money 1, or M1 (bank deposits and cash currencies) is now inadequate, and new rules have just been promulgated to include credit cards, which are now a form of money that disordered the earlier definitions. Money is now becoming so complex that there are not only M1, M2 (M1 plus commercial-bank time deposits other than $100,000-plus certificates of deposit); M3 (which includes also deposits at mutual savings banks, savings and loan shares, and credit union shares); not to mention the host of new problems of "stateless money," in corporate multinational banking. Eurodollars, etc., now make a mockery of efforts to "manage" a domestic economy. As mentioned in Chapter 2 (note 10), the monetary indicators were overhauled again in early 1980.

In Keynes's model, it is crucial that additional investment increase

employment and thereby aggregate income, and this additional demand for consumer goods adds to demand for labor and more manufacturing capacity and working capital, and so the system progressively utilizes its available resources (the multiplier effect). This is
what is called the trickle-down theory of investment and growth,
which I challenge in Chapters 9 and 10. However, Keynes *never said*
this process would culminate in full employment. Rather, it will
move the system in that direction, *or* peter out at some level of *under*employment *or* even go into reverse. It depends on a lot of metaphysical assumptions: the precise ratio between additional consumption and additional income that consumers marginally prefer: the
more of the new dollars they spend the more jobs and income; the
less they spend (i.e., the more they *save*), the sooner the expansion
will peter out unless new investment is pumped in. Thus the "habit"
of pump-priming, printing money, cutting taxes, easing credit, lowering interest rates, and seeking the "magic" multiplier (which eventually contradicts the laws of thermodynamics) is now being foreclosed, as I describe in *Creating Alternative Futures,* even though
Reagan-administration supply-side economists still believe in it.
We must now bring in *all* the "extrasystemic" variables into a larger
model and make Keynes a special case! This is necessary because
Keynes's model did not include multinational corporations or capital
flows, but portrayed an isolated domestic economy whose policies
were not constrained by global economic agreements and at the same
time one able to command cheap resources in a rigged world market. Neither did Keynes allow for political coalition-building between
labor, consumers, environmentalists, and unemployed; an energy
crisis; rising social or environmental costs; etc.

So Keynes found himself forced to a Marxian-type conclusion that
relates the time span of the cycles of boom and bust to "the average
durability of capital" (Keynes, p. 318). Even at best, Keynes's
model can only, under his ideal conditions, give us a set of *scenarios*
that are possible; it cannot be *predictive.* What it *appears* to confirm
is that since, on the average, consumers increase their consumption
as their income increases—but not by proportionally as much (marginal utility of income)—*this* leads, on the economy's upswing, to the
gap between income and consumption, which *widens* in the short run
with rising *employment,* and over longer periods, as *general wealth*
increases. This gap must be filled by increasing *investment* if the sys-

tem is not to periodically slump into underemployment. The post-Keynesians fall into this same methodological trap rather than examining what *kinds* of investment create *most* employment. Most economists *assume* that a "favorable" capital-labor ratio is one of greater, and increasing, capital intensity; see Chapter 10.

This *intra*systemic dilemma can only be resolved within a theory of capitalist growth that is as morphogenetic and biological, as those discussed in Chapter 11. It is simply the evolutionary formula "nothing fails like success."

But there is another aspect of Keynes's thought processes that is brilliantly dissected by Adolph Lowe (*On Economic Knowledge,* p. 222). Keynes hypothesized the supply of money as an essential determinant of the level of employment, output, and real income. But another wrinkle is added to challenge the orthodox equilibration processes that money mediates: *expectations,* i.e., consumers trying to second-guess future prices, investors, and the future yields on capital during various phases of the business cycle. But, inasmuch as these responses and the money that registers them in prices, demand, etc., are much more "frictionless" and rapid than the real-world processes, prices and money statistics become "decoupled" from the now set-in-concrete, large industrial mass-production processes they supposedly track and command. Thus production cannot respond to such rapid fluctuations in demand (what economists call "inelasticities of supply of industrial output"). So another destructive positive feedback loop pushes a sluggish economy into a downturn: people begin to wait for prices to fall (which they know will happen, because it has in the past; the flow of mass production cannot be shut off easily, and many companies find it easier to cut prices than to close down). Thus technological immobility now built into mature industrial economies prevents adjustment by *phaseout* of some production and its replacement by new. As the companies get even bigger, this inflexibility grows, as they have political power to *force* consumers not only to buy their products (via taxes and boondoggling publicly financed projects) but even to put up the investment capital (e.g., electric utilities and nuclear power plants and the bailing out of Chrysler).

Now we see the key role of *advertising* and its purpose for big companies in "managing" their *demand* in the marketplace. Consumers must not only keep increasing their spending, but they must do it *predictably,* for the system to work. At this point the classical capitalist model has almost been turned on its head! Today economists *create* business cycles: consumers are forced to be involuntary investors; the "market" is managed by business and government actions while economists refuse to recognize, as Christopher Lasch points out in *The Culture of Narcissism* (1979), that we live in a corporate state. Therefore, studying "consumer confidence" as a key economic variable *also* becomes essential (studied by the economists of the University of Chicago's "School of Rational Expectations" and those at the University of Michigan who survey consumer confidence). "Confidence" was identified as important for investment, by Marshall, "to touch all industries with her magic wand." But why, then, asks historian Routh, "did Marshall not lead the massed economists in a hue and cry after the Fairy Confidence . . . so as to capture her . . . and learn her secret powers?" (Routh, op. cit., p. 268). Because, according to Edwin Cannan, president of the Royal Economic Society, in his 1932 address, "General unemployment appears when asking too much becomes a general phenomenon . . . [the world] should learn to submit to declines of money income without squealing" (*Economic Journal,* Volume XLII, 1932, pp. 357–69). This is the classic statement of the "old-time religion"; i.e., the economy must take its unpleasant slump as medicine until profits are restored and capital investment starts a new upswing. Ironically, this is also Marx's view! But the issue concealed in all the abstractions about "the world" and "the economy" then and now is *who* is to take the medicine as Russell Baker's humor targets (Fig. 25, p. 158).

Thus the neoclassicists remained irresponsibly uninterested in the *political* problem of rising unemployment, and instead railed at the Keynesian remedies of pump-priming, public works, and easy credit (which had already been initiated by President Roosevelt in 1932) as "fiscally irresponsible" because they unbalanced the national budget and ran up deficits. The debate hasn't changed much since. By the 1960s, the mainstream was Keynesian, and President John F. Kennedy, with the help of his chief economist, Walter Heller, promised "to get the country moving again" with a general tax cut to stimulate it, because "a rising tide lifts all the boats." Thus, once more, the *structure* of the economy was ignored. The neoclassical economists teamed up with conservative politicians against the tax-cut stimulus. Finally, in the "Stagflation Seventies," the flaw in

Keynesianism became apparent to many: its too highly aggregated demand-stimulation policies never trickled down through the structure to mop up the unemployed: the stimulus passed through into higher prices and more *inflation,* multinational corporations used the supposedly investment-stimulating tax credits for their own—not society's—purposes, taxpayers worried about the future and *saved* their tax cuts instead of spending them, and so on.

Thus "inflation" is merely the system's expression of all the interacting variables that economists have tried to banish from their models. The mainstream Keynesians redoubled their efforts, while conservative neoclassicists *also* gave up their "objective observer" posture and intervened, but to reduce the money supply, cut the federal budget, and raise interest rates, thus throwing the system into reverse in the hope that a "reserve army" of unemployed would arrest inflation. In a new twist, Reagan conservatives now advocate huge tax cuts, more jobs, less inflation and increasing the military budget while reducing the federal budget—all the same conflicting goals of the Keynesians!

Another key methodological observation by Adolph Lowe (op. cit., pp. 235–37) is that Keynes's model did not fit the facts of the Great Depression in some key areas. Keynes was right about "marginal consumption" and the gap it created. Between 1923 and 1929, consumption rose but at a declining rate, while investment spurted just as consumption growth rates slackened. This raised aggregate employment and output to a crescendo, just as the Keynes model predicted. Then a sharp reduction in absolute investment followed, in mid-1929, and the boom crashed. *But,* there is no indication that personal *saving* rose during that upswing—the heart of the Keynesian hypothesis. Lowe states, "All in all, a picture emerges that is undecipherable in Keynesian terms: a falling rate of increase in consumption associated with rising real wages but *not* with rising personal savings; increasing profits combined with falling prices; a rapid increase in output accompanied by constant wage rates and constant, if not falling, marginal costs. But the riddle begins to resolve itself if we look at the *employment* figures. They remained practically constant over the whole period. . . . But stable employment by no means meant *full* employment. According to one estimate, unemployment during the period 1923–1929 never fell below 10% of the available supply of man-years." But how, then, was it possible for output to rise? "The answer is most dramatically illustrated by output in manufacture, which rose during the critical period by 20%, whereas employment fell by 5%—*an obvious case of labor-displacing technical progress.* There we have the crucial variable: a 'reserve army' of labor prevented money wages from rising, while the technologically-induced fall in prices concomitantly increased the buying power of the employed." Lowe adds in a footnote that the simultaneous revolution occurring in agriculture *aggravated* unemployment, and the rapidly expanding "service" industries *compensated* for the technological unemployment in industry. Thus Lowe concludes that Keynes's highly aggregated "underconsumption" was verified, but that it was not *voluntary* underconsumption or a rise in personal savings, but *forced* underconsumption of technologically displaced labor, and greater competitive pressure holding down the wages of those employed. Thus the Great Depression was a confirmation of certain features of Marx's model, while *dis*confirming some of the basic hypotheses of Keynes. It remained for Robert Theobald, a socioeconomist/futurist, Louis O. Kelso, a lawyer, myself, and others to lobby the issue of technological unemployment onto the political agenda through such coalitions as Environmentalists for Full Employment.

This brings us to the period when ecological limits impinge, and my own critique: involving mounting social costs treated in GNP as useful product, environmental destruction, diseconomies of scale, widening human alienation, and the new movements and coalitions for change, as well as the emergence of a nonmarketized "counter-economy."

However, the really new theorizing, which goes beyond Marx, Keynes, and all the rest, is that of Georgescu-Roegen's *The Entropy Law and the Economic Process,* which was informed more by physicist G. Helm and biologist Lotka than it was by Georgescu-Roegen's economist peers. Since I covered the entropy/economics problem in *Creating Alternative Futures,* I will not repeat it here, other than to note that thermodynamic models pin down the physical aspects of both production and distribution processes, making it possible to analyze the production process holistically—from agriculture to extraction to machinery to production to distribution to consumption to waste management—by the criteria of second-law-of-thermodynamics efficiencies, i.e., to measure their rate of entropy so that it can be reduced by redesign. Georgescu-Roegen's best student is Herman Daly, author of *Toward a Steady-State Economy* (1973); both build on the work of Frederick Soddy (1877–1956), an English chemist who shared the Nobel Prize with Rutherford for introducing isotopes into atomic theory. If economics can be rescued, Soddy and Georgescu-Roegen may be the founders of the next wave. Soddy decided that economists' dangerous drift into pseudo-scientific abstraction must be halted before they destroyed industrial societies, because their uninformed ideas contravened the first and second laws of thermodynamics. A paper Soddy gave at the London School of Economics in 1921 is entitled "Cartesian Economics." He asks rhetorically the question that economists are concerned with: "How do men live?" by asking what makes a railway train go. "In one sense or another the credit for the achievement may be claimed by the so-called engine-driver, the guard, the signalman, the manager, the capitalist, or the shareholder—or, again, by the scientific pioneers who discovered the nature of fire, by the inventors who harnessed it, by Labor, which built the railway and the train. The fact remains that all of them by their united efforts could not drive the train. *The real engine-driver is the coal.* So, in the present state of science, the answer to the question how men live, or how anything lives, or how inanimate nature lives, in the senses in which we speak of the life of a waterfall or of any other manifestation of continued liveliness, is, with few and unimportant exceptions, BY SUNSHINE." (Frederick Soddy, *Cartesian Economics,* Hendersons, 66 Charing Cross Road, London, 1922). Needless to say, Soddy was considered a crank and could not get his work published except in an unconventional way. Soddy does not use the word "Cartesian" in the somewhat pejorative sense of excessive empiricism. He is just pleading with economists for a bare minimum of empirical investigation of the natural world before leaping to their deductions.

Soddy's analysis was just the most accurate in a long list of valiant attempts through history to keep economics honest. Even the most worthy attempts from within the profession have derailed. For example, U.S. economist Wesley Mitchell (1874–1948) decided to use *only* empirical, statistical economic data for another try at reconstructing economics along more realistic lines, but he, too, failed. Mitchell taught at the University of Chicago and produced a huge statistical study, *Business Cycles,* in 1913. He then served as a U. S.

Government economist in the price section of the War Industries Board in World War I and after, and founded the National Bureau of Economic Research (NBER) to pursue a "positive" science of economics. Unfortunately, this heavily data-oriented research could not question the categories into which data had been slotted and patterned, and so it, too, fell into the reductionist style of today's bureaucratic economics. NBER was eventually absorbed by the U. S. Department of Commerce and was taken over by Arthur F. Burns, who later became Richard Nixon's chairman of the Federal Reserve Board and advised in President Reagan's election campaign.

I used to hope the future directions of economics would be influenced by the thermodynamic approach, and many other disciplines as well, since energy efficiency and the switch to renewable solar-based societies are necessary, though not sufficient. The cultural and social aspects of the shift and the distribution patterns are functions of the value system, which economics takes as a constant, pointed out by Benjamin Ward in *What's Wrong with Economics* (1972). Since values are also changing, they will lead to new theories of nonmonetary, resource allocation. We will have to redesign not only the social relationships to the means of production but also the means of production as well. Biologist Barry Commoner has probably done more than any economist to integrate the design criteria of ecological and technological systems of production and relate them to social and political strategies for the transition of industrial societies to the solar age. While I am dubious about his assertion that the United States should double gas production as the fuel during the transition to renewable resources, and about his emphasis on social planning and his rejection of most market mechanisms, Commoner has advanced the transition debate enormously in all his books, the latest of which is *The Politics of Energy* (Knopf, 1979). Similarly, physicist Amory Lovins, author of *Soft Energy Paths* (1977) and a prodigious output of brilliant policy papers (available in *Soft Energy Notes*—see Chapter 6), has done more than any economist to elucidate technologically, politically, and socially a sane transition path to the solar age. Most of the transition strategies and studies in other industrial countries have also been performed by physicists, engineers, and scientists in biology, zoology, chemistry, and ecology. Today, I expect that economics may simply peter out as it gets more absurd, as did the old science of alchemy, with its models of "earth," "air," "fire," and "water." Or economics may be subsumed by a new systems theory integrating biology, thermodynamics, information theory, psychology, and political philosophy. Once again, it is the material situation that is forcing the change; the planet Earth is speaking to us directly.

The challenges to economists' theories and their policy proposals provided by the events of 1979–80, as summarized in Part One, forced some rethinking. The two major lines of this agonizing reappraisal of economic theories involved resources and social choices and relationships. Even for the diehards, "free-market" theory became increasingly untenable. The main holdout was Milton Friedman and his University of Chicago school, and even so, Friedman felt the necessity of raising the pitch of his rhetoric by taking to the television airwaves in his popular series on taxpayer-supported public television (which would not exist, of course, in his "free-market" world). He and his wife, Rose Friedman, summarized the series in their polemic *Free to Choose: a Personal Statement* (Harcourt Brace Jovanovich, 1980), a sweeping set of assertions and generalizations signaling that Friedman had dropped the academic "objectivity" stance and joined in the political fray (presumably on Ronald Reagan's side). Friedman's approach to equality is illustrative:

> "Much of the moral fervor behind the drive for equality of outcome comes from the belief that it is not fair that some children should have a great advantage over others simply because they happen to have wealthy parents. Of course it is not fair. However, unfairness can take many forms. It can take the form of inheritance of property—stocks and bonds, houses, factories; it can also take the form of the inheritance of talent—musical ability, strength, mathematical genius. But from an ethical point of view, is there any difference between the two?"

The answer is, obviously, yes! We humans cannot prevent inherited talents and handicaps—but only try to ameliorate the latter. But we are responsible for *creating the societal inequities* by the various rules we choose to play our economic systems. As Lester Thurow points out in *The Zero-Sum Society* (Basic Books, 1980), "All economies are sets of rules—indeed, the so-called free-market economy is by definition a regulated economy; i.e., it is regulated by property rights," whereby it is *against the rules* for people to go around seizing others' property, and society spends a great deal of money in law enforcement to see that the rules are obeyed.

The resource/ecological perspective will necessarily *delimit* economics and the region of applicability of its methodology. Thus, we must deal not only with "appropriate technology" but also with "appropriate methodology" and "appropriate epistemology." This will require at least seven new approaches: 1) the systems approach; 2) an interdisciplinary approach; 3) a global view; 4) optimizing social and ecological flexibility as a key criterion—but one that we have little ability to *operationalize,* except in descriptive terms; 5) linearity of purpose replaced by acceptance of nonlinearity of complex systems and a new model of causality: i.e., mutual causality, as we shall discuss further in Chapter 11; 6) a focus on the phenomenon of exponential, "runaway" processes, which are modeled in cybernetics: using the concept of positive feedback loops, which *amplify* (rather than damp) rates of change or growth (see Fig. 11), and which can push the system into a new structural configuration (morphogenesis); 7) growth and decay models from biological theory and from thermodynamics (as in friction, heat loss, wear and tear, etc.); from information theory (accumulation of knowledge and its loss through "static" and in transmission); and from sociology and history (e.g., Sorokin's models of the rise and fall of cultural styles and value systems).

Much confusion arises because economics *inappropriately analogizes* from some of these models from the physical, social, and biological realms. For example, the best example of a "runaway" can be found in the hypothetical model that economists have imposed on the real world: compounded interest. Here, they have set up an *a priori,* positive feedback system (based on the value system of private property and its accumulation), in which the interest earned on a fixed quantity of money (capital) will be compounded and the next calculation of interest added on cumulatively. But this "runaway" accumulation process bears no relationship to the real world—only to the value system. However, it has profound real-world effects if enough people believe it is legitimate and employ lawyers, courts, etc., to enforce it! A similar "runaway" set up by economic theory is the real phenomenon of capital accumulation leading to more capital accumulation leading to larger corporations, more and more concentration of wealth, etc., until boundary conditions are encountered (e.g., revolution, ecological depletion, or the breakdown of human health or organizational structure). Thus economists believe in their own hypothetical runaway models but tend not to see all the real runaway situations in the physical world, so the rich get richer and the poor get poorer, and the same with rich and poor countries, until some real-world boundaries are encountered. To my knowledge, the only *economists* who have dealt with this are Georgescu-Roegen and Herman Daly. The Marxist economists have focused on the human/social boundaries but *missed* the biological and ecological ones.

This delimiting of all method and academic disciplines is now well underway, and the more generalized problem is now visible: that language itself, as well as scientific languages and methods, can *obscure* the nature of reality. So paradigm shifts are macro-cultural processes whereby we cleanse language and purge scientific disciplines, much as, on a micro scale, scientific experiment disproves bad theories.

But, apart from the theories of Georgescu-Roegen, Herman Daly, and others, such as Kenneth Boulding, who no longer even call themselves economists, no paradigm shift is occurring within economics—because no paradigm shift *can* occur within economics without exploding the discipline into a thousand pieces. However, there are some hopeful signs that some economists are trying to clarify the limitations of its models and methods (although this effort is hazardous, since economics is still a lucrative profession).

The future of resource-allocation theorizing and principles, as well as environmental policy, will increasingly be taken over by the physical sciences as we move to renewable, biological resource use by bioscientists and ecologists, already developing concepts of "carrying capacity" (of human activities) of various ecosystems. H. T. Odum is still doing good work in this area at the University of Florida, comparing the bio-productivity, say of marshlands (in fish-spawning, maintaining wildlife, preventing flooding, etc.), versus other supposedly "productive" economic uses. On resource allocation and sharing within human societies, the sociologists, anthropologists, political scientists, and theologians will move into the field, along with the systems analysts, decision theorists, information theorists, and psychologists. This process is already visible in recent books on subjects formerly the domain of economists. Some innovative books were written by economists who had transcended the discipline, including: Barbara Ward's *Progress for a Small Planet* (1979), a detailed global wrap-up of all the alternative citizen movements and local initiatives, and Nicholas Georgescue-Roegen's *The Entropy Law and the Economic Process* (1971).

Finally, it is essential to alert readers to the *least* helpful reformulations that economists have attempted in the area of resource and environmental protection, which unfortunately is one of the fastest-growing new fields academic economists are developing for themselves.

Environmental and resource economics have become real problem areas, since thousands of newly graduated economists in these fields are now invading the policy process, taking models not from biology, ecology, or thermodynamics, but from their own abstract and seemingly all-purpose models and imposing them on nature, valuing a marshland, for example, by people's "willingness to pay" to keep it "un-used." The environment is reduced to a "supply of services" or a "set of preferences" for clean air, water, or open space, revealed by people's willingness to pay. Then the idiocy is extended using welfare economics, Pareto Optimality, etc., to model social choice, simply derived from inappropriately aggregating individual marketplace choice.

The concept is that the environment is an extension of human property rights: "common property," which can then be used to devise more models of how to "manage common property resources," bringing along all the baggage of equilibrium supply/demand, pricing, and the "compensation principle," which says, "How much money will you pay me to stop polluting your environment or harming you in some way?" Typical of this literature is *Economics and the Environment,* by Matthew Edel (Prentice-Hall, 1973), replete with supply/demand curves depicting dollars per unit of "desired levels of cleanliness"; pleas for "effluent taxes" (see my critique in Chapter 9); and even a chapter entitled "Environmental Fine-Tuning," symptomatic of the inappropriate mapping of a larger system with a subsystem map, i.e., order of magnitude errors, etc. In the same genre is *Environmental Improvement: the Economic Incentives,* by Anderson, Kneese, Reed, Taylor, and Stevenson (Resources for the Future/Johns Hopkins Press, 1977). In the past, I referred to the work of several Resources for the Future economists which, at the time (1969–73), looked promising. Since then, Kneese has focused almost exclusively on effluent taxes and "managing common prop-

erty," and this book is basically a repetition of all these "market" solutions and further citing the feasibility of compensation, as in Japan's
Law for the Compensation of Pollution-Related Health Damage,
passed in 1973. Better than nothing, one supposes! Other titles similarly reveal economists' assumptions in trying to map larger systems:
Urban and Environmental Management, Berry and Horton, eds.,
Prentice-Hall, 1974; *Economics of the Environment,* R. Dorfman
and N. Dorfman, eds., Norton, 1972; "Air Pollution Abatement:
Economic Rationality and Reality," by Azriel Teller, in "America's
Changing Environment," *Daedalus,* 1967, p. 1082, where it is argued
that since we cannot quantify precise thresholds of human illness related to auto pollution, we must use an *economic* rationale. This is
still argued today by all the "environmental economists" now hired
by corporations. Some of them are questioning the effluent tax pre-
scription which grew out of the earlier work on externalities by
A. C. Pigou, *Wealth and Welfare* (1913). Talbot Page does so in *The
Economics of Involuntary Transfers* (Springer-Verlag, 1973), and
others review refinements of Pigou's concept of externality: Tibor
Scitovsky, in *Two Concepts of External Economies* (1954), Bator in
Anatomy of a Market Failure (1958), and Coase in *The Problem of
Social Costs* (1960), finding them all narrowly applicable only.
Coase even argued against the minimal acknowledgment of social
and environmental constraint of taxing effluents. In his "neutrality
theorem," he held that it would be optimal if polluter and victim simply bargained with each other without state intervention, since to re-
strain the polluter would harm him as much as the victim! Page
covers much more in his later book *Conservation and Economic
Efficiency* (1977), adding a critique of economists' excessive discounting of the future but falling short of shifting out of the paradigm
of economics, although he is one of the more honest of the "environmental management" school.

Mathematical economist Tjalling Koopmans, in *Three Essays on
the State of Economic Science* (McGraw Hill, 1957), reviews the
limitations of economics, noting that the most important insights
achieved by economic analysis, such as "the efficiency of resource-
allocation by competitive markets in a predictable world in which
technology permits perfect competition" does *not* require a high degree of technical training (p. 131). Note the restricted nature of his
proposition! The case for delimitation of economics was more elo-

quently made by Robert Warnbach, associate professor of forestry at
the University of Montana, in "An Economist's View of the Environmental Crisis," in *Ecology Economy Environment* (eds. Behan and
Weddle, University of Montana, Missoula, 1971). He begins with
proper humility: "I am going to talk about the environmental crisis.
But I want it understood from the outset that I do not pretend to be
an expert in this field." He notes the differences between economic
and ecological points of view: ". . . the ecologist tends to view man
[sic] as a part of nature—the philosophy of the economists, on the
other hand, is clearly man-centered . . . believes that man is dominant . . . in the extreme the economist is a cornucopean optimist
and the world essentially capable of providing an unending supply
of resources" (p. 188).

This is why the only legitimate new role for economics is that of,
as far as possible, quantifying social costs—to taxpayers, consumers,
and even other producers—of treating polluted water, cleaning and re-
pairing pollution damage, collecting throwaway containers, building
new community services for incoming factories, police, schools, access roads, fire-protection, sanitation services, etc. All can be
quantified or well estimated. The external-costs approach is built on
K. W. Kapp's *Social Costs of Private Enterprise* (1950).

The social-cost approach is that of Ralph Nader, the Council on
Economic Priorities in New York, and most public-interest research
(see the Council's studies, e.g., *The Price of Power* [1972], *Paper
Profits* [1972], and *Cracking Down* [1975] all available from 84
Fifth Avenue, New York City 10011). The best current examples
include Lester Lave's and E. B. Seskin's studies for the Environmental Protection Administration, mentioned in Chapter 10.

Within the field of economics, some change is visible, brought
about, I suspect, by the intense pressure economists are now under
vis-à-vis their inability to manage the "economy," let alone explain
what is happening. A group of dissidents from orthodoxy, calling
themselves the Association for Social Economics, put out a small
journal and try to get attention for the concept of limiting economic
methodology at the American Economic Association.

A huge jolt was given to the neoclassical, mathematical, econometric school when, in 1978, a Nobel Memorial Prize was given to Herbert Simon, who has specifically stated that he no longer uses economics as his framework and has moved to information and organiza-

tion theory and general systems models. His lecture to the American Economic Association "Rationality as a Process and Product of Thought" turned the logic inside out. Simon now holds a joint appointment, at the University of Pittsburgh, in psychology and political science. He rejected the "maximizing theory" of human behavior, on which economics rests.

Here again, much better models come out of other disciplines, such as philosopher Sidney Hook's editing of a symposium in which he extracted papers from a group of economists on the bases of value judgments in economics: *Human Values and Economic Policy* (1967). Some of the contributors, e.g., Kenneth Boulding, are frank about the value base of economics, but most hedge and obfuscate. Two other books on ethics, altruism, and economics that are much more forthright are produced by religious groups: *People/Profits, the Ethics of Investment,* ed. Charles Powers (Council on Religion in International Affairs, 1972), which covers the responsibility and control of corporations, and *Economics and the Gospel* by Richard K. Taylor (United Church Press, Philadelphia, 1973), as well as *In Search of a Third Way: Is a Morally Principled Political Economy Possible?* by Tom Settle (McClelland & Stewart, Toronto, 1976).

Therefore, the paradigm shift in economics is the *end* of economics as the predominant policy tool for industrial countries (or any country) and the recognition of its proper range of applicability (i.e., for accounting purposes between firms and keeping cash records for individuals and small enterprises, etc.), all corrected by internalizing, to the fullest extent, socially and environmentally necessary regulatory costs *within* the enterprises' accounts and reflected in the prices of products (full-cost pricing). This formula could be applied to alternative businesses and cooperatives, collectives, etc., where the models of maximizing self-interest and competition are relaxed. Referring to my diagram (Plate 18), its parameters would be set by expanded models of dynamic equilibrium and much longer time scales of the relevant ecological matrix, i.e., economics would be the mapping of a *sub*system (human production/consumption processes) operating within larger chemical-exchange/energy-flow/material-transformation processes operating in nature.

Even this reformulation of economics bounded as a subsystem, as in the steady-state economics of Georgescu-Roegen and Daly, must be seen, like Newtonian physics, as limited to "middle-range phenomena." It cannot be expected to work at the macro levels any more than Newtonian physics can account for astrophysical phenomena. Similarly, it is because economics' coordinate systems are in error (at least, limited to special cases), e.g., technology, resources, etc., are treated as coordinates, rather than variables, while the money coefficient cannot express qualitative differences, creating the absurdly high levels of aggregation that have rendered macroeconomics useless.

I recommended kilocalories as a more accurate coefficient to measure the efficiency of physical production/extraction/distribution/recycling, as Amory Lovins expanded on in *Soft Energy Paths,* calling for second-law efficiency standards for energy use, following H. T. Odum and Barry Commoner. However, I also cautioned that all this contextual mapping for economic subprocesses and use of kilocalories instead of money was only a correction of the physical-process models of economics. But while correcting it parametrically, they can say nothing about human values, purposes, and social arrangements for sharing and utilizing production: who will do which jobs, with what status; how leisure is used; etc. The great temptation for humans is always to try to derive such moral and ethical issues from *"data"* (thus avoiding responsibility for our actions, blaming God or an "invisible hand," and rationalizing power structures).

Continuing the analogy with Newtonian physics, economics is also inappropriate to map the equivalent of the "sub-quantum level," i.e., the interpersonal, subjective levels of human functioning and interaction. Boulding drew attention to this limitation of economic models in *Beyond Economics* (University of Michigan, 1968). Economics' absurd model of human motivation was summed up by psychologist David McClelland: "Economists use a totally outdated model of human motivation. They haven't even discovered Freud, let alone Abraham Maslow" (personal communication at the Conference on Steady-State Economics, Johnson Foundation, Racine, Wisconsin, 1970). Thus the current trend of economics, trying to actually colonize new areas such as "environmental management," urban planning, site-location studies for factories, new towns, etc., and invading the political area with their spurious "social-choice theories," which reduce the political process of democracy to another "free market" for competing lobbies, is disastrous.

We need only review the increasing list of books dealing with the real-world crises we face, to see that most of them are by *non*economists. The limits of economics are clear. The era of posteconomic policy making has begun.

NOTES – CHAPTER 8

The following list of books represents some of the best reformulations of economic theory by economists.

Herman Daly, *Toward a Steady-State Economy* (W. H. Freeman, 1973), follows and extends the work of Georgescu-Roegen. *Steady State Economics* (W. H. Freeman, 1977), is a basic textbook.

Tibor Scitovsky, *The Joyless Economy* (Oxford University Press, 1977), is a very useful wrap-up of the problems of using consumer dissatisfaction and "keeping up with the Joneses" as the basic flywheel of an economy.

Fred Hirsch, *The Social Limits to Growth* (Harvard University Press, 1976), is an indispensable thesis showing the limitations of the "keeping up with the Joneses" game [which culminates in the fruitless competition for status, style, and amenities he calls "positional goods," i.e., a house in the country, trips to exotic places, etc., all of which become less desirable and possible as more people aspire to them (the familiar problem of the "tragedy of the commons" in a new form)]. Hirsch demonstrates that, in complex societies, many *individual* desires now require *collective* action.

Mark A. Lutz and Kenneth Lux, *The Challenge of Humanistic Economics* (Benjamin Cummings Publishing Company, 1979). Economist Lutz and psychologist Lux offer one of the best new textbooks, setting out the contradictions and dilemmas of traditional, neoclassical, Keynesian, monetarist, and post-Keynesian economics.

Jaroslav Vanek, *The General Theory of Labor-managed Market Economics* (Cornell University Press, 1970), a basic theory of the hybrid, worker-self-managed economy of Yugoslavia, which is becoming an ever more important model of a "middle way" between state-corporate capitalism, as in the U.S.A., and the state socialism of the U.S.S.R. *The Participatory Economy* (Cornell University Press, 1971) lays out the philosophy of worker-self-managed economies, which rely on markets *specifically created* where feasible. It offers a convincing argument that such economies create greater citizen and worker participation, increase general education in self-government (on and off the job), and lead to capital-saving and greater labor- and skills-intensive, decentralized economies.

Joan Robinson, *An Introduction to Modern Economics* (with John Eatwell, Cambridge University Press, 1973). Robinson is the towering figure of post-Keynesian economic theory, which in spite of its shortcomings is an advance over traditional, mainstream economics, facing up to the uncertainty, market power, and institutional structure of industrial economies. *The Evolution of Economic Ideas* (coedited with Phyllis Deane, Modern Cambridge Economics, 1978) is an excellent discussion and comparison of major economic ideas, culminating in today's impasse and the efforts of the post-Keynesians to resolve the problems of late-stage industrialism.

Alfred Eichner (ed.), *A Guide to Post-Keynesian Economics* (Foreword by Joan Robinson, M. E. Sharpe, 1979). Summarized in Chapter 4. A useful synopsis of what we can and cannot expect from the post-Keynesians.

Martin Carnoy and Derek Shearer, *Economic Democracy* (M. E. Sharpe, 1980). The best general exposition of how a Yugoslav-inspired, worker-self-managed, market-oriented economy might be adapted to U.S. problems of inflation and unemployment. Discusses in detail what we can and cannot learn from experiences of the drive for greater economic democracy in Britain, France, Germany, and Sweden. This book may become a bible of the "third-party forces" in the U.S.A., trying to break out of the Tweedledum-Tweedledee choices of the major parties.

Robert Hamrin, *Managing Growth in the 80's: Toward a New Economics*, (Praeger, 1980). Hamrin, the former staff economist for the Joint Economic Committee, organized a series of hearings in 1976–77 on the problems and prospects for economic growth in the future U.S. economy. Drawing on much of the testimony presented at these hearings, Hamrin has argued for the introduction of many new variables to overhaul the old economic models and proposes some of the needed shifts to an "economics of quality" as well as to a "total employment economy." The book relies almost entirely on summarizing the positions of others, and makes most of its arguments from within the economic paradigm. However, it is a useful documentation of all the issues that traditional economics now fails to address realistically.

Lester Thurow, *The Zero-Sum Society: Distribution and the Possibilities of Economic Change* (Basic Books, 1980). An innovative view of the U.S. economic dilemma as a political problem of our government leaders being unable to deal with the issues of distribution and equity (i.e., all political and technological choices create different groups of winners and losers). American politics is based on "progress," "growth," and the time-honored idea that all citizens can be winners. Thurow argues convincingly that until we can stop kidding ourselves that everyone can win we will be paralyzed. All groups have veto power over choices that will cause them loss, while any positive policy to address a national problem (e.g., energy) will create winners—who will lobby for such policy—but can be stopped by losers (he cites our inability to impose high taxes on gasoline because most people want to drive, whereas in European democracies, their parliamentary systems can act to curb gas consumption in spite of its unpopularity). In spite of his still myopic thinking on environment (typical of most economists), he bites the bullet in urging that our politicians can no longer avoid "specifying equity" rather than pretending that the "market" takes care of the issue *à la* Milton Friedman. His main point is that of Part One of this volume, i.e., that "our society has reached a point where it must begin to make explicit equity decisions if it is to advance." Symptomatically, economist Thurow has relied on game theory for his central thesis.

Edmund Phelps (ed.), *Altruism, Morality, and Economic Theory* (Russell Sage Foundation, 1975). A valiant effort to get some of his fellow economists to focus more creatively on the hidden values in economic theory. Sadly, most of these halfhearted essays attempting to address the crippling model of individual self-interest as the main motivation of humans fall far short of clarifying why, in fact, humans behave altruistically in many situations and societies. Phelps almost proves the bankruptcy of economics, because he has gone to the preeminent figures in the field: Kenneth Arrow, William Baumol, Bruce Bolnick, James Buchanan, Thomas Nagel, Amartya Sen, Burton Weisbrod, Peter Hammond, and others. There is much wallowing within the limited paradigms of economics, arguing about Pareto Optimality and whether charity is just "cooperative egoism." The book is disparaging of other creative efforts, such as Arrow's contribution, which does little more than tear down the brilliant work of Richard Titmuss, *The Gift Relationship* (Allen & Unwin, 1971), which showed that the British tradition of voluntary donation of blood for transfusions was more socially and economically efficient, as well as safer for the recipient, than the American "free market" in commercial blood collections and transfers.

David Collard, *Altruism and Economy: a Study in Non-Selfish Economics* (Oxford University Press, 1978). An excellent review of the history of nonselfish economics, theory of cooperatives, communes, and utopian thought, with an extensive bibliography of further sources. The book discusses the problems of collective choice and the superiority of cooperative action, reviewing the many cases in society in which altruism plays a part, including voluntary wage restraint, charitable donations, income transfers, cooperation in disasters, choices that favor future generations, the relationship of altruism and sympathy in political action, and the possibilities of a good society.

Roefie Hueting, *New Scarcity and Economic Growth: More Welfare Through Less Production* (North-Holland, 1980). Hueting is a Dutch statistical economist who has taken the useful approach of quantifying to a considerable degree the social and environmental costs of continuing current forms of industrial growth. Hueting, by factoring estimates of these growing social costs into national income accounting, has taken economics a good way further toward a correction of its errors in assessing what constitutes human welfare. This is the essential and legitimate approach toward expanding economic models, which I have advocated in this volume and in *Creating Alternative Futures*.

Jan Tinbergen, *Reshaping the International Order: a Report to the Club of Rome* (E. P. Dutton, 1976). Tinbergen, a celebrated European economist, has transcended economic models and coordinated an important political/economic review of the explosive issues of economic imbalance and inequality between nations, and the efforts toward creating a New International Economic Order.

Jonathan David Aronson, *Debt and the Less Developed Countries* (Westview Press, 1979). An important review of the main factors involved in the crucial set of issues surrounding the staggering debt loads of poor countries and the instabilities they are causing, with contributions by other experts, including Susan Strange, Hyman Minsky, Barbara Stallings, Clark Reynolds, and others. Aronson explores the *political* role of international banking and financial institutions, often obscured by reliance on "economic" justifications, as reviewed in his earlier book *Money and Power: Banks and the World Monetary System* (Sage, 1978).

Irma Adelman and Cynthia Taft Morris, *Economic Growth and Social Equity in Developing Countries* (Stanford University Press, 1973). A ground-breaking book and still one of the most cogent studies showing how economic growth, as currently pursued and measured by economists, leads to *greater* social inequality.

Nandini Joshi, *The Challenge of Poverty* (Arnold Heinmann, New Delhi, 1978). A well-buttressed argument by an Indian economist of the potential for less developed countries to break out of the control of economic policy makers in the industrialized countries—by more regional cooperation with each other and with more indigenous forms of development. She points out that interdependence is now the inevitable new framework, and thus cooperation, not competition, must be the new moderating force in any future world economic order.

John Kenneth Galbraith, *The Nature of Mass Poverty* (Harvard University

Press, 1979). One must compare invidiously this slim polemic with Nandini Joshi's, with the latter's much richer set of alternatives. Galbraith's usual creativity and insight fail him in this latest volume, which resorts to witticisms, rhetoric, and an unwarranted fixation on the migration of individuals as the best way to escape the conditions of poverty—surely a much too Eurocentric solution, based on emphasizing individual salvation at the expense of the group.

Michael Harrington, *Decade of Decision: the Crisis of the American System* (Simon & Schuster, 1980). Another insightful book by America's leading exponent of democratic socialism, that examines the worsening stagflation syndrome and among other causes (such as inequality) pinpoints the basic instability of the erratic business cycle itself. Making the nation whole via national planning of the social investments that are now made inadvertently is seen as a key, together with coherent policies on welfare, transfers, taxes, and health, rather than today's crazyquilt of mutually inconsistent, costly, and self-defeating government policies.

Orio Giarini and Henri Louberge, *The Diminishing Returns to Technology* (Pergamon, 1979). A crucial argument documenting the belief that technology can no longer serve as the mainspring of economic growth.

Orio Giarini, *Dialogue on Wealth and Welfare* (Pergamon, 1980). A groundbreaking alternative view of world capital formation, focusing on the nonmonetized sector of the global economy.

E. J. Mishan, *The Economic Growth Debate* (Allen & Unwin, 1977). Mishan's *Costs of Economic Growth* (1976) was one of the earliest critiques of economic theory's inability to deal with environmental destruction and resource depletion. This book is an excellent summary of the debate it helped to create. Mishan details the inadequacy of responses in economic theorizing to the challenges of the new limits to old-style economic growth.

E. F. Schumacher, *Small Is Beautiful* (Harper & Row, 1973), is the now-classic statement by an economist of what is wrong with economics. *A Guide for the Perplexed* (Harper & Row, 1978) delineates Schumacher's underlying philosophy and thoughts on the needed metaphysical reconstruction of much of Western scientific reductionism. *Good Work* (Harper & Row, 1979), published posthumously, contains many of Schumacher's best lectures in the U.S.A.

Lester R. Brown, *The Twenty-ninth Day* (W. W. Norton, 1978). Brown, who began as an agricultural economist, has summarized in this book much of the most crucial evidence of breakdown in the natural systems on which economic processes rely—carefully documented by his brainchild, the Worldwatch Institute. Brown makes a compelling case that human societies do not have too much time left to begin grappling with these resource crises so long obscured by myopic politicians and inadequate economic theories. Worldwatch has published a steady stream of reports on the Earth's systems that are approaching critical stress: from population pressures, desertification, and energy, to food, malnutrition, global employment, inflation, the arms race, and new definitions of national security. Worldwatch Papers are available from 1776 Massachusetts Avenue, N.W., Washington, D.C., 20036.

Nake M. Kamrany and Richard H. Day, *Economic Issues for the Eighties* (Johns Hopkins University Press, 1979). Another praiseworthy attempt by two economists to elicit a set of relevant papers from economists addressing the complex of global issues that lie beyond economics' scope. Day, an agricultural economist, reviews the history of the development of industrialism and its technology in epochal time frames as a more realistic context for understanding our brief, two-hundred-year period of economic growth. Kamrany reviews the new debate about basic human needs as criteria for development, rather than gross national product, and argues for international distribution of rights and resources, rather than competition between the northern and southern hemispheres. Significantly, noneconomists such as systems theorist Jay Forrester make contributions (Forrester shows in his models of the real economic behavior of the U.S.A. how economic theories diverge from what actually occurs).

The following list of books represents a clear contrast, for comparison, by authors addressing global, economic, social, and resource issues from noneconomic viewpoints and disciplines:

Andrew Levison, *The Full Employment Alternative* (Coward, McCann & Geoghegan, 1980). An excellent overview of the stagflation syndrome and the crises of mature industrial economies, with more emphasis on what government can do and has done in other countries. Contains a good critique of macroeconomic policy approaches, and points to the integrative, stabilizing role labor unions have played in the economies of Sweden and Germany.

Norman Furniss and Timothy Tilton, *The Case for the Welfare State* (University of Indiana Press, 1977). Two political scientists deftly classify the major debates and critiques of the welfare state as it has emerged in Sweden, Britain, and the U.S.A.

W. Van Dieren and M. G. Hummelinck, *Nature's Price: the Economics of Mother Earth* (Marion Boyars, 1979). The collaboration of a former businessman and a science policy analyst, this book (first published in Dutch) shows the needed direction for restructuring economics so as to take into account the many delayed and displaced social and environmental costs that industrial societies have swept under the rug for so long.

Kirkpatrick Sale, *Human Scale* (Coward, McCann & Geoghegan, 1980). Another insightful book from the author of *Power Shift* (1975), which first drew attention to the growing power of the Sunbelt in the U.S.A., foreseeing the election of a southern president. In *Human Scale* the author zeroes in on the very scale of all industrial institutions, big government, big business, big *everything* as a source of the crises we face in governing complex, incomprehensible industrial societies. Sale is perhaps our foremost theorist of decentralization and how to accomplish it.

Mildred Loomis, *Decentralism: Where It Came From. Where Is It Going?* (1980). The Grandmother of American Decentralism sums up its theory and practice. (Available from the School of Living Press, York, PA 17402)

Jeremy Rifkin and Randy Barber, *The North Will Rise Again* (Beacon Press, 1978). An ingenious thesis concerning the unused power possessed by labor unions, and the outcome *if* they were to use the economic leverage they have over large banks and financial groups by virtue of their enormous pension fund investments.

Richard Barnet, *The Lean Years: Politics in the Age of Scarcity* (Simon & Schuster, 1980). The coauthor of the widely influential *Global Reach* assesses the new strains on the U.S.A. and other industrial societies due to the source crunch. Barnet correctly points out that it is not an absolute crunch and could be ameliorated if the world order and domestic societies were restructured so as to mitigate the exploitative role of multinational corporations (both "private" and state-owned), together with the effects of an inequitable world trade system controlled by the rich countries.

Alvin Toffler, *The Third Wave* (William Morrow, 1980). Toffler again demonstrates his creative touch in turning all the complex, confusing issues and crises of waning industrial societies into popular, even enthralling reading. As in *Future Shock*, the reader is given a framework in which to grasp the swirling changes in society and is introduced to many new ways of thinking about them.

Bertram Gross, *Friendly Fascism: the New Face of Power in America* (M. Evans, 1980). A timely warning concerning the creeping takeover of the Big Brother corporate state via advertising propaganda, managed news, the control of research agendas, and the machinery of representative government.

Marcus G. Raskin, *The Politics of National Security* (Transaction Books, 1979). Raskin rethinks the meaning of "national security" in much more realistic terms of dealing with our domestic and economic crises, rather than adding to them by escalating military expenditures in a world where the U.S.A. can no longer be "the world's policeman" or maintain its multinational empire.

Charles E. Lindblom, *Politics and Markets: the World's Political-Economic Systems* (Basic Books, 1977). A comprehensive reassessment of democracies and the extent to which they are now controlled by corporations; democracies are actually "polyarchies" with veto power possessed by corporations that control *economic* power that operates as indirect political power. This book has been as highly praised in business and conservative circles as it has in liberal reviews.

Edward Wenk, Jr., *Margins for Survival: Overcoming Political Limits in Steering Technology* (Pergamon, 1979). A veteran science policy maker follows up on his monumental study *The Politics of the Oceans* (1972), urging greater citizen participation in scientific and technological policy decisions.

Denis Hayes, *Rays of Hope: the Transition to a Post-Petroleum World* (W. W. Norton, 1977). A leading theorist of solar energy development sets forth his proposals for a peaceful transition to the Solar Age.

Henry W. Kendall and Steven J. Nadis (eds.), *Energy Strategies: Toward a Solar Future* (Ballinger, 1980). A meticulous review of the best of the wide range of technological choices for safe, renewable solar energy for our future.

Jeremy Rifkin with Ted Howard, *The Emerging Order* (G. P. Putnam's Sons, 1979). A fascinating thesis concerning the potential benefits (and the problems) of the rise of the forty-million-strong movement of evangelical Christians.

Jeremy Rifkin with Ted Howard and Noreen Banks, *Entropy: a New Worldview* (Viking Press, 1980) is likely to be a key book for the 1980s, since it expands on the insights into the entropy law provided by Nicholas Georgescu-Roegen and draws important social and political conclusions. Economists will no longer be able to avoid the debate with thermodynamicists. Georgescu-Roegen, whose own book of essays *Energy and Economic Myths* (1976) is highly recommended, has written the Foreword.

James Robertson, *The Sane Alternative* (1979) and *Power, Money, and Sex* (1976). Two highly recommended books by Britain's imaginative futurist.

The following books are highly recommended reading:

Rachel Carson, *Silent Spring* (1962); Jay Forrester, *World Dynamics* (1971); Ralph Nader, Mark Green, and Joel Seligman, *Taming the Giant Corporation* (1976); Warren Johnson, *Muddling Toward Frugality* (1978); Theodore Roszak,

Person/Planet (1977); Philip Slater, *Earthwalk* (1974) and *Wealth Addiction* (1980); Barry Commoner, *The Closing Circle* (1971) and *The Poverty of Power* (1976); Amory Lovins, *Soft Energy Paths* (1977); Dennis Pirages, *Global Eco-Politics* (1978); Lewis Perlman, *Global Mind* (1976); Sydney Hook, *Human Values and Economic Policy* (1967); Frances Moore Lappé, *Food First* (1977); Paul and Anna Ehrlich, *Human Ecology* (1973); John and Magda McHale, *Basic Human Needs* (1978); Donella Meadows with Dennis Meadows, *The Limits to Growth* (1972); Howard T. Odum, *Environment, Power and Society* (1971); René Thom, *Structural Stability and Morphogenesis* (1976); Garrett Hardin, *Exploring New Ethics for Survival* (1977); Leopold Kohr, *The Breakdown of Nations* (1956); Gregory Bateson, *Steps to an Ecology of Mind* (1971) and *Mind in Nature* (1980); and Robert Socolow, *Patient Earth* (1971).

Turn Your Investment $$ Into
Human Development

The Southern Cooperative Development Fund, Inc. is a financial institution seeking funds in the form of long term loans. Since being incorporated in September of 1969 under the laws of Delaware, SCDF has been able to obtain over 2 million dollars in long term investments. Investors are paid five per cent on their investments.

Since inception SCDF has made or approved loans totaling over $2,500,000.00 to 35 rural low-income cooperative groups needing economic and technical assistance located through 8 Southern States. Cooperatives receiving loans from SCDF are largely engaged in farming, marketing of produce, fishing, retail sale of groceries and sewing.

Traditional financial institutions consider many of the cooperative organizations served by SCDF ineligible for financing on conventional terms. Existing conventional sources of credit have not met the needs of the cooperative organizations, which are primarily composed of low-income persons and are unlikely to do so in the near future. SCDF, through its loans operate to help cooperatives reach the point where they can enter the main stream of commercial credit resources.

Here is how you can really make your investment count; make a long term "Creative Loan" to SCDF. For further information write:

Plate 25

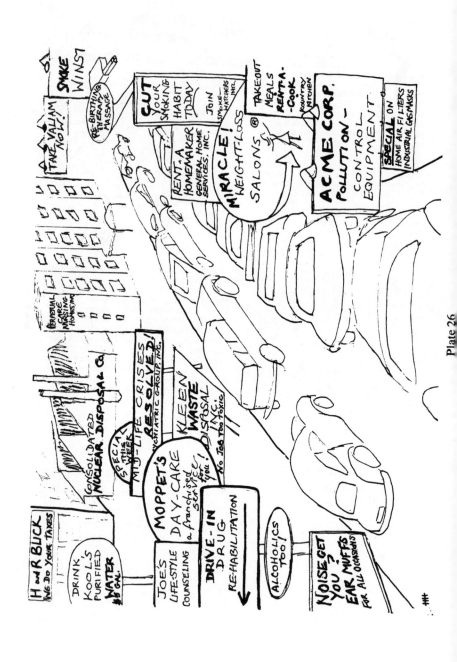

Plate 26

For the past ten years, Action on Smoking and Health has been the only national organization dedicated to legal action on matters of smoking and health. Through legal action ASH has been instrumental in getting anti-smoking messages on radio and television; ending broadcast cigarette advertising; and getting no-smoking sections on planes, trains, buses, and in other public places.

Today ASH is working to preserve and extend these gains. In numerous other ways we are working to protect the nonsmoker from the harm and discomfort of tobacco smoke, both on the job and in public places.

ASH also offers posters, T-shirts, buttons and other items designed to help the nonsmoker stand up and speak out for the right to breathe air unpolluted by tobacco smoke.

ASH is nonprofit and depends entirely on voluntary, tax deductible contributions. We need your support to carry on our efforts on your behalf.

Please send your gift today; or send the coupon below for additional information, including an order form for our nonsmokers' rights materials.

☐ Please send further information about ASH.
☐ Here is my tax deductible gift of _____
   to the work of ASH.

Name

Street

City   State   Zip

ACTION ON SMOKING AND HEALTH
2013 H STREET, N.W.
WASHINGTON, D.C. 20006

Plate 27

One Cure for Cancer

Cancer kills about one thousand Americans a day, almost 400,000 a year. To combat this killer, thousands of scientists are working hard to understand the exact mechanism of cancer and develop cures.

Laboratory research is vital, but we must fight cancer on other fronts, as well. After all, many causes of cancer are well known: cigarettes, air and water pollutants, radioactive materials, certain food additives. In these cases, the cancer battle is really a political one, one of regulating industries that market hazardous products (and that have great political influence).

The ultimate cure for our cancer problem is to *prevent* cancer from developing, rather than *curing* it after it has found a victim. Considering the size and power of the cancer-promoting industries, prevention is certain to be as much a political effort as a scientific one.

Son of Red 2

When the FDA outlawed Red dye No. 2 two years ago, companies turned to another coal tar dye, Red No. 40. The new dye is used in soda pop, gelatin desserts, candy, and most other artificially colored red and orange foods. We (and our pets) eat over one million pounds of this synthetic dye every year.

Unfortunately, Red No. 40 was not adequately tested when it was approved by FDA in 1971. Because it was so poorly tested, the Center for Science in the Public Interest (CSPI) asked FDA in 1976 to ban the dye until it was proven safe. Since then, two tests on mice indicate that the dye promotes cancer.

One FDA pathologist has concluded, according to an internal FDA memo, that "Red No. 40 has behaved as a carcinogen." But higher-ups at FDA are maintaining that the dye is safe. Until it is banned, consumers should make a special effort to avoid red and orange dyed foods.

Red No. 40 is a classic example of FDA approving a poorly tested chemical, then permitting its continued use even after tests indicate a risk. One cure for Red No. 40 and other cancer-promoting additives is strong citizens' groups, which can awaken the bureaucrats and stand up to industry.

Fat, Too?

Steak, bacon, butter, and other foods that are rich in saturated fat have been known for years to contribute to heart disease. The latest new studies indicate that a diet rich in fat (any kind of fat) also increases the risk of bowel and breast cancer, two of the three major cancers. The American diet is one of the fattiest in the world. Judging from the past, we can safely predict that the meat, dairy, and vegetable oil industries will never admit the fat-cancer link, but will demand more and more studies . . . as the tobacco industry has done with lung disease. When profits are at stake, there is never enough evidence.

Concerned citizens can improve their health by eating a diet low in fat. CSPI will help by arming people with information and by pressing the government to require more informative food labels and ads to make the shopper's job easier.

Does Everything Cause Cancer?

What with DDT, Red No. 2, PCBs, asbestos, and other widely used chemicals having been found to cause cancer, it sometimes seems that *everything* causes cancer.

In fact, though, *most* chemicals do not cause cancer. You can feed enormous amounts of most chemicals to animals and they will not develop cancer. This even includes most pesticides and food additives.

Industry has marketed thousands of new chemicals in recent decades. Most of the chemicals are safe. However, a number of cancer-causing products are widely used. The task now before society is to restrict the use of these products and prevent new carcinogens from being marketed. Alarms about newly identified carcinogens should be cause for hope and optimism, not helpless despair, because they indicate that private watchdog groups and government agencies are finally tracking down the culprits.

One Cure . . . CSPI

The non-profit Center for Science in the Public Interest (CSPI) is fighting cancer, heart disease, and other health problems, as well as irresponsibility in government.

In recent months, CSPI's scientific and legal experts have urged the FDA to ban Red No. 40 dye and other unsafe additives . . . petitioned the FTC to ban ads for sugary foods on children's TV . . . pressed FDA to inform pregnant women that caffeine may cause miscarriages or birth defects. . . and pushed for food labeling that says how much fat and sugar are actually in the food. CSPI publishes factual, attractive books, posters, and a magazine about food, the food industry, and the government regulation of food. CSPI also originated and sponsored National Food Day for three years and disclosed links between nutrition professors and the food industry.

If you are tired of having your stomach and your government controlled by the giant foodmakers, won't you join our fight to build better food and health policies?

Plate 28

PART THREE

Coming Home:
From Redoubling Old Efforts to
Reconceptualizing Our "Problems"

CHAPTER 9

Workers and Environmentalists:
The Common Cause

The notion of "declining productivity" of American workers is a bum rap. Rather, it is the declining productivity of *capital investments* in our mature economy, due to its rising social costs, the transaction costs of its complexity, the declining quality of its resource base (which yields less and less return on capital invested in extraction processes, as well as declining yields in net energy), and the fact that, overall, the entire industrial production structure is overly energy-, resource-, and capital-intensive. Thus I conclude that the Phillips Curve interpretation of inflation—as a trade-off with unemployment—does not adequately account for these new conditions, not to mention the new problems of an interlinked global economy, the legacy of inflation due to the "guns and butter" policies of the Vietnam War, and the flood of dollars swamping the world's currency markets. Yet election debates in most mature, industrial democracies continue to feature discussion of the Phillips Curve trade-off, although even A. W. Phillips, the economist after whom the so-called trade-off is named, never postulated its existence in his 1958 study of the British economy of the early-twentieth century. The facts that unemployment remains high during inflation and that prices remain high in recessionary times of tight money and expensive credit indicate clearly that these problems are now built into the structure of mature industrial economies and their new pattern of ever-more-inflationary recessions. The U.S. economy is laboring under the replay of this now familiar cycle, after the recession policies of 1980 and the failure of the Federal Reserve Board's swing to monetarism. Neither trying to control the money supply directly nor setting high interest rates seems to have made much difference.

Blaming workers for inflation has always been a convenient tactic of management and conservative, business-oriented politicians. It can no longer be justified by the facts, for wage costs are by no means the dominant factor in prices in today's industrial societies. The rising costs of energy and raw materials and the declining productivity of capital investments are now making labor the more efficient factor of production in many excessively automated, capital-intensive processes.

Further proof is available in the work of Dale Jorgenson, of Harvard University, whose data support my earlier contention that energy and capital are substitutable for labor. Because of this substitutability, we need to examine how our present tax code favors capital intensity and energy intensity and discourages employment (see Fig. 5). Jorgenson cited the impact of energy price rises since 1974, while conservation measures to cut manufacturing costs have also had the effect of decreasing capital intensity and therefore have resulted in a widespread substitution of labor for capital (*Business Week,* October 1, 1979). Jerome A. Mark, assistant commissioner for productivity and technology at the U. S. Bureau of Labor Statistics, also pointed to increasing energy prices, while Seymour Zucker's commentary in *Business Week* (October 29, 1979) concurred that energy prices were the largest chunk of the new inflation headache. Zucker estimated that half of the United States' 13 percent inflation rate at that time was attributable to oil price increases in 1979 alone and went on to criticize the Administration's fixation on monetary restraint, which he believed worked only in a "demand-push" type of inflation but could not affect the new situation of inflation, in which energy prices constitute 10 percent of the consumer price index and 15 percent of the producer price index, not to mention the multiplier effect I had predicted as these costs surged through the economy. Naturally, such price increases soon force workers into demanding wage increases to keep up, since real incomes (adjusted for inflation) are now dropping in relation to prices. In September 1979, personal income rose $12.1 billion, representing an annual "rate of advance" of 7.5 percent, but prices rose at almost twice that rate. Economists, rather than rethink their model, see this as evidence that the Phillips Curve trade-off is "getting worse." For example, Phillip Wachter, of the University of Pennsylvania, shared the general dismay that "pushing unemployment below 5.5 percent solely by stimulating the economy would send inflation soaring" (*Business Week*, July 31, 1978, p. 93). In the same vein, former Treasury Secretary Michael Blumenthal noted in June 1979 (quoted in *Dollars and Sense*, July–August 1979), when the economy was peaking, "We need a slowdown and we're getting it in the right way," explaining that consumers were going to have to cut their spending, while business was not. Similarly, business leaders applauded Federal Reserve Chairman Volcker's tight-money policy at their October 1979 meeting of the Business Roundtable, where their economic forecast called for a "moderate" recession with unemployment peaking at 7.5 percent. The general business sentiment—that workers, once again, will have to pay for wringing out inflation by job losses—was summed up in a Freudian slip by Chairman Volcker's characterization of the central bank's chief economic problem as dealing with an employment trend that is *too strong*. (*Business Week*, October 29, 1979, p. 41) Yet as the Fed's tight-money policy took hold, most private economists expected that joblessness would hover around 8 percent through 1980.

Thus, even though it is increasingly clear that the interests of workers and the interests of their employers and of investors and bankers are diverging rapidly, inexplicably labor unions are still willing to fall into the trap of allowing traditional economic interpretations that inflation is due to "declining productivity" of labor to go unchallenged. Few labor economists counter with the increasingly validated argument that it is the declining productivity of *capital* and of *management* that must be pinpointed, even though this became clearer in the many investment and managerial errors at Chrysler. One of the reasons for this unwillingness to raise the issue may be that the largest unions in the United States are precisely those whose members are currently employed in the most capital- and energy-dependent industries and that their members have enjoyed the long period of growth based on cheap energy and materials and have become somewhat "symbiotic" with such older corporations in the now unsustainable industrial sectors.[1] In spite of union rethinking on energy issues, the pattern emerging is that scenario of lame-duck corporations in a *de facto* coalition with lame-duck unions. For example, we have witnessed the United Auto Workers taking on the dubious enterprise of bailing out Chrysler as a last resort and the New York City unions committing their members' pension funds to buy the New York Municipal Assistance Corporation (MAC) bonds. One fears that such understandable short-term commitments of their members' hard-won pension funds to such "broken vases" may lead to tragedy in the long run. We shouldn't expect cities to be profitable, but companies are *not* expected to be parasitic on the economy, and savvy taxpayers increasingly shun bailing them out. Yet these older, more powerful unions, the AFL-CIO and the United Auto Workers, for example, still predominate as the voice of the workers in general, even though they represent a minority of workers in the U.S. economy. For example, the AFL-CIO's membership is 13.5 million workers in a total labor force of 96.4 million as of the second quarter of 1979 (Joint Economic Committee, *Mid-Year Review,* 1979). In Britain, by contrast, labor unions organize about half the total work force.[2] Thus it is understandable that these older unions, representing workers in the most troubled areas of the U.S. economy—the auto, steel, construction, electrical, and utility industries—often accept their managements' definition of economic problems, as well as supporting, in many cases, their employers' fights against greater plant safety and environmental protection, since this, too, has been defined by economists (even "liberal" economists such as long-time Kennedy adviser Walter Heller) as "unproductive"—a definition challenged in Chapter 10.

By now it should be obvious to the reader why business prefers to shift the onus for "declining productivity" onto workers, and other scapegoats such as environmentalists and "welfare chiselers," rather than admit that it is capital productivity that is declining. What is not so obvious is why labor unions go along so passively.[3] It is clearly to business's and investors' advantage if they can continue the consensus that has been maintained over the previous years of economic expansion: that labor and management had much more to gain by cooperating to raise "productivity" than by fighting each other—European-style—by keeping their separate sets of interests clear by class-conscious analyses. Yet the warnings of British-based economists Joan Robinson and Karl Polanyi are now clearly confirmed in all the mature, mixed economies of the West: the forces of capital (money) and the forces of labor and consumers (votes) are moving into a deadlock, with governments forced into stop-go policies, toppling with greater frequency, and reaching the stalemate we see today in France, Britain, the United States, Sweden, Canada, and even that spectacular latecomer Japan.

I believe that the current haggling of economic policy makers in both parties across the very narrow spectrum of concepts generated by the Golden Goose model will continue until they realize 1) that all mature industrial societies have reached a new and different stage and 2) that their ideas have been shaped without regard to problems of distribution of wealth and income or of recognition of the stage at which the quality of life begins to decline and heavy social costs are incurred. These social costs include:

1. Unemployment, due to increasingly automated, capital-intensive production and the greater centralization of economic power it requires.
2. Increasing maldistribution of wealth and income, as narrow criteria of corporate efficiency and "labor productivity" lead to larger-scale, more capital-intensive technologies and the organization scale required to develop and manage them.
3. The increasing human costs of technological complexity and organizational giantism and the unanticipated side effects of high-technology production and consumption. These include widespread occupational hazards to health and safety, the effects on workers of continued stress, effects on families and communities of plant relocations and the growing hazards of much consumption (see Figs. 9 and 10).
4. The increasing public costs of maintaining environmental quality, clean drinking water, and breathable air. We face the cost of cleaning up mountains of waste and the additional cost of necessary government regulation to prevent more pollution. Indeed many of our most thoughtful scholars now believe that this type of industrial society is creating problems faster than it is able to devise cures for them. Thus I have contended that the only fraction of the gross national product that is growing is this social-cost factor.

I believe that this is the much larger crisis of industrial societies and that this crisis faces the centrally planned as well as the market-oriented societies. It is not merely a problem, as socialists thought, of who owns the means of production, but a problem of our choice of the means of production. If taken too far, this choice can exploit human beings as well as our environment, turning both into cogs in a vast technocratic, bureaucratic system. Karl Polanyi was right in predicting, in *The Great Transformation,* that a system of allocation emphasizing only market transactions would simply dislocate the social, human, and environmental components of society and that it would require continual government intervention and ever-larger income transfers to keep it going. These chickens are now coming home to roost in the mature industrial societies.

Economies now must conserve materials and energy, distribute the fruits of their production more equitably, and be managed for sustained-yield, long-term productivity. As described in Chapter 8, the more limited economics for the 1980s will be neither capitalistic, socialistic, nor communistic, but will move on from all these nineteenth-century ideologies. It will need to incorporate both the knowledge of how to design regenerative production systems based on renewable resources and the knowledge developed by humanistic psychologists on the almost unlimited potential of human beings as our greatest natural resource, in which our investments will yield the largest returns. The time has come to heed Walter Mondale's repeated call, while he was in the U. S. Senate, for an interdisciplinary council of social advisers to look at the broad problems of society. As is now clear, leaving that task to the Council of Economic Advisers, in charge of economic planning, would be to continue emphasizing narrow economic goals at the expense of human and social goals and the general quality of life.

James Robertson points out in *The Sane Alternative* (1979) that a society's currency will remain stable and noninflationary only when all its members have faith that the society is fair. This corresponds to the game-theory concept that players in a game who suspect that others can get away with cheating will respond by cutting as many corners as they can. In his earlier book *Profit or People: the New Social Role of Money* (Marion Boyars, London, 1974), Robertson, a former government systems analyst in Britain, approaches his subject with a holistic and radical analysis of the role of money as a vital common calculus of value and as a scoring system that permits individuals to reconcile their interests and agree on collective choices. By using the tools of game theory and operations research, Robertson feels, we could investigate how to better manage the money system as a social choice and as a scoring mechanism.

Money can function properly in this role only when it is viewed by all as honest and unmanipulated by powerful interest groups. To prevent its inflation and distortion by such manipulation will require, Robertson believes, an end to profit maximization as a principle of business activity. To this end, he also outlines changes needed in the tax system, and a series of other reforms in financial administration. Corporations, he claims, must be managed on a cash-flow basis, rather than one of profit maximizing, with money flows being directed to "stakeholders" rather than stockholders, by boards of directors with all such stakeholders—management, employees, stockholders, government, and customers—represented. Capital could still be raised in Robertson's reconceptualized corporations by bond-type securities offering fair returns to stockholders.

Robertson's last chapter, "Money Science and Money Metaphysics," dismantles the current absurdities of economics in one of the most elegant and lucid analyses I have yet encountered. Keynesian demand management is pinpointed as one of the destabilizing factors of the current system, and market concepts are reinterpreted —in systems-analysis language—as systems requirements necessary for decentralizing, countercyclical decisions. Robertson combines meaty pragmatism with clarifying reconceptualization of the cybernetic requirements for operating interdependent economies on a finite planet. One of many such operating principles is stated by Robertson as follows: "An honest money system will only be restored in a society which is seen by all its members as being just and fair." Read it again, slowly. It is not primarily idealism; it is a correct axiom from general systems theory. Economists, take note or prepare to be outflanked.

Yet old-fashioned inflation remedies continue to be applied. Lester Thurow, of M.I.T., reminds us that during the 1973–75 recession and inflation in the United States, while the conventional economic tools were vigorously applied, from 17 percent to 29 percent of manufacturing capacity was idle. Similar problems of overcapacity were evident in the 1980 recession. Yet business has lobbied relentlessly for additional investment tax credits to create still more plants and, under Ronald Reagan's economic approaches, these business tax credits and cuts continue as the new conventional wisdom of "supply-side" economics. In the 1973-75 recession, hardly a dent was made in inflation, and unemployment officially ranged between 7 percent and 9 percent. During the 1980 recession, unemployment reached the same range, while inflation was higher. The Congressional Budget Office, in 1975, estimated that each 1 percent of unemployment represents some $13.7 billion of lost tax revenues and costs the government $5.6 billion in unemployment, welfare, food stamps, etc., not to mention the heavy economic and human toll of family disruption, stress diseases, and crime. By 1980, *Business Week*'s May 26 editorial updated these cost estimates to $20 billion lost taxes and $7 billion increase in spending. We must conclude that such policies are not only ineffective but staggeringly costly. The budget-balancing flurry of spring 1980 was replaced by official estimates that these costs of recession would instead produce the same $60-billion deficit in 1980.

Not only is unemployment costly in money terms, but medical sociologist Dr. Harvey Brenner, of Johns Hopkins University, in his report to Congress *Estimating the Social Costs of National Economic Policy* (U. S. Government Printing Office, Washington, D.C., 1976), noted that a 1 percent increase in the unemployment rate in the United States sustained over a period of six years has been associated with an increase of approximately 36,887 deaths. The most direct effects on health of involuntary unemployment are determined by following the workers from the trauma of being told that they are to lose their jobs through either their eventual reemployment or the despair that they experience if they cannot find another job. In addition to the severe psychological effects (notably depression), high blood pressure and other indications of stress persist until the workers have settled down in the new job, according to medical researchers (S. Kasl and S. Cobb, *International Journal of Epidemiology*, 1972, Volume 1, p. 111). Therefore, one might well ask, why, in any case, is a fall in labor productivity such a bad thing, when it leads to greater energy productivity (i.e., conservation) and higher levels of total employment?

Even in mid-1978, economists were expressing "concern" at the strong labor market and that the economy, rebounding from the 1973-75 recession, was creating new jobs at a record rate, and that by June 1978 the unemployment rate had fallen to 5.7 percent (*Business Week*, July 31, 1978, p. 93). But even though at least two maverick economists, Harvard's Dale Jorgenson and New York-based consultant Gary Schilling, had both pinpointed the cause—that companies were finding it cheaper to hire more workers than to buy more machines and use more energy—the old Phillips Curve model won out again. For example, the view of Carnegie-Mellon University economist Arnold R. Weber was typical, in that it was merely that the Phillips Curve trade-off had gotten worse: ". . . the failure of *wage demands even to moderate* in the face of the realities of the demand for labor," even when the economy was slowing. Some glimmers of light came from George Perry, of the Brookings Institution, who noted that an extra percentage point of unemployment today lowers the inflation rate by only 0.3 percentage point in a year and only 0.7 percentage point in three years: "The anti-inflation gains from restraining aggregate demand are disappointingly small" (*Brookings Papers,* Summer 1978). Thus it was that economists began to see that the lever they had used to "manage aggregate demand" had stripped its gears. But, unfortunately for the rest of us, the only other ball game economics then provided was to try the Reagan approach: to force-feed the investment of capital in the hopes of increasing "supply" and sustaining a now unsustainable rate of increase of machine-assisted per-capita labor productivity—as if the trend of the past hundred years of industrial growth based on labor-saving automation could continue indefinitely.

Rather than redouble these old efforts, it seems more sensible to noneconomists to reconceptualize the situation and see that declining labor productivity is the wave of the future and, in fact, that it is good news for workers, consumers, environmentalists, blacks, women, and all minorities, as well as for future generations. In fact, the only groups likely to be upset about it are the investors, managers, and bankers trying to keep the old capital-intensive and energy-intensive industrial companies afloat—if necessary by even becoming parasitic on the economy and society as a whole.

Unfortunately, several million workers are caught in the middle and are naturally fearful that they will be expected to help sustain the profitability of these old companies until all their capital is fully amortized by layoffs, short hours, plant closings, and job losses; not to mention efforts to cut corners by reducing in-plant safety and increasing plant pollution of the neighboring environment, where workers and their families must live.

An example of the tragedies that can occur in these increasingly zero-sum games workers are forced by management to play is the story of shop steward Richard Ostrowski, of the Utility Workers Local 1–2, of the AFL-CIO, at the Indian Point nuclear power plant, operated by Con Edison, of New York. Ostrowski had become alarmed about the secrecy of management concerning the exposure to doses of low-level radiation of the members of Local 1–2 and had sought medical information from the local anti-nuclear coalition, the SHAD Alliance. Ostrowski invited Dr. Thomas Najarian, of the Boston VA Hospital, to talk to some of his fellow welders who volunteer to work inside high-radiation areas of the plant. Najarian discussed the results of his study published in *Technology Review* (November 1978), showing the health effects of low-level radiation on workers in the Portsmouth (New Hampshire) Naval Shipyard: nuclear workers had nearly double the cancer death rates (and roughly five times the leukemia deaths) of both the general U.S. population and the other workers at the same facility not exposed to nuclear radiation.

Sadly, the story, written by Susan Jaffe in *The Village Voice* (October 8, 1979) was unfairly titled "The Tyranny of the Working Class," since it concerned the tragic but understandable reactions of the other workers of Local 1–2 at Indian Point. They turned against their own co-worker and, accusing him of willfully harming the union by questioning management's secrecy about plant safety, suspended Ostrowski from his union position for fourteen months, creating (in the fear that if Ostrowski dug any deeper the nuclear plant might be closed down) a sad spectacle of blind and misdirected anger. Similar tales of management callousness leading to this type of disarray among the ranks of workers, as well as the fomenting of confrontations between environmentalists, are becoming commonplace. Even in the many cases in which management actually plans to close down a facility because it is no longer efficient or profitable, there is an added advantage to blaming environmental regulations and "environmentalism"—an easy new scapegoat.

The files of the U. S. Environmental Protection Agency are full of such accounts, and I am reminded of my own experience when addressing the annual meeting of the New Jersey State Council of Machinists in June 1976 as one of the founders of Environmentalists for Full Employment. I had talked about the shifting energy and resource situation, which was making the older industries and their workers in the United States so vulnerable, and outlined the possibilities of forming coalitions with local environmental groups over issues such as in-plant and neighboring pollution, rather than allowing management of such companies to continue manipulating the situation by using "divide and conquer" strategies. One unionist rose and spoke to the general Catch-22 situation with which workers in such industries are faced. "We don't enjoy being *for* everything our children are against: pollution, waste, loss of recreation areas and the like," he said, "but, you know, management tells us, when there is a local air-pollution control law being debated in Trenton, that we are free to take the day off and go down there and lobby against it, *if we know what's good for us.*"

Threats of this nature, raising fears of plant closings, are documented in most union files, as well as evidence in the suit brought by the International Association of Machinists and Aerospace Workers charging that U.S. corporations are violating election laws by using threats and intimidation to get employees to contribute to corporate-run political action committees. The Machinists' complaint, filed with the Federal Election Commission in October 1979, charged that employee donations (used to further the political aims of management and investors) to the proliferating corporate political action committees, were extracted through the use of "psychological threats" to "force the employees to give for political activities against their own wishes." The companies named in the complaint included Dart Industries, Eaton Corporation, General Electric, General Motors, International Paper, Standard Oil of Indiana, Union Camp, Union Oil, United Technologies, Winn-Dixie, and the Grumman Corporation (Washington *Star,* October 9, 1979). The suit contends that "most employees solicited to contribute to the company's political cause are not really free to refuse, because they have no union or contractual job protection, nor anonymity protection if they decline to give."

But last-ditch attempts to prevent a reexamination of the "declining labor productivity" bum rap are being fought, by those corporations with vested interest in that definition, along many fronts, particularly the excoriation of government regulation of health, safety, and environmental standards. This type of corporate political offensive will be dissected in the next chapter, but meanwhile, politicians are scurrying to join what they see as general voter sentiment against government, which corporations are riding.[4] Thus it became conventional wisdom, even for liberals, to pronounce themselves in favor of the "free market" and for a rollback of regulations, as well as providing "incentives," as if the taxpayers could afford to bribe every company that pollutes or violates health and safety standards, etc., rather than simply legislating them to desist.

Another twist in this kind of absurd "economic" remedy for polluters was proposed by a key adviser to Edward Kennedy, Harvard law professor Stephen Breyer. Writing in the *Harvard Law Review* (January, 1979), he outlined his idea of regulatory reform: that we should abandon the simplistic idea that "imperfect" markets call for the introduction of "perfect" government regulation. Instead, Breyer opted for using the tax code (i.e., more "incentives" to behave decently), enforcement of anti-trust laws, bargaining between parties at interest, and full public disclosure of corporate wrongdoing. Thus we should simply redouble our efforts to do more of the same. Breyer even noted that there was a market-based alternative to a "pollution tax" or "effluent taxation" (which I critique in the next chapter). Why not, asked Breyer, have the regulatory agency set an absolute limit on the amount of pollution that could be emitted in a given area and then sell "marketable rights" to pollute the area up to that level? This appallingly ignorant concept (apart from its callousness in terms of the people affected by such bureaucratic deals with corporations and its misunderstanding of biological productivity) did not acknowledge the concentration of economic and political power that would cut out of the "pollution market" all but the very largest multinational corporations.

However, Breyer was overtaken by events in any case. In 1979, an ad placed in *The Wall Street Journal* ran as follows:

> For Sale: Substantial Hydrocarbon Emission Offset in the Chicago area. For details, contact Box EZ-300.

Mother Jones, a corporate watchdog magazine, followed up and discovered that the "seller" of this "license" to pollute Chicago's air with hydrocarbon emissions was the Abitibi Corporation, which is moving out of the area, which the Environmental Protection Agency has designated a "nonattainment" area, i.e., an area that already has substandard air, where new industrial sources of pollution are not permitted. But the EPA, under constant attack from industries, has begun making such deals with new polluters to "offset" existing pollution) for example, allowing a new polluter to buy control equipment for a belching utility plant in the area in exchange for its right to dirty the air back up to the same, substandard level of smog as before! Thus we find ourselves, as voters, taxpayers, and breathers, witnessing the spectacle of having the very air we inhale "bought and sold" out from under us by large corporations and government bureaucrats (*Mother Jones,* November 1979, p. 12).

Another set of anomalies due to this excessive corporate power to manipulate workers and taxpayers and to define the country's "problems" in terms of their own problems is the new call for "work sharing," which sounds good and of course is better than being laid off. Increasingly workers are finding themselves with such narrow choices. Here again, government "incentives" are sought by companies that, for humanitarian reasons, decide to put their workers on short hours, rather than fire them. Many resigned unionists and social analysts view the situation as the best that can be expected (given the capital-intensive structure of the highly concentrated U.S. economy). Thus a new "consensus" between labor and management seems to be shaping up over work sharing, permanent part-time workers, job sharing, and the move toward a four-day work week.

The demands for shorter work weeks have long been made by European unions, but not, as in the United States, with cuts in pay. The European unions point to the increasing capital intensity of production and the march of automation and demand that management more equitably share the fruits of this per-capita productivity of an ever smaller work force, not by cutting the numbers of employees or reduction of pay and hours, but by simply sharing the profits with workers as extra leisure time. However, in the United States, business-press stories look upon shorter hours with docked pay with approval and even as signs of corporate "enlightened self-interest."

Armco, Inc., is one of the corporations that tried the plan now in effect in California, which allows for state government subsidies to a company that puts its work force on short hours and short pay. The Armco plant manager noted that this allowed the company to avoid the cost of retraining green workers, by keeping on the more experienced ones, with California's taxpayers picking up part of their wages. But since the state would have had to pay them unemployment benefits anyway, the companies have the taxpayers coming and going, and in California's case, the new work-sharing program was estimated to have cost some $150,000 in subsidies in 1979. But employment officials noted that if 20 percent of the seventy-six hundred workers who have received these payments had been laid off for ten weeks and had received the average unemployment check of seventy-seven dollars per week, the total cost would have been $1.2 million.

Such stop-gap measures to avoid grosser indignities of job loss may be all that we can hope for *if* we continue to allow corporations to define our social issues for us. Labor-union leaders express some muted fears that, without safeguards, companies will be able to figure ways to get five days of production out of an employee in four days, rather than go back to full time and full pay when business improves. Workers who have participated in the California program tend to be resigned to the situation. As one worker put it, "It's a lot better than being laid off."

The U. S. Labor Department is studying the California experience, and Wilbur J. Cohen, chairman of the National Commission on Unemployment Compensation, says, "The issue will become even more pressing three or four years from now, when we get more women into the work force and more pressure for accommodating part-time workers, and as the energy crisis begins to affect employment patterns" (*Business Week,* October 29, 1979). Meanwhile, during 1980 youth unemployment remained at 14 percent overall and at 33 percent for blacks and other minorities, while the Carter administration paid $150 million lip service to it in the fiscal 1981 budget.

Thus, if traditional economists' and business definitions of our inflation and "declining productivity" problems are unchallenged, the writing is clearly on the wall. Already it is becoming clear that the increase in total employment—that worrying tendency of the economy to maintain a level of employment "too high" for the comfort of economists, bankers, and business executives—is due to the involuntary sharing of available work as more and more workers are subemployed or can find only permanent part-time work. According to a 1979 survey by the Dartnell Institute of Business Research, 84 percent of the responding companies have part-time or temporary employees on their payrolls, and 81 percent of the total surveyed had permanent, part-time employees working somewhere between twenty and thirty hours a week.

In addition, some people are trying to drop out of the rat race, preferring to work as few hours a week as they can for their cash income needs and to live in a more communal, reciprocal, sharing lifestyle for their nonmonetary needs by bartering and cooperating with their neighbors, rather than competing for ever-less-valuable dollars in the cash economy. Others have learned how to create their own jobs, and fully 10 percent of the rise in the employment level in 1978 was accounted for by the rise in the numbers of self-employed people. Similarly, part-time work appeals to many people for lifestyle reasons, such as when a husband and wife want to both experience the joys of parenting, and split the household work and monetized work between them, i.e., share both jobs.[5]

But although this new, part-time phenomenon is not all bad news, it merely means that the same number of formerly full-time jobs are now, in many cases, being counted as two jobs. Thus I contend that this is another reason why labor productivity appears to be declining —even though many of the people represented by the statistics may be enjoying richer life experiences and more satisfaction. Yet the more ominous side of work sharing is that it is a portent of a *new trend:* the increasing speed of automation of all kinds of processes, from engineering, drafting, and manufacturing to office services, due to the onrushing revolution in microprocessor technology.

Over a decade ago, Louis Kelso, archcapitalist author of *The Capitalist Manifesto,* pointed out that if industrial societies did not deal realistically with the labor-saving effects of their ever-more-capital-intensive technologies and automation, we would have to distribute purchasing power to citizens who did not own capital, via transfers: "workfare," "warfare," and "welfare." Kelso's remedy was simply: "As the machine takes over your job, you must push for corporate and social policies that enable you to buy a piece of the machine." Kelso's remedies were published in *How to Turn Eighty Million Workers into Capitalists on Borrowed Money,* coauthored with Patricia Hetter (Vintage, 1967). Kelso, a brilliant corporate lawyer, showed how the tax code might be used to help workers buy stock in their own corporations and thus derive a second income from capital stock ownership. Kelso contended that if we think capitalism is a good thing, then why are we not working to make every American a capitalist? The Employee Stock Ownership Trusts (ESOTS), which he lobbied into law with the help of Senator Russell Long, of Louisiana, chairman of the Senate Finance Committee, underwent some serious distortion in the process, in that they were also shaped into a corporate boondoggle. However, the issues Kelso raised and the concept of worker ownership of company stock and worker self-management are slowly being accepted in the United States, as documented, for example, in Daniel Zwerdling's *Democracy at Work* (1978), published by the Washington-based Association for Self-Management. Similarly, with the new success stories surveyed by Zwerdling has come the New School for Democratic Management, in San Francisco (see Plate 24). Another success story from Youngstown, Ohio, is that of the worker-owned Republic Hose Manufacturing Company reborn from the shutdown of Aeroquip Corp. in 1978. The new company is profitable, and productivity is up 40 percent (*Time*, December 24, 1979, p. 65).

It is therefore no wonder that labor and management find their mutual suspicions deepening as the energy/inflation crunch continues to worsen and the mature industrial economies' policy makers turn back to rigid monetarism and the old-time orthodoxies of imposing "inflation remedies" that not only make matters worse in the new conditions but still fall heaviest on those least able to bear them: workers and the poor. The National Urban League's 1980 study showed that black Americans had lost ground in the 1970s.[6]

No wonder that social strife and labor unrest are beginning to emerge in all mature industrial countries trying to solve their new problems with old economic nostrums. Britain's Conservative government of Margaret Thatcher has been plagued by strikes, many of which involved the new demands to attack increasing levels of structural unemployment by phasing in shorter work weeks at the same pay, with the target of a thirty-five-hour week by 1982. Britain's economy contracted under the same squeeze of high interest rates, to the point where firms cannot borrow or hope to be bailed out, as formerly, by public money. Similarly, the austerity of France's Premier Raymond Barre caused similar political discontent, and even the strong West German economy fell under the tight monetarist grip, so that as recession took hold, German workers, formerly cushioned by sending home "guest workers" in a downturn, now bear the full brunt of layoffs. In the United States, labor leaders reluctantly agreed to serve on the tripartite Pay Advisory Committee, set up in late 1979, where power will at least be more evenly divided between labor, management, and public members than the group under Nixon and Ford. Thus wage and price guidelines played another round at arbitrating the ancient but lately submerged conflict between labor and capital.[7]

Fears abounded on both sides: business feared that President Carter wanted to use the Council to blame business for inflation and pinpoint their oligopolistic power to raise prices, while labor leaders worried that their members would be on the short end of it by being forced to accept a continued real decline in their purchasing power by some arbitrary new formula devised by economists related to their "productivity."

By fall 1979, President Carter, seeking reelection support from labor in spite of the policies of his hand-picked Federal Reserve chairman, was caught out on both sides: promising business more tax cuts to "spur capital investment" while promising construction workers that the tight-money regime would not be used to destroy their jobs—even as the construction sector was crumbling (*The Christian Science Monitor,* October 16, 1979). His administration's chief economist, Charles Schultze, reassured worried Americans in October 1979 that although admittedly there would be a recession and higher unemployment, "we" (or, rather, the poor and those laid off) should take the medicine and look forward to a brighter future in the long run. Thus the evils of excessive aggregation of data come home again. How long can we expect the same segments of society to continue to swallow the painful purging medicine in order to keep the rest of us comfortable?

Even the usually hailed statistics to prove that all Americans were gradually moving up the ladder, i.e., that between 1959 and 1976 the percentage of Americans living below the "poverty line" had dropped from 22.4 percent to 11.8 percent, were challenged. Sheldon Danziger, in the *Wharton Quarterly* (Fall 1979), of the University of Pennsylvania, pointed out a statistical flaw of vast implications: the figures on the rising incomes of the poor not only included wages and salaries but added on all transfer payments, welfare, Social Security, unemployment benefits, and food stamps; thus, there had been no actual change in the incidence of poverty: the pre-transfer level was 21.3 percent in 1965 and 21.0 percent in 1976.

To be sure, economic misery and hunger had been relieved, but the economy itself had not created jobs, nor the government the social and training programs, to bring this permanent underclass into the society.

When we add to these dismal statistics the new worries over the revelation that corporate pension funds are often financially precarious and carry enormous unfunded liabilities, and that even Social Security is in trouble as fewer and fewer working-age people are paying in to fund the current benefits of more and more retirees, one may ask how much longer it will take for workers, minority-group members, women, poor people, consumers, environmentalists, and all the ordinary, modest citizens in this country for whom the old corporate economy is *not* working to join together in a new coalition? Such a coalition, which the Citizens Party, launched in 1980, failed to weld together, would, of course, constitute a clear majority of the electorate. The Democratic Party faithful used to contend that any third-party efforts would just take votes away from good Democrats. But no one is sure any more who is a Democrat and who is a Republican and what those terms mean anyway. More than half the registered voters were so turned off at the last presidential election that they didn't bother to vote at all. Thus the hopes of a new coalition party growing around the new awareness of "the rest of us" could change everything dramatically by bringing the nonvoting majority surging back into the process and creating one of the watershed periods in all democratic societies, when a new consensus emerges out of the splintering of the old alliances and out of a redefinition of the whole situation. The John Anderson campaign channeled much of the frustration in 1980 but failed to chart a clear, new course.

Thus workers who cling to the old corporate power game out of fear of immediate job losses or because their union leaders are simply unable to rethink the situation may find themselves holding onto a sinking ship. Even their corporate pensions (in which they now, at least, have vested rights) are not much of a hope for security, even if they happen to be securely financed. Inflation will erode such money to a mere shadow if current rates continue, and they will continue if we do not change the direction and structure of our economy toward a renewable-resource base designed for long-term optimal productivity[8] of labor *and* capital *and* energy *and* the health of the nonmonetized sector *and* our social and environmental values. The writing is on the wall in terms that are very immediate: the decline of the American Dream, for example of home ownership; some decades ago, seven out of ten Americans could expect to own their own home, while by 1979 only four out of ten could. Even automobile ownership is beginning to be limited by the price not only of the vehicle but of insurance. Today many cars are on the road uninsured, because their owners cannot afford to insure them. Our auto industry, former crowning glory of the U.S. economy, is in deep trouble, unable, it seems, to produce the kind of fuel-efficient cars Americans now want to buy, thus allowing economical foreign cars to capture 25 percent of the market. The Japanese became the scapegoat of Detroit and Washington. Rather than facing the facts, import limits and U.S.-based production of Japanese cars were promoted. Predictably, the highly automated Japanese car production would yield few additional U.S. jobs (*The Christian Science Monitor,* August 4, 1980). Detroit's declining productivity was not the fault of workers, but of the managers who are supposed to make decisions about investment, production, design, and marketing.

We know now that the Keynesian policies of pumping up consumer demand for cars, boats, refrigerators, and appliances, and giving tax cuts and credits to the middle class, who can afford to buy all these things, will not be able to keep the system going and all of us employed, at least not for very much longer. First, there are not enough consumers in that system to keep all of us employed. Secondly, the products themselves are pricing themselves out of the market. The old way of baking that economic pie in order to share out the slices and keep everybody happy is never going to give us full employment. Thus it is now better strategy for workers and environmentalists to work together to help create the new job opportunities in the renewable-resources sector, as Minorities Organized for Renewable Energy (MORE), of Washington, D.C., is doing.

In addition, I believe that the inflation level is *higher* than the official consumer price index tells us, in spite of economists' contentions that this index *over*states inflation by overstating housing costs. Yet it doesn't include local taxes, one of the fastest-rising components of the cost of living. Neither does the C.P.I. reflect the much higher inflation of necessities: energy, housing, food, and health, as shown by Leslie Nulty for the National Center for Economic Alternatives.

But there are other ways of fudging the inflation figures, such as comparing different time phases so that one can make it look as if the rate is less than it is. Then there is what I call the "vanishing candy bar" inflation syndrome, in which everything is getting a little smaller and a little shoddier. That's another form of inflation. We have the same form of inflation in our public services: the post-office service is declining in quality and the garbage-removal service is declining, but the taxes keep going up, much higher than the official statistics recognize.

Likewise, the unemployment levels the Bureau of Labor Statistics owns up to are probably tremendously understated because of all those people who stopped looking, although economists often take the opposite view, that many job seekers are "only women" looking for pin money. For especially disadvantaged groups, such as black teenagers, the rate is 33 percent. I remember Hubert Humphrey remarking, "Where is this 8 percent? I go to Minneapolis and it's 9 percent. I go to Philadelphia and it's 11 percent. I go to Los Angeles and it's 11 percent. Where is this mythical 8 percent?" Or, one may ask, *where* is it 2 percent? Similarly, we now need measures of energy productivity and of capital productivity. The Bureau of Labor Statistics does not track these phenomena well and often uses projections of labor costs as if they were still in the same relationship to energy and materials as in the past, causing them to predict that labor costs are going up at a linear rate. Capital, energy, and materials costs now are rising more steeply.

Thus the "declining labor productivity" rap surely cannot stick for much longer, any more than the prevailing view that wage costs are the biggest villain in inflation. It is possible to show that this is not true to both business people and government leaders who might be willing to listen. It is only necessary to say, "All right, suppose you beat the workers right back into the Stone Age. Suppose that they are your slaves. Suppose you have no wage costs at all! You find that you've *still* got double-digit inflation (from energy, imported raw materials and capital costs, as well as social costs), so you'd better start looking in these other places.

In Britain the case is clearest. That little island has exceeded its carrying capacity by a factor of two: i.e., the land can carry only twenty-five million people, but they have over fifty million. Therefore, a large part of their inflation is caused by the fact that they have to import half of their food supply every day. Next, they no longer have an empire whose raw materials they can use as cheap inputs for their production processes. What is hitting Britain is an entirely inappropriate set of production facilities, designed for producing goods that can no longer be made at a profit because the import of raw materials and energy for them costs too much. They also have not even begun to reconceptualize their situation, a failure of political leadership and management, hardly the fault of the working people.

This squeeze in the older industrial countries fuels the militancy of Third World countries who are demanding a new economic order, and with good reason. All they want is to raise their share of the gross world product from 7 percent to 25 percent between now and the year 2000, which I think is very modest. They realize that we are energy-and-resource junkies and that industrial economies are now very vulnerable. One thing they're doing—quite properly for their self-interest—is nationalizing their resources, making it tougher for us to get them at the kind of knockdown prices we're used to and on which we have built our industrial growth. We have to realize that natural resources and energy inputs into our economy are never again going to be cheap.

Of course, there are other reasons why corporations are making very poor decisions about their mix of labor and capital in new production facilities. One of those is the tax code, which allows all those tax credits for capital investments, and the continued substitution of more and more capital for labor even though it is the labor that is plentiful and the capital that is in short supply (see Fig. 5). We have to remember why those credits for capital investment were put into our tax code. Capital investment was going to create jobs. This is the Keynesian-trickle-down theory of job creation discussed in Chapter 8. The idea was that you poured capital in at the top of a corporation and it came out the bottom as jobs.

But we're beginning to notice that this doesn't always happen. Sometimes corporations export the capital to Taiwan, where there is a docile cheap labor force and few pollution laws. Sometimes they use the capital to *dis*employ people. The most awesome example now looming is the microprocessor revolution, which horrified European unionists are researching. The implications are staggering, since the new wave of microprocessor automation will decimate employment in the service industries and other white-collar employment—the very "tertiary services sector" that Daniel Bell expected to be able to absorb the disemployed in agriculture, extraction, and manufacturing.

But first a little history. In the small recession around 1960, the Eisenhower administration initiated an advertising campaign with the slogan "You Auto Buy Now," urging people to their "patriotic duty" to buy an automobile and hype up the economy. When President Kennedy came in, he changed this and proposed a tax cut instead. The tax cut was supposed to be "the rising tide that would lift all the boats." That is Keynesian economics: the simple hydraulics of managing total levels of demand. They view the economy as a bathtub. If you want to hype up activity, you open up the faucet and pump in some credit—print some money—and that is supposed to fill the bathtub. One hopes to stop before the bathtub starts overflowing. When the bathtub overflows, that's inflation. The economists who used this simple, equilibrium model of supply and demand seemed to believe that this bathtub was empty and had no structure in it but was full of a thin soup of tiny producers and consumers, like little atoms, all buying and selling and trading with each other. What they hadn't noticed was that over the past hundred fifty years of industrial development, this bathtub had accumulated all kinds of watertight compartments in it. This is the structure that has developed around complex technology, which requires tremendous organization, linkages, interdependencies, and regulatory agencies. There is no such thing anymore as the "free-market" business cycle as if it were a natural thing that was created by God. Blaming these mysterious "market forces" is a cop-out that allows us to avoid taking responsibility for *human* interventions.

Economists themselves create business cycles by tinkering with the faucet. Each new batch of economists comes in and tries to correct the mistakes of the previous batch. That is aggregate-demand management—pumping up consumer demand using tax cuts, or deflating it by tightening up credit, and supply-side economics is just the other side of the same coin. Unfortunately economists don't notice that there is a structure and compartments in that big bathtub, so the danger is that they flood the front end, which overflows as inflation, while nothing flows to the back end at all—where the unemployed are. This is basically the problem we have with the simple hydraulics of supply

To Restore the Lands, Protect the Seas, And Inform the Earth's Stewards

The New Alchemy Institute is a small, international organization for research and education on behalf of humanity and the planet. We seek solutions that can be used by individuals or small groups who are trying to create a greener, kinder world. It is our belief that ecological and social transformations must take place at the lowest functional levels of society if people are to direct their course towards a saner tomorrow.

Among our major tasks is the creation of ecologically derived forms of energy, agriculture, aquaculture, housing and landscapes, that will permit a revitalization and repopulation of the countryside. The Institute has centers existing, or planned, for a wide range of climates in several countries, in order that our research and experience can be used by large numbers of people in diverse regions of the world.

The Institute is non-profit and tax-exempt, and derives its support from private contributions and research grants. Because we are concerned with ecological and social tools useful to small groups or individuals, many orthodox channels of support are not available. The success of the Institute will depend upon our ability to address ourselves to the genuine needs of people working on behalf of themselves and the earth, and to the realization by all our friends that financial support of our research is necessary if the task ahead is to be realized.

The New Alchemy Institute has an Associate Membership ($25.00 per annum, tax-deductible) which is available to anyone with an interest in our goals. Upon joining, associates receive the most recent Journal, dealing with theoretical and practical aspects of new world planning. Over the years the support of our associates has been critical to the continuance of the Institute and its work.

Those wishing to have their membership payment qualify as a deductible contribution under the tax regulations of Canada should make checks payable to The New Alchemy Institute (P. E. I.) Inc.

ASSOCIATE MEMBERSHIP: $25 per annum

Contributions of larger amounts are very much needed and, if you can afford more, that would be beautiful.

SUSTAINING MEMBERSHIP: $100 or greater

PATRONS OF THE INSTITUTE: $1,000 or greater

We invite you to join us as members of The New Alchemy Institute. A company of individuals, addressing themselves to the future can, perhaps, make a difference during these years when there is waning reason to have hope in the course of human history.

THE NEW ALCHEMY INSTITUTE
P. O. Box 432
Woods Hole, Massachusetts 02543 U. S. A.

PRICE PER COPY $6.00

Plate 29

PRINCIPLES OF THE CONSCIENTIZED INTERNATIONAL EXPERT

Caveat lector

Project participants

1. Participant representativity :

I will only participate in meetings or projects in which the distribution of participants (a) by **age**, (b) by **creed**, (c) by **race**, (d) by **ideology** and (e) by **nationality**, reflects, in each case (within say 15 %) that for the world population, or at least for that sub-set with which the project is concerned.

Any imbalance toward a predominance of one age-group, creed, race, ideology, or nationality, or any coalition thereof, runs a considerable risk of restricting the validity of any conclusions (to the degree of that imbalance) for a society in which the interests of each are improperly guaranteed and protected by the others.

2. Sexism :

I will only participate in meetings or projects in which the difference in the numbers of males or females is not greater than (say) 15 %.

Any imbalance runs a considerable risk of restricting the validity of any conclusions (to the degree of that imbalance) for a society composed of approximately equal numbers of each sex. The dominance of one perspective is not tolerable.

3. Discipline representation :

I will only participate in meetings or projects in which the disciplines represented reflect the complete spectrum of those which consider that the topic under discussion is related to their concerns and current activities.

In a socity bedevilled by complex problems it cannot be expected that any international issue can be appropriately and responsibly discussed in the absence of the insight of disciplines which may provide clues to useful action. It is intolerable that discussion should be dominated by views emerging from one discipline or group of disciplines. It has been remarked that few of the 1800 disciplines would acknowledge that their expertise is not relevant to any complex problem.

4. Qualifications :

I will only participate in meetings or projects in which special care is taken to ensure that those participating are appropriately qualified to guarantee that the discussion and conclusions are of quality commensurate with the best available in institutions around the world and are not influenced by superficial considerations.

DIPLOMA
+ IQ +
EXPERIENCE

The participation of people who have devoted insufficient attention to the topic under discussion, as recognized by their peers, is irresponsible. If the project is to be of any significance, every effort must be made to protect the project from inappropriate inputs and inadequate insight.

5. Community representation :

I will only participate in meetings or projects in which representatives of the community to be affected participate on an equal footing with outside experts.

It is intolerable, in the light of past experiences of the misuse of expertise, that « experts » should make plans for a community without involving those to be affected right from the start. In doing so, care should be taken to ensure that all factions and minority groups are appropriately and genuinely represented at that level.

6. Economic and social equality :

I will not participate in meetings or projects in which participants are supposed to participate as equals but in which their annual salaries (including financial privileges) differ by a significant amount (e.g. 15 %).

It is unreasonable to engage in projects to reduce economic and social inequalities when the participants in such projects constitute a living and unselfcritical demonstration of such inequalities.

7. Family participation

I will only participate in meetings or projects in which arrangements are made to facilitate the presence and participation of family members including children.

It is unrealistic to expect that a balanced discussion concerning the problems and future of any community can be achieved when persons representing important dimensions of that community are excluded or allocated a purely decorative role. This procedure further reinforces adult arrogance (usually male) and does much to perpetuate the unenlightened sterility of such proceedings. It constitutes a demonstration of the insidious dynamics whose prevalence in society is of so much concern. And, in so far as it is towards improving the world in which children will have to live, it is only natural that such meetings should be more responsive to childrens' views of the future – rather than designing their world with eyes blinkered by the views and methods of the preceding generation.

Project facilities

8. Location.

I will not participate in meetings or projects based in cities to which a disproportionate percentage of resources is already allocated at the expense of other locations whose underprivileged condition is consequently aggravated.

It is intolerable that projects to reduce the unequal allocation of resources should constitute a basic demonstration of the mentality which reinforces the continuation of such practices.

9. Transportation :

I will only participate in meetings or projects which do not require use of vehicles constructed by aerospace and automotive enterprises intimitately associated with the manufacture of arms and military equipment.

Such enterprises are at the heart of the military-industrial complex and contribute directly to the arms race, to misallocation of resources, and the aggravation of world problems. Each time such vehicles are used to transport participants the position of such enterprises within the international system is reinforced.

10. Handicapped:

I will only participate in meetings or projects which make provision for the participation on an equal footing of those who may be physically handicapped, whether confined to wheel chairs, blind, deaf or with a speech impediment.

A physical handicap should not constitute an excuse for disqualifying a participant, particularly when this is reinforced by the design of the conference building, furniture, toilet and audio-visual facilities, and the arrangements made for transport and accommodation. This is especially the case when the project in concerned with problems which may give rise to the handicap by which the person is afflicted and to which that person is uniquely qualified to bear witness.

11. Equipment:

I will only participate in meetings or projects which do not make use of technical equipment (e.g. typewriters, computers, telephones, desk calculators, interpretation equipment, etc.) manufactured or maintained by multinational enterprises.

Aside from making use of scarce resources, and displacing individuals from jobs, such products are often partly manufactured in developing countries under exploitative labour arrangements and are then sold under monopolistic conditions.

12. Ergonomic factors:

I will only participate in meetings or projects which involve use of furniture and time schedules which respect the physiological and psychological needs of participants (as determined by such disciplines as ergonomics).

It is unreasonable to expect a group to generate useful results for a wider society when its members do violence to their posture, health and general well-being by the manner is which they are obliged to work. Furthermore, solutions to this difficulty should reflect the needs of individuals from cultures where both the actual furniture and the organization and pace of interpersonal interaction may be very different.

Working arrangements

13. Holidays:

I will only participate in meetings or projects which respect the national holidays (independence days, etc.) of the countries of the world and the weekly and other holy days of the religions of the world.

It is insulting to a people or a creed to arrange international meeting or project schedules in terms of the practices in one nation or culture whilst ignoring the practice in another which may be less favourably presented. This can only aggravate international misunderstanding as well as reinforce the insensitivity of those from dominant nations and cultures.

Forms of address:

I will only participate in meetings or projects which take account of the appropriate forms of greeting and address to which people from different cultures are accustomed.

It is insensitive and possibly insulting, for example, to use the first (or given) name of a person who does not wish to be automatically coopted into the western fraternity which favours this practice.

15. Group size:

I will only participate in meetings or projects which call for interaction in groups in numbers not significantly greater than the « critical group size », namely 7-12 participants.

Groups of larger size inhibit the freedom of expression of the less articulate and less aggressive participants whose views may well be both specially relevant and less distorted by needs for personal aggrandisement. Larger groups are generally less effective. in this mode of operation and therefore absorb considerable person-hours when those involved could better employ their attention on other activities.

16. Participant expression:

I will not participate in meetings or projects in which the speaking is done mainly by a small clique (representing less than 30 % of the participants) whilst the remainder, for whatever reason, seldom speak, if at all.

This is an extremely inefficient use of human resources in which many travel long distances without having the opportunity to express their views and priorities. The more aggressive individuals or those speaking the congress « jargon » fluently, dominate such events which thus stand as models of the oppressive communication practices in the wider society.

Project environment and resource use

17. Resource conservation:

I will only participate in meetings or projects which make deliberate efforts to counter the wastage of natural resources normally associated with such activities.

It is only appropriate that an international project should make use of recycled paper and reuseable envelopes, for example.

18. Meeting budget:

I will only participate in meetings or projects in which the total budget for the preparatory meetings (including travel, accommodation, and cost of work lost) does not exceed a reasonable percentage (say 33 %) of the total annual operational budget (i.e. excluding administrative overheads) of the project.

It is counter-productive to assemble people to engage in projects whose operational budgets are of the same order of magnitude as the cost of their preparatory meetings, and to waste time discussing the details and viability of such budgets.

19. Smoking:

I will only participate in meetings or projects in which smoking in any form is not tolerated.

Smoking fouls the air. It is inconsiderate of non-smoking participants, is bad for the health of smokers and non-smokers, and reinforces the attitudes which ensure continued environmental pollution. It is especially inappropriate in meetings about environmental questions.

20. Alcohol consumption:

I will only participate in meetings or projects enlivened by receptions, banquets or entertainment in which alcoholic beverages are not tolerated.

Alcohol dulls the mental faculties, diminishing the value of participant contributions and thus of the exercise. In addition, it is against the principles of some major cultures and therefore offensive to them. It constitutes a significant waste of scarce resources at a time when budgetary restrictions are increasing and many viable projects go unfunded. Furthermore, it is unseemly that particpants should engage in such activity, which often amounts to conspicuous overconsumption, at a time when two-thirds of the world's population is short of food or starving.

21. Meat consumption:

I will only participate in meetings or projects in which the meals provided, available or recommended, do not involve consumption of animal flesh.

Such consumption is offensive to the principles of some cultures and as such is not appropriate to an international project. Delegation of the (often painful) slaughter of such animals, by people who find it distasteful and yet hope to benefit thereby, reinforces the split-minded attitude whereby the slaughter of human beings is tolerated by those who would supposedly prefer not to engage in it themselves.

Plate 30

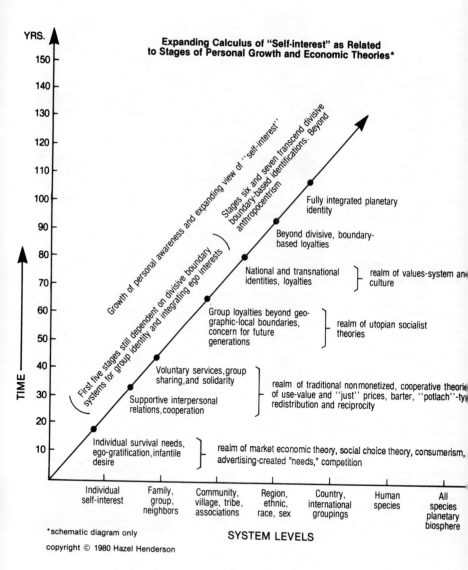

Expanding Calculus of "Self-interest" as Related to Stages of Personal Growth and Economic Theories*

YRS.

Growth of personal awareness and expanding view of "self-interest"

Stages six and seven transcend divisive boundary-based identifications. Beyond anthropocentrism

First five stages still dependent on divisive boundary systems for group identity and integrating ego interests

Fully integrated planetary identity

Beyond divisive, boundary-based loyalties

National and transnational identities, loyalties } realm of values-system and culture

Group loyalties beyond geographic-local boundaries, concern for future generations } realm of utopian socialist theories

Voluntary services, group sharing, and solidarity

Supportive interpersonal relations, cooperation } realm of traditional nonmonetized, cooperative theories of use-value and "just" prices, barter, "potlach"-type redistribution and reciprocity

Individual survival needs, ego-gratification, infantile desire } realm of market economic theory, social choice theory, consumerism, advertising-created "needs," competition

TIME

| Individual self-interest | Family, group, neighbors | Community, village, tribe, associations | Region, ethnic, race, sex | Country, international groupings | Human species | All species planetary biosphere |

SYSTEM LEVELS

*schematic diagram only

Plate 31

and demand. It doesn't take into account the fact that the structure has changed.

By 1966, the tax-cut plan had not lifted all the boats and we had a debate about structural unemployment. That was when we found that there were all those people who were "unemployables." What people didn't focus on was the *other* possibility, that the economy was not providing enough jobs for everybody. It was just pretty unfortunate if you happened to be one of those people who were designated "unemployable" because they had been read out of our increasingly automated, mechanized production system.

In 1966, this whole question of structural unemployment became such an important issue that we set up a President's Commission on Technology, Automation, and Economic Progress to study the whole question of automation. The Commission reported, in essence, that automation does create some unemployment, but we need not worry about it. They made some assumptions that, looked at with hindsight, we know were not very bright: Assumption number 1 was that we had "perfect labor markets," i.e., the economists' notion assuming, for example, that if you're an unemployed checker in a Newark, New Jersey, supermarket you can magically become a computer programmer in an aerospace firm in California, with no dislocation or transaction costs. In fact, these were the words they used in the report: "We view our U.S. economy as one gigantic shape-up hall." (*Technology and the American Economy,* Volume 1, U. S. Government Printing Office, Washington, D.C., 1966) Assumption number 2 was that, even though unemployment was unavoidable, economic growth was going to absorb all the new workers that came into the system.

By 1972, we took another tack. If the problem was the increased productivity created by technologically advanced mass production, then the question was how to distribute the fruits more fairly. An answer was to legislate a guaranteed income, but that idea foundered on the "Puritan work ethic." By the time President Nixon had finished with it, it had all kinds of means tests, shame, and degradation attached to it, and Robert Theobald, who invented the guaranteed income, almost cursed the day he ever thought of it. We were still hung up on the "no-workee-no-eatee" ethic. But the absolutely imperative corollary to that ethic is an economy in which everybody has the right to a job. If having a job is going to be the only way that one can be attached to an income (unless you own capital), then we must design that kind of economy: an economy of full employment.

That is why I think it is very important today for all environmentalists to assert the right of every person to a job, since we have decided as a society that this is the way we want it. Having made the decision, we have to focus on the structural design principles. The principles were embodied in the Employment Act of 1946, but we never delivered on it, because it was easier to pretend that economic growth would solve the design and distribution problems by papering them over. The passage of the Humphrey-Hawkins Full Employment Act of 1978 at least reminded Americans that they have not yet faced up to these issues. We have to remember that a full-employment economy is by definition a more environmentally benign economy. Although the Humphrey-Hawkins Act reasserts the right of every person to a job, we cannot leave it to the old bathtub economists to design full employment for us, because all they try to do is hype up the old, failing system and bail out more corporations. They'll try to freeze the economy into this old pattern, rather than see the *new* economy that we need to develop. We know from the sad British experience of trying to shore up their economy by using tax funds to bail out lame-duck companies in order to save jobs that you can't do it. What you *can* do is make sure that the individuals who were unfortunate enough to be trapped in one of these obsolete sectors of the economy are helped with whatever extended benefits are needed, as well as retraining, to get them into the growing sector of the economy.

The question is, how are we going to get through the 1980s? Tax cuts proposed by the supply-siders are the least effective way, and they are certainly unjustified in view of the current federal deficit. It is much more efficient to target the unemployment with a rifle shot, in public-service employment, especially in urban areas, where it can be matched to real needs, than to puff up the whole system, hoping some of it will trickle down. We have tried that before. If we use the tax approach, then a better way would be to institute the negative income tax. What we need is to get the purchasing power into the hands of the people who can't even afford to buy the things they need every day. They will *spend* that money on basics rather than save it. If you give tax cuts to middle-class people, they're likely to save it or buy speculatively, as a hedge against inflation. Tax cuts should put the money in the hands of the people who need to spend it immediately on the most basic things in life, not give more breaks to rich people and corporations on the erroneous assumption of the new "supply-side" economists that they will invest it productively. Bayard Rustin once said, "In America, we have welfare for the rich and rugged individualism for the poor."

If we give tax credits to private industry, we shouldn't give them for capital investment, we should give them on the basis of *new jobs* they create; that is, credit to the employer for every new person employed. This principle of targeting the actual unemployed, for example, with public-service jobs such as the CETA job programs, is still contested by business, which prefers, naturally, subsidies to private employers. Such subsidies can be justified for small businesses, but not for huge corporations.

We need public-works projects, but of the kind that really meet new needs. Let's not pour more concrete for the automobile, because the next system we need in an energy-efficient society is mass transit. Mass transit will take just as much concrete, just as much steel, but in a different design. The only way to get a new design, to get this shift in priorities, is to stop allowing corporations and old government agencies, who have vested interests in old projects, to define the new needs. We have to get together in our own communities and say, "We don't want *those, old* kinds of projects, we want *these, new* kinds of projects." Such new initiatives at all levels may see us through the transition period and buy us some time to design the new productive system that we need to create: the regenerative economy. A step in the right direction, the National Consumer Cooperative Bank was, by July 1980, making $300 million available for loans in new neighborhood projects such as food and housing co-ops (*The Christian Science Monitor*, July 18, 1980).

One of the things we at Environmentalists for Full Employment developed was the concept of an employment impact statement (see Plate 23). Public officials or business people often try to justify a project (however mismatched with the community's needs, however costly, or however insane—and some of them are insane, such as the Mx Missile—just because it will create jobs. Employment Impact Statements force them to account by asking the following questions:

1) Since we are talking about tax money: jobs creating what, producing what, at the expense of what other public priorities?

COMPARISONS OF EMPLOYMENT FOR SEVERAL TECHNOLOGICAL OPTIONS
worker years based on a 1000 MW (E) plant or equal

Technology	I Processing & Manufacturing (cumulative worker years)	II Construction (cumulative worker years)	III Plant Operation (continuous) (workers each year)	IV Fuel Supply (continuous) (workers each year)	V 30-year plant operation and fuel supply— (worker years)	VI Jobs 1st 30 years Total cumulative worker years
Nuclear	3000 to 5000	5000	120	15	4100	12,000 to 14,000
Coal	2000 to 3000	4000	120	65	5500	12,000 to 13,000.
Wind w/o storage	3000 to 4000	6000	300	0	9000	18,000 to 19,000
Wind with storage	4000 to 8000	8000	350	0	10,500	22,000 to 26,000
Wood from existing forests	2500 to 3500	5000	150	1800 to 2300	59,000 to 73,000	65,000 to 80,000
Wood from fuel plantations	2500 to 3500	5000	150	1600	52,000	60,000
Conservation	1000	10,000	25	0	750	12,000
Conservation	2000	20,000	50	0	1500	24,000
Conservation	3000	35,000	100	0	3000	41,000

Technology	Capital Costs $/Kw
Nuclear	1000
Coal	700
Wind w/o storage	600
Wind with storage	1500
Wood from existing forests	900
Wood from fuel plantations	900
Conservation	300 Avg. Cost
Conservation	600 Avg. Cost
Conservation	1000 Avg. Cost

Fig. 6

2) Jobs at how much capital per workplace? There are all kinds of price tags for jobs. For instance, jobs in nuclear power are incredibly expensive to create with the taxpayer's money, whereas jobs in the construction industry to retrofit buildings to conserve energy, in which you end up with the same number of BTUs as if you had built a power plant, give you a much bigger bang for your buck, as the Council on Economic Priorities' study of energy jobs on Long Island shows (see Figs. 7 and 8).

NATIONAL EMPLOYMENT PER DOLLAR EXPENDITURE ON ENERGY CONSUMPTION, PCEs* AND CONSERVATION/SOLAR
(labor years per million dollars of expenditure)

Energy Production & Supply

Industry	Domestic Fuel Oil	Fuel Oil**	Elec-tricity	Natural Gas	Mixed ***	PCEs	Conser-vation/ Solar
Agriculture	------- less than 0.05 -------					2.3	0.2
Mining	1.9	1.1	1.4	1.7	1.3	0.3	0.4
Construction	0.7	0.4	1.4	0.9	0.9	0.7	19.0
Manufacturing	4.6	4.4	1.3	1.2	2.6	12.0	18.2
Transportation	1.7	1.2	1.2	0.3	1.1	2.1	1.2
Communications	0.2	0.2	0.2	0.1	0.2	1.0	0.3
Public Utilities	0.3	0.3	10.7	7.4	5.2	0.6	0.2
Wholesale & Retail	3.0	3.0	0.8	0.5	1.9	15.2	3.3
Personal & Pro-fessional Services	3.2	3.2	1.9	2.0	2.5	14.8	5.7
Gov't Enterprises	0.3	0.3	4.8	1.0	2.2	1.1	0.4
Total	16.1	14.1	24.0	15.5	17.7	50.3	48.8

Note: All employment is on-site plus multiplier.

* Personal consumption expenditures.
** Adjusted to reflect 51.4% importation from foreign countries to the Nassau/Suffolk region.
*** Represents a mix of fuel oil (49.6%), electricity (39.3%) and natural gas (11.1%) expenditures. This is the mix of energy resources which would be conserved through implementation of the Conservation Scenario.

Fig. 7

CONSERVATION SCENARIO
NET NATIONAL EMPLOYMENT BENEFITS
(labor years)

Industry	Scenario Implementation*	Reduced Energy Consumption**	Increased Discretionary Spending	Net Employment
Agriculture	1,000	***	17,000 to 24,000	18,000 to 25,000
Mining	1,000	−14,000 to −18,000	2,000 to 3,000	−11,000 to −14,000
Construction	77,000	−7,000 to −12,000	5,000 to 7,000	75,000 to 72,000
Manufacturing	73,000	−36,000 to −40,000	89,000 to 128,000	126,000 to 161,000
Transportation	5,000	−12,000 to −16,000	16,000 to 22,000	9,000 to 11,000
Communications	1,000	−2,000	7,000 to 11,000	6,000 to 10,000
Public Utilities	1,000	−33,000 to −67,000	4,000 to 6,000	−28,000 to −60,000
Wholesale & Retail Trade	13,000	−27,000 to −29,000	113,000 to 162,000	99,000 to 146,000
Personal & Professional Services	23,000	−31,000 to −37,000	110,000 to 157,000	102,000 to 143,000
Government Enterprises****	2,000	−12,000 to −28,000	8,000 to 12,000	−2,000 to −15,000
Total	197,000	−172,000 to −249,000	374,000 to 535,000	399,000 to 483,000

Note: Totals do not add due to rounding. Where figures are given as a range, the first figure in the range was calculated assuming high fixed utility costs, and the second figure was calculated assuming low fixed utility costs.
* On-site plus multiplier employment.
** The negative employment effects of reduced energy expenditures totalling $11.48 to $14.67 billion ($16.27 billion in energy savings less $1.60 to $4.79 billion to cover possible fixed utility costs).
*** Negligible.
**** Includes such utilities as the Tennessee Valley Authority and the Bonneville Power Administration.

Fig. 8

3) Are these just temporary jobs or are they permanent jobs? What happens to these people after the jobs are finished? Are they going to go on the local welfare rolls?

In addition, we have to get government officials to prepare *job-potential studies.* The Sheet Metal Workers did a study projecting a $2-billion industry by 1990 in the solar heating and cooling business. Why should public-interest groups do such studies when we're paying the bureaucrats? If they can't estimate what the alternative job potential of this plan versus that plan is, then they haven't done their homework, any more than if they can't tell you how many jobs are going to be lost if a particular technology is discontinued. In other words, we have to force all our government agencies to do job-impact studies before the fact.

However, at the urging of many public-interest researchers, including myself, some agencies in Washington are beginning to add Employment Impact Statements to their policy analyses. The Office of Technology Assessment was an early convert; its 1978 study on the potential for residential energy conservation, OTA-E-92, available from the U. S. Congress Office of Technology Assessment, Washington, D.C. 20515, included data to show that home energy saving in insulation and improved design was a much more labor-intensive economic activity than energy production or importing fuel (which destroys jobs by sending dollars abroad). In July 1980, with the housing industry in a slump, the home insulation industry was operating at full capacity and the $4-billion-a-year home rehabilitation industry was growing, as the trend to reclaim old city neighborhoods accelerated (*The Christian Science Monitor,* July 18, 1980). Thus a shift toward conservation would create more jobs per dollar invested than construction, manufacturing household appliances, average investments in fixed capital, general maintenance and repair, the production of natural gas, coal, fuel oil, or electricity, as well as general personal consumption. One government study, that of the New York State Legislative Commission on Energy Studies, chaired by Daniel Haley and entitled *Operation Bootstrap* (1976), provided an early model. It plotted the course of an alternative set of energy futures for New York State and included the job impacts, comparatively, of all the options. Following the model of the Lucas Aerospace Company workers in Britain, whose alternative management study I described in *Creating Alternative Futures* (p. 213), other job-alternative studies of how to unhook workers from military production to peaceful, useful production have emerged, for example, *Creating Solar Jobs* (1978), by the Mid-Peninsula Conversion Project, in California, funded by labor-union, environmentalist, church, and community resources.

In fact, as companies become more shrill and defensive, many of us are learning that we might be able to fight inflation simply by getting their management reorganized for greater management productivity. A case in point is the growing consensus of management consultants, state utility commissions, and consumer advocates regarding the poor managerial performance of electric-utility executives. Among factors cited for increasing electric rates were 1) poor fuel management, partly due to the now discredited fuel adjustment clauses lobbied onto consumers' bills by utilities, which provides no incentive to buy fuel cheaply, 2) inefficient scheduling of personnel, and 3) expensive debt financing. Many utilities are overly dependent on costly, short-term debt financing, when less costly, long-term financing has been available. Dennis J. Callahan, of Theodore Barry and Associates, a Los Angeles-based consulting firm, says that it is not unusual for a utility to save 19 percent of its payroll costs by better work-force management, while in 1977 an Arizona state commission reported that Tucson Gas and Electric Company was able to cut $5 million from its costs by changes in management and plant efficiency (*The Christian Science Monitor,* October 16, 1979). It is clear that companies and business economists are beginning to feel cornered. Paul McCracken, former Nixon administration chairman of the Council of Economic Advisers, said of the economy of 1979 that the nation did not experience a recession but, rather, a series of "dislocations" caused by government bungling and demagoguery in energy programs, which caused the shrinkage of auto sales and tourism.

Even the Joint Economic Committee, in its *Mid-Year Review of the Economy* for 1979, felt constrained to point out that there was some fallacious reasoning afoot, wherein the much-sought-after "improvement" traditionalists looked for in the capital–labor ratio (i.e., more capital intensity) was viewed as a problem for labor (since it was so often accompanied by slower growth of the total number of jobs available). It firmly squashed this heretical idea: "Because a falling capital–labor ratio is consistent with a rapidly growing labor force growth, some economists conclude *fallaciously* that improvements in productivity will not be accompanied by a reduction in unemployment" (p. 146, italics added here and in later quotations from the *Mid-Year Review*). Rather than deal with the *evidence* that this has, in fact, been the case in the past, the report simply *asserts* that the argument is fallacious and repeats itself: "The fallacy in this reasoning lies in the assumption that high productivity growth must be accompanied by slower labor force growth. But there is no reason why this must be so." The report then adds the familiar litany of wishful thinking about programs designed for retraining of unskilled workers and accelerating the rate of capital accumulation, adding hopefully, "productivity improvements *should* follow. Moreover, an increase in labor force growth *could* also result. Assuming steady economic growth, the unemployment rate *could* decline to about 4% by 1989 by this approach." All one can add is that hell might freeze over too.

Yet the data now accumulating on the shattering employment impacts of the latest wave of "improvement" in the capital–labor ratio may be sufficient to shock even the Joint Economic Committee out of its dream world. Computer-industry spokespeople have always maintained that computers are no different from any other labor-saving machine and that the problems created by dislocation of the work force through their introduction would always eventually iron themselves out by creating new jobs in other parts of the economy and generally increase social development and economic growth. What is often overlooked is that the period in which the computer was introduced and diffused throughout industrial economies also coincided with the long postwar period of fossil-fueled, unprecedented levels of economic growth. Even so, its impacts on workers were sufficient to create the fears of widespread unemployment that emerged in the United States in the early 1960s, with calls for, among other things, a guaranteed income for all citizens.

In an important new view of the evolution of the computer and microprocessor industry presented at the Berlin meeting of the Club of Rome in 1979, Dr. Juan F. Rada, a Chilean economist, drew some devastating conclusions.[9] Rada pointed out that the microprocessor revolution represented not only a qualitative change in the production process (through the new power of microprocessors to control entire manufacturing processes) but that it would produce other global qualitative changes as well. First, Rada concluded, the microprocessor had clearly repealed the labor theory of value, since now production was occurring, quite untouched by any human hand, not only in automated factories but in the microprocessors' increasing ability to *reproduce themselves* by their ability to program their own production. Thus what such guaranteed-income advocates as Robert Theobald predicted in the 1960s has now occurred: the link between a job and an income has been broken, and new ways will thus have to be found to justify placing an income at the disposal of consumers for whom society can find no role as producers in the monetized economy. Rada adds that for the Third World the impact is even greater, since they were told, in the traditional economic development theory, that comparative advantage in world trade would gradually advance the economies of all nations. Thus the "less developed" countries must find their own particular comparative advantage: learn to produce a few things well, and sell them by joining in the world marketplace. Lately it has become clear that this theory of classical economics has become a cruel hoax.

Even *The Wall Street Journal* reported in an article "Mixed Blessing: Do Multi-Nationals Really Create Jobs in the Third World" (September 25, 1979) that foreign investments of $70 billion have created fewer than 4 million jobs (of the 680 million needed). As Michael Harrington points out in *The Vast Majority* (Simon & Schuster, 1977) the comparative-advantage, trickle-down model was generalized from the one-time historical experience of the development pattern of European colonial nations. It won't work today because now that these same nations have climbed the ladder, their very presence at the top is the new situation that changes the model; i.e., when the European nations began their climb, the top of the ladder was empty. Therefore, the dilemma of Third World nations is that in the existing world trade game, still based on comparative advantage, the only commodity they can sell in the world market with clear comparative advantage is their cheap labor force.

The microprocessor revolution takes even this from them, as it is already cheaper, in many manufacturing processes, to simply automate the whole factory than send it offshore, as has been the practice where the labor force was cheaper and the environment could still be exploited with impunity. Today, in the interim period, we see the anomalies of "space-age sweatshops" in Third World countries, where microprocessor-based products such as citizens-band radios, watches, and computer toys are assembled by women whose pay is often about three cents an hour and whose eyesight is jeopardized. Their production is then shipped back to the industrial country, where it can be sold even more cheaply than if assembled by U.S. workers. But as Rada points out, this game can continue for only a few more years, at the rate of innovation of the microprocessors since they are moving so fast toward total automation of such assembly and even their own reproduction. Thus Third World countries, after giving tax holidays to companies to build these assembly operations, will soon be left with idle factories and "holding the bag" of displaced workers. Rada also shows how there is no way out for Third World countries, since the colossal capital intensity of the microprocessor industry *requires* a world market, and no national economy can sustain this technology domestically (a point I shall explore further in the next chapter). Also, the naïve hope that computers and information technology will be used to decentralize industrial economies must be soberly reevaluated, in that this technology has even greater potential for centralizing economic and political control, as can be clearly seen in the new corporate plan of William Norris, president of Control Data, of Minneapolis, to create computer-based "small farm franchises" in the United States, which can only be described as computerized sharecropping. Furthermore, Norris is seeking government funds to demonstrate the feasibilty of this gruesome parody of Jeffersonian agrarianism and E. F. Schumacher's concept of humanly scaled, more appropriate farm technology.

Meanwhile, European unionists are a decade ahead of their American counterparts in examining the impacts of microprocessors on the economies of the industrial nations.[10] While the issue was swept under the rug in the United States by the 1966 Commission on Technology Automation, and Economic Progress, the Europeans have been facing up to it, as, for example, in the study of the future economic trend toward "jobless growth," released by the OECD, mentioned earlier. In industrial countries, the only area left to achieve increased labor productivity is precisely in the "services sector," the clerical, white collar, information-handling activities that we were promised would expand to absorb all the disemployed in other, older, manufacturing sectors and farming. Thus it is clearly admitted that economic growth can no longer be achieved by the increase of manufacturing and agricultural labor productivity alone. Rada quotes the managing director of Olivetti Corporation, of Italy, on the new needs that the microprocessor revolution will fill. "The Taylorization of the first factories, developed as the answer to competition between companies, is a 'digitalization' of the productive process. . . . It enabled the labor force to be controlled and was the necessary prerequisite to the subsequent mechanization and automation of the productive process. . . . Data processing is a continuation of the story which began in the industrial revolution." He added that this Taylorization must now move into the office and to white-collar workers, as a technology of coordination and control over this labor force, which up to now has not been rationalized. As we shall explore in the next chapter, the information needs of coordination of excessively complex, late-stage industrial societies (which I refer to as "the entropy state") push inevitably in this direction—even though the effort will be abortive, since it runs counter to the second law of thermodynamics.

Nevertheless, the corporate sector will continue with the tragic drama until it becomes clear to them that the task is impossible, since perfect information systems to run complex societies as Orwellian computerized technocracies are not feasible. The rest of us will realize this much sooner, since working people and taxpayers will feel the impacts first. At the same Club of Rome meeting, German analyst Günter Friedrichs presented a paper entitled "Micro-Electronics: a New Dimension of Technological Change and Automation." Pointing out that microprocessors in specific cases can be either labor-saving or capital-saving, and sometimes equally labor-and-capital-saving, thus yielding real benefits in more efficient processes, he raised a key issue: "The bill for these benefits has been paid by workers, and in the next decade it is now clear that industrial societies can no longer cushion these hardships of lay-off and dislocation by hyping growth rates. . . . Thus even higher rates of unemployment than heretofore can be expected." Friedrichs cited statistics from West German industry for the 1970–77 period. In forty-five sectors of manufacturing and mining, data-processing machines were able to realize the third-highest production increase. But in terms of employment this was accompanied by the lay-off of 20,600 people. In 1976 the production index in the German computer economic sector was still not too much larger than for general manufacturing. But in 1977 it increased 27.8 percent. In spite of this huge production explosion, the number of people employed in this industry fell again, by 2,300, or 4 percent. Thus Friedrichs shows that even unusually high growth rates in productivity cannot prevent job losses. He cites the study of the Siemens Corporation entitled "Office 1990" (still classified), which reports that a high percentage of normal office work can be automated. Out of some 2.7 million typical office jobs, 43 percent could be standardized, and between 25 percent and 30 percent more could be automated. Thus, a possible 73 percent of existing typists and clerks had better brace themselves for finding alternative kinds of work.

Friedrichs continues by examining the potential for replacing drafting, engineering, banking, and public-administration personnel and predicts that in all these areas, as well as in transportation, where it is already evident, employment will continue to decrease. Friedrichs also punctures the facile argument that the microprocessor revolution will decentralize the economies in question and allow people to work, more to their individual liking, at home. For a lucky few, this may be so. But while he confirms that such decentralization of work locations may indeed occur, with this will come greater centralization of control. A better hope is worker ownership; a study, *Workplace Democracy and Productivity,* by Karl Frieden, found worker-owned companies were 1.5 times as profitable as conventional firms (*World of Work Report,* May 1980).

Thus it behooves American labor unions to study these issues far more closely than heretofore and break with the glib "productivity" formulas of traditional economics. It also makes possible a coalition of workers, environmentalists, and all those for whom the microprocessor revolution will mean not profits and "productivity" but wrenching dislocation and economic hardship. Unions are beginning to break with tradition as they see their old power bases eroding and new opportunities for organizing among women and white-collar workers faced with office-work automation. The Teamsters have organized thousands of clerical workers at Blue Cross/Blue Shield and at the University of Chicago. The Communications Workers of America is focusing a membership drive to "organize all the unorganized"—clerical, retail, service, and blue-collar workers. Unions have belatedly realized that women, overlooked as were blacks until recently, have risen from a work force of 18 million in 1950 to 42 million in 1978, and, in fact, a financial "role reversal" is now occurring, with women now exceeding the numbers of men in the work force, while male workers drop out at a growing rate. Far from the economists' stereotype of "untrained" women lowering overall "productivity," Caroline Bird notes in *The Two-Paycheck Marriage* (1979), it was the women who worked while their men were unemployed that in the 1970s saved the faltering U.S. economy from falling into deeper recession. A new, winning coalition is clearly emerging. The issues are what *kind* of productivity, and *for whom?*

NOTES—CHAPTER 9

[1] However, the increased anti-union lobbying of these very corporations and their push to repeal worker safety legislation has forced a reappraisal of the energy policies once endorsed by most big unions. The AFL-CIO adopted a strong resolution supporting environmental protection legislation and expanded programs for resource recovery and alternative energy sources (Resolution on the Environment No. 13, passed by the Thirteenth Convention of the AFL-CIO, November 15, 1979). The United Auto Workers in Michigan opposed the dumping of radioactive wastes in Michigan as "simply too hazardous," as Vice-President Irving Bluestone testified before the state House of Representatives in February 1978 (press release from UAW, February 23, 1978). Robert A. Georgine, president of the Building and Construction Trades Department of the AFL-CIO, circulated a letter he wrote on July 18, 1979, to the board of directors of the Edison Electric Institute (the trade association of the investor-owned electric utility companies) and to Carl Walske, president of the Atomic Industrial Forum (the nuclear-industry trade association), noting that his union's traditional support for these industries' positions on energy was being reevaluated in the light of many of these companies' attacks on the labor movement in lobbying for "right to work" and union-busting legislation and repeal of safety legislation. Environmentalists for Full Employment gathered up all the evidence of new thinking on these issues by labor unions and issued a joint statement of support and a pamphlet, *Working People*, identifying the growing numbers of issues relating to workers and environmentalists on which they could present a common coalition, including "nuclear blackmail" and jobs in renewable-energy alternatives. They demanded: 1) an end to job and nuclear blackmail, 2) an end to nuclear construction until all danger and waste problems have been solved, 3) guaranteed new, safe jobs at full union wages and with paid retraining to all now employed in nuclear and nuclear-related industries, 4) a huge shift of investment, resources, and worker power to conservation, clean and safe coal, and solar energy, and 5) local, democratic control of energy sources, production, and distribution. (The pamphlet is available in quantity from Environ-

mentalists for Full Employment, 1536 16th St., N.W., 1st fl., Washington, D.C. 20036.)

2 William Rees-Mogg, editor of *The Times*, of London, in *The Christian Science Monitor*, November 14, 1979.

3 Even the very useful journal *World of Work Report*, covering the growing concern for quality of working life, worker self-management, job sharing, flextime, shortening work weeks, and other new issues of labor unions internationally, reports "productivity" issues and economic studies without basically challenging their assumptions (see *World of Work Report*, monthly, 700 White Plains Road, Scarsdale, New York 10583). Again the key problem is that "productivity" statistics are in a shambles, yet another symptom of the end of macroeconomic policy usefulness. For example, there is currently hot debate about whether declines in "productivity" in construction (long thought to be leading the general "decline in productivity") were real or statistical "garbage"—a word used by the Bureau of Labor Statistics (see "A Productivity Drop That No One Believes," *Business Week*, February 25, 1980, pp. 77–80).

4 Much of the 1980 press analysis of this so-called new conservatism and the perception of a general shift to the right was, I believe, a misreading of the new disgust with the growing absurdities and abstractions of *all* centralized policy making, both in Washington by government *and* by large, unaccountable corporations. An example was the article by Peter C. Stuart, "Senate Shifting to the Right Even Without Being Pushed" (*The Christian Science Monitor*, March 3, 1980). The mass media, too, must begin the task of reconceptualization —perhaps the most urgent one in our democracy.

5 While the regrettable competition for jobs in the institutionalized, formal economy increased between white men and other groups (blacks, youth, and women), the reconceptualization of entitlement to the formal wages of the traditional "breadwinner" began. Following legislation in Sweden, marriage was slowly being recognized as an economic partnership (although one partner might be paid in cash and the other not), as pointed out by Martha Keys, special adviser to the Secretary of the former Health, Education, and Welfare Department, now of Health and Human Resources, and that both partners have entitlements to these cash wages. This view was taken by Stanford G. Ross, a former Social Security commissioner, who suggested some form of earnings sharing between husbands and wives—a reform endorsed also by the Advisory Council on Social Security (*World of Work Report*, January 1980).

6 The report showed that, a decade ago, black family income averaged about 61 percent that of whites. By 1978, it had dropped to 59 percent. In real purchasing power, the median family income of blacks rose only 3.1 percent, to $10,879, between 1970 and 1978, whereas whites saw theirs increase by 6.8 percent, to $18,368. Black unemployment, of 11.3 percent, remained more than twice as high as white joblessness (5.1 percent) and is actually higher than it was at the start of the 1970s (8.2 percent). The gap showed in other areas too. The U. S. Civil Rights Commission labeled 1979 "the year of the drift" in civil rights, since half of all minority-group schoolchildren were still in racially isolated schools and housing discrimination remains a pattern in the United States (*The Christian Science Monitor*, January 24, 1980). It was also clear in the energy/inflation squeeze, that low-income people were bearing the brunt of the agonizing adjustments and that the cost of heating alone was taking 50 percent of the incomes of many of the poor and elderly. An advisory report to the Department of Energy's Office of Consumer Affairs warned of an immediate

and urgent need for a $3.2-billion direct-assistance program—twice the amount of the program passed by Congress in 1979. Tina Hobson, the Department of Energy's tireless director of consumer affairs, fought hard for more recognition of the energy inequities in the existing situation and pointed out that those on low incomes are already conserving as much as they can—because they can hardly afford to keep warm (*The Christian Science Monitor*, March 3, 1980). The National Center for Appropriate Technology of Butte, Montana, funded by the Community Services Administration, came up with an energy plan, *Energy and the Poor* (1979), focusing on conservation and using CETA employees to insulate old and inner-city housing that would save 1.7 million barrels of oil a day and create jobs.

[7] In January 1980, the Pay Advisory Committee raised its wage-increase guidelines from the old 7 percent to a range from 7.5 percent to 9.5 percent. Even though this was puny compared with the inflation rate, then at 18 percent, business was alarmed and tried to use the new guidelines to lever price increases (*Business Week*, February 4, 1980). Environmentalists for Full Employment published a useful review of the whole subject of the shift to a full-employment, resource-conserving society in a paper by Leonard Rodberg, with Gail Daneker and Richard Grossman, which surveyed the methodologies that were failing in economics and some of the studies of actual job impacts and effects of various energy technologies. Entitled *Energy and Employment: A Review and Commentary*, February 1980, it is available from Environmentalists for Full Employment, 1536 16th St., N.W., Washington, D.C. 20036.

[8] For example, the Conference on Long-Term Energy Resources, a UN Meeting held in Montreal in December 1979 and run by the United Nations Institute for Training and Research, brought together a diverse international group, with many contributions of renewable and other alternative energy/materials research by countries of the southern hemisphere. The UN Conference on New and Renewable Energy Sources, in Nairobi in 1981, will expand these new dialogues on technology-*sharing* rather than the old model of technology transfer. The industrial nations will have to learn from the rest of the world for the rest of this century.

[9] Juan F. Rada, "Microelectronics and Information Technology: a Challenge for Research in the Social Sciences," *Social Science Information* (Sage, London and Beverly Hills), 19, 2 (1980), pp. 435–65.

[10] See, for example, the comprehensive report prepared by the European Trade Union Institute, *The Impact of Micro-electronics on Employment in Western Europe in the 1980s* (published by Gunter Kopke, Boulevard de l'Impératrice, 66; 1000 Brussels, Belgium), as well as the many valuable studies by the International Institute for Labor Studies, in Geneva, the research arm of the International Labor Organization (ILO). The United States, after its piqued withdrawal from the ILO, in 1977, has now rejoined. This reconsideration makes good sense, since the ILO is an important meeting ground for new ideas on the quality of work life for all workers, and its research is of great benefit to the United States, as well as providing a meeting place for discussing the issues of how to reshape industrial societies and help build a more equitable world order in an ever more polarized world.

CHAPTER 10

Dissecting the "Declining Productivity" Flap

As the 1980 presidential election approached, the issue of the "declining productivity" of the U.S. economy became predominant. The fashionable new slogan, "Reindustrialization," a suitably catchall proposition advanced by sociologist and White House adviser Amitai Etzioni, allowed all candidates to reflect their own images of a concerted government initiative toward a new industrial policy.[1] The very idea is a major heresy in terms of the dominant economic ideology of the free market—laissez-faire—currently enjoying a revival by Milton Friedman and the Reagan-led Republicans, since it is assumed that the "invisible hand" guides investment and innovation to produce a Panglossian best possible set of technological outcomes. The catch in "Reindustrialization" is, of course, the *kind* of industrial policy. Policy makers of all stripes profess horror of the kind of government strategies pursued so disastrously in Britain, of bailing out the lame-duck industries at the behest of their symbiotic lame-duck unions. Yet this becomes the line of least resistance for politicians on the firing line, who can rarely take a longer view and rethink such issues and are urged to *do anything,* as long as it is something. Thus, they continue backing into the future, eyes firmly fixed on the rearview mirror, as revealed by the very term *"re*industrialization." It smacks of the New Deal and the Depression—indeed, some influential voices, such as that of investment banker Felix Rohatyn, adviser to independent candidate John Anderson, called precisely for a revival of the Roosevelt administration's Reconstruction Finance Corporation—a straight bail-out operation. Jimmy Carter committed himself to the past by bailing out Chrysler, providing further aid to the ailing auto industry, repealing air-pollution and safety standards, and generally blaming the Japanese (for building fuel-efficient cars), rather than the recession engineered by his administration's economic policies. Ronald Reagan's view of "Reindustrialization" is simply to turn back the clock, deregulate the economy, remove the shackles from toiling corporations, and reinstate Adam Smith and his invisible hand. Both major parties scrambled to offer the voters the most delectable set of tax cuts, leaving many voters who still bother to exercise their franchise (the lowest participation rate of all the democracies) wondering whether there is a two-party system at all, or a single party: the "Republicrats."

Yet it is not only American politicians who have cornered the market on nostalgia and policies of restoration of a simpler past. The detour back through monetarism continued in Britain, Germany, France, and other stressed industrial, resource-dependent countries. All these remedies rely on similar diagnoses, which must first be evaluated critically to judge the likely efficacy of their prescriptions. To give Professor Etzioni his due, the underlying diagnosis—a decade of overconsumption and underinvestment—has some merit. The U.S.A. has certainly overconsumed, beginning with the still-to-be-paid-for war in Vietnam; massive continued military expenditures (added to the GNP and then used to "prove" that, as a *percentage* of this overinflated GNP, these military expenditures have decreased!); colossal oil import bills; and a rising, unaccounted backlog of social costs and environmental bills now coming due. But Etzioni's broadly shared underinvestment thesis is much less convincing; it begs many important redefinitions and will be challenged herewith.

The conventional range of remedies is familiar; they include tax credits and other incentives to spur private capital investment; modernization and automation of plant and equipment; similar policies to speed up research and development and innovation efforts; faster depreciation of capital equipment; deregulation of business; airlines, railroads, trucking, etc.; dismantling legislation to protect workers, consumers, and the environment and "cutting red tape" (as the U. S. Energy Mobilization Board permits by overriding various states'-rights and due-process guarantees in siting energy facilities); holding down wages; repealing "unproductive" work rules; challenging labor unionizing by extending "right to work" laws; and opposing increases in the minimum wage and resisting equal-employment-opportunity legislation and on-the-job training of "hard-core unemployables" unless further government subsidies are forthcoming.[2] By now, readers will have no difficulty identifying the assumptions underlying these prescriptions and the view of the economy they typify, i.e., the Golden Goose, linear, either/or model of the GNP half of the society, which is then divided into a public sector and a private sector. The private sector is hypothesized as being the only part that is "productive," while all the uncounted costs of production that fall on consumers, taxpayers, the environment, and future generations are simply "externalized," resisted, and—where possible—pushed into the "public sector," where they can be excoriated as the general growth of "big government." The other conventional remedies to "declining productivity," rarely discussed in public, include the diverse demands of industrial producers to have the government assume costs and socialize risks, for example by insuring increasingly hazardous technologies such as nuclear power and liquefied-natural-gas transport or pay for clean-up costs of chemical dumps, estimated as high as $50 billion, with $1.3 billion earmarked to begin the task in the 1980 superfund mandated by Congress.

But, as we have seen, there is increasing evidence that this Golden Goose model is becoming threadbare, and the conventional array of remedies for the Goose's new headache—"declining productivity"—have been advocated for at least a decade, and applied with some regularity in the past five years even as the headache has worsened into a full-blown disease. Economists, still searching in all the wrong places, are (not surprisingly) still mystified as to where the holes in the productivity bucket are. In fact, the United States' star productivity experts, Edward F. Denison, of the U. S. Commerce Department, William Kendrick, former chief economist for the U. S. Chamber of Commerce, and Murray Weidenbaum, who follows the vanishing-productivity mystery for the corporate-funded American Enterprise Institute, all bear witness to the disease's progression, as does the 1979 study by the prestigious Business Roundtable. All agree on the general list of contributory causes: big, free-spending, overregulating government; "less-productive" workers; finicky, paranoid consumers; effete environmentalists; lazy, overly fertile welfare recipients; a sinister "new class" of disgruntled intellectuals and a gutless generation of undisciplined, less-honest citizens, who expect too much of life and avoid taking risks, while spending and borrowing like drunken sailors compared with their thrifty, upright, sober forebears.[3] Yet even such exhaustive definition of the problem no longer accounts, it seems, for the new holes appearing in the productivity bucket that defy identification and are even harder to plug. For example, Denison, mentioned earlier, made an expensive but credible stab in the dark and came up with some $40 billion (about 2 percent of the GNP) of costs to *producers* (reducing their productivity in terms of output per unit of input) due to government regulation of the workplace and environment, and private costs of crime control (e.g., shoplifting). However, Denison didn't notice any of the additional costs consumers and taxpayers bear, since these are *social* costs.[4] Surprisingly, *Business Week*'s analysis of the problem included the faults of managers who avoid risk-taking and myopic focus on short-term profit-maximizing at the expense of the future.[5]

But Denison kept up his search, and after vast additional research expenditures, concluded, in the November 1979 Commerce Department *Survey of Current Business,* that "there is no explanation for the productivity slump." Denison found that the much vilified government regulations he studied initially do not alone account for the productivity decline the United States has experienced since 1974. Furthermore, of the two dozen possible causes—ranging from the falloff in research and development spending to the effects of higher tax rates on the incentives to work and save, to the general toll of inflation on efficiency—none can be *singled out* (italics added) to explain the overall decline by an annual rate of 0.5 percent between 1973 and 1976. The simplistic style of economists' reasoning is becoming more transparent. The simple either/or, public/private model, with its linear assumptions, leads to one-at-a-time focusing on causes that are "singled out" for case-by-case study.

However, recently some progress has been made in understanding that linear, static economic models cannot map a dynamic, nonlinear society. Important modifications of the traditional general prescription for continued "growth" and "productivity" have been made in response to the persistence of the "stagflation" syndrome. During the recession of 1973–74, it at last became clear that the Keynesian notion that had been so widely accepted—that industrial countries could keep stimulating aggregate demand and manage to consume their way out of recessions—had become counterproductive. In fact, policy makers became more aware that this type of stimulus actually contributes to inflation faster than it increases economic activity and reduces unemployment, just as deflationary policies do little to reduce inflation, while cutting into production and employment. In *Creating Alternative Futures,* these counterproductive policies were compared to burning down the house to bake a loaf of bread. I warned that such linear-based policy models, if continually applied to complex, nonlinear industrial societies, would only increase levels of both inflation and unemployment—as has occurred since. But, as discussed earlier in this book, the new realization that hyping aggregate demand (now often called "naïve Keynesianism") inflames the fever and puts off the day of reckoning has not led to any reconceptualizing of the basic model. The either/or logic is still visible in that economists now at least see that increasing demand can only worsen what they persist in calling "cost-push inflation." But moving their focus between the two poles of their linear model of either "demand-pull" or "cost-push" inflation is no substitute for modeling inflation as a systemic phenomenon, which I have described as the last, "entropy state" of industrialism, characterized by no-longer-avoidable social and "transaction" costs creeping back onto the monetized economy's balance sheet (see Plate 26). In other words, the economists' creative-accounting game is up, and it is getting harder to "externalize" costs from producers' books, load them onto consumers and taxpayers, and extract them from the nonmonetized sector than it used to be, when they were smaller and less noticeable. Today, one could just as easily view an industrial society as one vast accumulation of social costs, which, when accounted for, may be merely a mirror image of the production we have already taken credit for prematurely. Instead of reconceptualizing the old linear, supply-demand, input-output mental traps, the economists have simply shifted their focus from "demand" back to "supply." Even the post-Keynesians who insist on a more accurate view of the structural configuration of mature, industrial economies still hew to many of the same mental traps of the economic method itself, as described earlier. To capture both the nonlinearity and the dynamism and structural evolution of industrial societies will require wholly new systemic models, discussed in Chapter 12, familiar to biologists, thermodynamicists, and general systems theorists but largely ignored by economists, who tout supply-side economics as a new breakthrough.

Thus, within traditional economic methodology, the search for the vanishing productivity is likely to become ever more esoteric. While economists are now trying to deal with the highly interactive, nonlinear world economy by defining more and more perturbations, such as OPEC price increases, as "exogenous" shocks to a basically unchanged system, they are in spite of themselves being forced into ever-more-evanescent explanations for the productivity leaks, including "uncertainty about the future" affecting investment decisions; "consumer expectations and confidence," which has produced the new "school of rational expectations" view that both fiscal and monetary policies are ineffective, because market participants discount and counteract them. While realistic about the limits of economics, the school is hardly new, but traces back to 1938 with G. S. L. Shackle's *Expectations, Investment and Income.*[6] To quote Dr. Donald Michael, planner, psychologist, and author of *Learning to Plan and Planning to Learn* (Jossey-Bass, 1973), "We are awash with unproven and probably unprovable theories about the nature of society and the dynamics of social change, from economics, psychology, history and political science. Without a validating theory, the meaning of data becomes highly uncertain. Decision on *what* data to gather and how to interpret them, both depend on a theoretical context. But both the data about the state of society and the theory for interpreting these data are so fragmentary and often dubious, that theory can seldom adjudicate among contending data and vice versa." In other words, what you see depends on where you stand and what you decide to look at. In a very real sense, reality is what you pay attention to, but we are seldom conscious of this truth until the social system in which we are embedded undergoes a rapid transformation, such as industrial societies are experiencing today.

With this caveat, we will now attempt to unravel the short-circuited logic of economic theory that obstructs the development of a new view of what is occurring, prevents a clear public debate, and thwarts the discussion of alternative policies and theories. Since the social systems we are trying to examine anew are multidimensional and best represented as spherical (see Fig. 3), we must first assume that they cannot be understood via any single "cut" along any particular axis with any single methodological tool. Thus, our approach must be pragmatic, focusing on the phenomenon (in this case defined as "declining productivity") in an open-ended way, with an "ecumenical" view concerning what disciplinary spectacles may be most fruitful and what methodologies may be appropriate. We should even withhold judgment as to whether the phenomenon is a "problem" at all, or whether it has just been stated from within a conventional view as one.

Here the social and policy "sciences" must learn from the frontier experience of physics research, as outlined by Fritjof Capra in *The Tao of Physics* (1975) and Gary Zukav in *The Dancing Wu Li Masters* (1979), that an experiment now yields the observer a set of phenomena that can be interpreted via a number of theoretical constructs, depending on the particular world view, or epistemology, chosen by the physicist. A similar state of affairs exists on the policy level, which I have observed firsthand in methods of technology assessment; i.e., in any good technology assessment there are many diverse interpretations of the phenomena under investigation, and it is the interactions among all these "biases" that determine what parts of a study are highlighted and how the data are interpreted, both scientifically and politically. In extreme cases, the clash of views leads to such power-wielding games as withholding data, burying reports with "bad news" for some powerful interest group, and similar shenanigans. However, in a time when one-dimensional, linear approaches employing single-discipline methods are failing, we must plunge ahead with the multidisciplinary mapping of the phenomena we seek to understand. This may involve, as in technology assessments, environmental and social-impact statements and futures studies, the overlay of many different methods, each, as it were, a sheet of different-colored cellophane highlighting a different pattern. In addition, one must pay attention to differing macro-level and micro-level data and the eternal problem of when to aggregate and when to disaggregate, as well as keep in mind that unquantifiable data must not get lost but must be stated clearly up front of the study as a set of disclaimers and areas of uncertainty.

Thus, with such a humbler but wiser approach of accepting indeterminacy and embracing the possibility of error and ignorance, let us examine the methods of traditional economics, as typified by Denison's studies on "productivity," which were hailed by the business community when published, in January 1978, and which I critiqued in a privately circulated "Memorandum to Colleagues" as follows, and later in an article for *Spectrum,* the journal of the Institute of Electrical and Electronics Engineers, October 1978.

MEMORANDUM TO COLLEAGUES

From: *Hazel Henderson*   *March 24, 1978*

Subject: Study by Edward F. Denison, of the Brookings Institution, "Effects of Selected Changes in the Institutional and Human Environment upon Output per Unit of Input," *Survey of Current Business,* U. S. Dept. of Commerce, Volume 58, Number 1, January 1978

This study is a classic example of traditional concepts in economics, i.e., "productivity" as measured by output per unit of input, applied inappropriately to the analysis of a nonlinear system: the U.S. socioeconomy, its institutions, and the human environment. It was lauded in the *National Journal* and the Washington *Post* by Robert J. Samuelson (February 1978) as the work of "an economist with no axe to grind" and covered extensively in *Forbes* (March 6, 1978) and elsewhere. On the contrary, it is clear from some of Denison's statements in the *Forbes* interview that although his purpose "is not to judge the wisdom of government programs, which have benefits as well as costs," he shares the economists' bias in favor of the price system. For example, Denison states, "Whether it's worth it or not, we go about it in a hell of an inefficient way. We do not use the price system." Denison favors effluent taxes [taxes on pollution], which "could be set high enough to achieve any desired reduction in pollution," over a "rigid regulatory approach." The purpose of this memo is to clarify the often unconsciously value-laden approaches of economic analysis such as Denison's in this study, which serves, however inadvertently, to buttress the standard business viewpoint that our economy is being strangled by government regulation, red tape, unnecessary costs, etc. An alternative view can be found in "Forcing the Hand of the De-Regulators," Chapter 17 of my *Creating Alternative Futures,* pp. 297–302.

Denison's study must be viewed with caution, firstly because, as is made clear, it is only "part of a comprehensive study of U.S. economic growth" in which the author is engaged (financed by the National Science Foundation for the Brookings Institution). Such a partial analysis, when published out of its context, is subject to all manner of misinterpretation, as has already occurred in the reactions cited above. The study focuses, according to its Summary, on the phenomenon in the past decade that

> the institutional and human environment within which business must operate has changed in several ways that adversely affect output per unit of input. This article examines the effects of three such changes:
>
> 1. new requirements to protect the physical environment against pollution;
> 2. increased requirements to protect the safety and health of employed persons;
> 3. a rise in dishonesty and crime.

The common characteristic of the changes is that they have reduced the measured output that is produced by any given measure of input. By measured output, I mean national income, or net national product as defined by the Bureau of Economic Analysis. By 1975, the last year for which this article provides estimates, output per unit of input in the nonresidential business sector of the economy was 1.8% smaller than it would have been if business had operated under 1967 conditions. Of this amount, 1.0% is ascribable to health programs and to the increase in dishonesty and crime. The reductions had been small in 1968–70 but were rising rapidly in the 1970s.

The first cautionary note is that the focus on these three selected classes of social costs, which Denison estimates, quite credibly, at approximately $40 billion or almost 2 percent of GNP, detracts attention from the systemic array of social and environmental costs that our particular type of ecologically and socially incompatible technologies have produced thus far. Presumably, Denison is now in the process of studying this much wider array of social costs, many of which are larger, for example, those borne largely by the public due to smoking and alcohol abuse, recently estimated by health economists Schweitzer and Luce at approximately $60 billion (*New England Journal of Medicine,* March 1978), or those assessed by economist Lester Lave in a study for EPA that found that reducing sulfate and particulate levels in the air by 50 percent would save $7 billion due to decreases in pollution-related sickness and death, or similar monetary and nonmonetary costs of work-related hazards, diseases, and death. In addition, as ecologists, physicians, and political scientists know well, the price-system solution of effluent taxation cannot address many problems, including toxic substances where prohibition is required, nor can they deal with the political power of corporations to lobby to set taxes artificially low so that they become merely licenses to pollute.

It is certainly a useful endeavor, in my view, to begin looking at what percentage of GNP should now be classified as the "social cost fraction" and subtracted from totals, rather than added (i.e., not differentiated), as at present. In my own analysis of the conceptual problem in economics ("The Entropy State," *Planning Review,* April 1974), I suggested that this burgeoning "social cost fraction" might be the only part of the GNP in mature industrial societies that was growing, and that a point of "maximum entropy" would be reached when these societies were generating social and transaction costs (i.e., unanticipated second-order consequences, social and environmental disruption, growth of necessary government regulation) that would exceed real productivity. At such a stage, they would drift to a soft landing in an accidental "steady state" and inflation would mask their declining condition. The Morgan Guaranty Trust Company of New York's January 1978 monthly letter, cited earlier, bears out my thesis quite well, not to mention the steady rise in inflation rates. Another definitional error is the confusion of *rates* and *levels.* For example, "declining productivity" should be more accurately defined as decline in the *rate* of productivity *increase,* since this clarifies that this *rate* of increase is calculated from a base. If the base of productivity is already large, then *rates* of increase related to it cannot go on rising.[7] Similar confusions exist in expectations that the rate of GNP growth could continue in relation to the enormous *base* of the U.S. economy.

Therefore it is now necessary to examine as quantitatively as possible this growing "social cost fraction," which Denison sees, quite correctly, as leading to a pervasive decline in overall productivity. However he might better have first constructed a model arraying the entire U.S. economy in a general systems perspective and assessing the social costs arising from *all* major sectors, i.e., publicly borne costs of the tobacco and alcohol industries, rising drug-abuse costs attributable to drug companies and their advertising, health costs of tooth decay and poor nutrition due to promotion of oversweetened cereals and snacks, clean-up costs of polluted waters, costs of cancer related to environmental carcinogens, etc. Instead, Denison's logical leap from this systemic, macroeconomic problem to the initial selection of three particular micro areas of special concern to *producers* (rather than consumers or taxpayers) reflects his choice of a narrow conceptual tool, i.e., the "productivity" favored of economic analysis (but often rejected by biologists, physicians, ecologists, psychologists, sociologists, and others) and his own weightings and value judgments and priorities. If he had approached the problem with paradigmatic rigor, rather than with the preconceived conceptual tools his discipline offers, he might have plotted on the nonlinear U.S. socioeconomic system the various types of social costs I have mentioned, the business sectors responsible for them, and plotted the pathways these social costs were taking as they diffused spatially and temporally through the system, as well as how they were allocated and borne, not just by producers and at the regulatory level but at all other levels in society, including those I have noted, and the many new classes of costs, such as the socializing of risks considered unacceptable to private insurance [see Chapter 11], and socialized investments such as the charges on many consumer utility bills for "construction work in progress (CWIP) by utility companies, as well as those being pushed forward onto future generations such as the costs of decommissioning obsolete nuclear reactors.

Lastly, some specific methodological problems with the study include:

1. Measuring "productivity" in terms of output per unit of input. Productivity measures such as this are simplistic and based on historical ability of producers to externalize costs to others (taxpayers, municipalities, consumers). A major study to overhaul traditional economic productivity measures is currently underway, sponsored by the National Academy of Sciences and the National Commission on Productivity, chaired by Dr. Albert Rees, of Princeton University. Viewed from a futurist's perspective, all Mr. Denison is measuring is the extent to which, in the past decade, producers have been forced to internalize these costs through new laws and regulations, rather than continue passing them on to

others as both monetary and nonmonetary costs, whether municipal refuse collection or cleaning costs, ill health, stress, and environmental degradation, as well as the considerable increase in costs to taxpayers of the needed regulatory bodies to enforce the new laws *without which* these social costs would have been even higher.

2. Economists' traditional "productivity" measures have simply *overstated* productivity gains for decades, and now the social and environmental bills are coming due. By using his input-output productivity measure, Denison focuses on costs to producers and opportunity costs to capital and costs of regulation borne by producers. He could equally well have elected not to use the "productivity" approach and focused on taxpayer and consumer costs.

Therefore, the correct view of this study is as one small module in an expected forthcoming comprehensive model of the U.S. economy. We should watch carefully to see whether this forthcoming model is, in reality, a comprehensive model of the full range of social costs that industrial societies such as our own are now generating, i.e., a mirror image, as it were, of the GNP. Its implications will be profound, for they may show that the very structure of our capital-energy-materials-intensive, technological society is itself generating social costs faster than real production. Mr. Denison's initial exploration, if pursued to its logical conclusion, may demonstrate this situation, which may be much further than he or any economist would like to explore. It may demonstrate, for the first time, what those from many other disciplines, including biology, ecology, sociology, thermodynamics, as well as economists such as Nicholas Georgescu-Roegen and Herman Daly, and futurists such as myself have pointed out for more than a decade: that the gross national product must now be replaced by one of the more holistic, "quality-of-life"-type indicators, for example, Japan's new net national welfare, Tobin and Nordhaus's measure of economic welfare, the more recent PQLI (physical quality of life indicator, developed by the Overseas Development Council [see Fig. 18]) and other, similar efforts that still languish behind the walls of academe. It may also demonstrate the ways in which the inadequacy of simple linear and equilibrium models that still drive much of traditional economics are now almost useless for mapping nonlinear, rapidly changing, often morphogenetic systems, such as late-stage industrial societies including those of Western Europe, Japan, and the U.S.A. These societies are now so complex that they can be adequately understood only in multidisciplinary terms, rather than as abstractions referred to by economists as "economies," using narrow, linear definitions of "productivity" measured as units of "output" related to given "inputs."

Given these multidimensional problems we face today of energy and materials constraints, social disruption, and environmental depletion, economists must engage in intensive dialogue with analysts of global problems from other disciplines. I have always welcomed the chance to debate and dialogue with economists on such issues, and hoped to have a similar opportunity vis-à-vis Mr. Denison, whose later 1979 study for the Commerce Department left my questions and many others unanswered. However, the opportunity never materialized, so I took up another aspect of the growing debate; the policy implication that grew out of Denison's work and the increasing number of business-funded studies of growing businesses' costs due to regulation (in other words, costs society was now demanding that businesses "internalize"). The next round of the battle was the demand by industry that these costly regulations be submitted to administration review and "rationalization" (a worthy goal but one that concealed a wealth of interpretations, depending on whether one was a worker, or a consumer whose environmental hazards were reduced, or a producer who had to clean up), and the introduction of what was billed as a major innovation in economic theory: that government regulations must be subject to an "inflation impact statement," spelling out their economic costs and economic benefits! Thus coming full circle, the economic view was reimposed, and those concerned with social-welfare, environment, or future impacts on succeeding generations were forced back into the economists' ballpark, where the discussion could be "managed" by economic "experts."

One attempt to disentangle the new Gordian knot of chicken-and-egg reasoning behind the new "inflation impact statements" was a joint symposium, held on April 19, 1979, by the Office of Technology Assessment and the Library of Congress, focused around the Carter administration bill calling for such "inflation impact statements" to assess the economic costs and benefits of federal regulations.[8] The disastrous bias of the economic method became clear, reinforced by economists themselves, who had pushed economics the furthest in terms of quantifying social costs. For example, Lester Lave (whose studies of the social costs of air pollution and benefit of control had pioneered the quantifying of social costs producers have hitherto "externalized") pointed out the limitations of this approach, since it can proceed only on the strength of medical and epidemiological studies proving cause-effect relationships between specific pollutants and disease incidence.

In my own contribution to the symposium, I pointed out that this dilemma signaled the end of this type of Cartesian, analytical approach of trying to relate causally one specific pollutant to the incidence of a particular disease and its costs to the victim. To repeat, no such linear cause-effect relationships exist in a nonlinear, complex, dynamic social system, since multiple causes in diverse environmental circumstances prohibit the usual neat "laboratory approach" of controlling some variables while others are explored. Even if direct cause-effect relationships existed (itself highly unlikely), there would never be enough research funding, or enough specialists in ever-more-narrow specific fields of toxicology, epidemiology, meteorology, pathology, etc., let alone enough computer time to track and document the evidence to "prove" that factory X had emitted a toxic pollutant that had made citizen Y ill and inflicted Z dollars of social costs and medical expenses.

An interesting footnote to this discussion of the exhaustion of linear cost/benefit and social-costs analysis is the past resistance of regulators of pollutants to attempt to pinpoint actual sources of wastes. Given the power of companies to affect legislation, such a simple, commonsense approach has almost never been tried, since directly metering smokestack gases, etc., would indeed affix liability quite simply and cheaply. The less controversial approach of setting up costly monitoring programs to track pollutants after their release (thus increasing the difficulty of identifying their sources and proving the regulators' cases against polluters) won out over common sense and the public interest. Needless to say, this capitulation to industries that pollute has not only proved enormously costly to taxpayers and workers in terms of uncompensated ill health and other costs of pollution damage but also in terms of burgeoning regulatory agencies, which must hire batteries of lawyers and toxicologists and construct costly computer "pollution-diffusion" models in order to prove their cases before endless hearings and in court battles.

But what is left of the market system sometimes works in ironic ways; for example, Reliance Steel Company installed pollution-control equipment on its Cucamonga, California, plant and discovered that it also cut energy use by 45 percent and saved some thirty-three thousand dollars per month, since the pollutants were burned as fuel. This would surprise an economist but not a thermodynamicist (*The Christian Science Monitor,* April 11, 1980). Recently a Pennsylvania-based computer company, Materials Consultant Laboratories, Inc., announced an innovative electron-beam control system, Suspended Particulate Evaluation, which can analyze air samples in a polluted area and instantly "fingerprint" pollution sources by the size, shape, and composition of dust particles, thus providing evidence to nail the polluter.[9] The general sentiment of the symposium was summed up by one legislative assistant charged with promoting President Carter's bill on submitting federal regulations to "inflation impact statements" or any other economic cost/benefit analysis method: "The snag is that we have quantified the costs [to producers], but we know very little about the social costs and benefits." So Americans are forced to rely on the volunteer efforts to quantify them of such public-spirited groups as Accountants in the Public Interest, which has local affiliates in most cities and is headquartered at 19 West 44th Street, Room 1608, New York City 10036. The same kind of general bias of the economic method was clear in that only *some* regulations were targeted for their "inflationary impact." No mention was made of such huge, costly regulatory systems as the Securities and Exchange Commission, the Federal Reserve Bank, the Federal Deposit Insurance Corporation, and other vital government regulators without which an orderly capital-market and private-enterprise system could not function. Thus it became clear that definitions of what was to be considered a "government regulation" were just as arbitrary as definitions of what was a "cost" and what was a "benefit." At the boundary of a logical system's applicability, we are forced back to arguing the old issues of power: costs and benefits to whom and for what purposes?

The final absurdity of the economic method is the macabre necessity of assessing the monetary value of a human life. Moral repugnance as well as gut reaction are already forcing the necessary recon-

General disease or risk condition by economic category	Examples	Specific diseases, problems of risk
PRODUCTION		
i. Conditions primarily related to the *nature* or *organization* of production	the use of various chemical and other toxic materials in mining, industry, and agriculture	occupational diseases and injuries, e.g., asbestos diseases; numerous skin, lung, bladder, and other cancers; radiation diseases
	the careless use of capital-intensive productive methods	industrial injuries and deaths; capital-labor substitution leading to unemployment and thereby to anxiety states, depression, alcoholism, and the cigarette diseases such as bronchitis and lung cancer
	increased use of human beings in passive, repetitive or machine-like roles	obesity; industrial accidents; cigarette diseases; alcoholism; boredom and stress-related diseases and conditions
	industrial pollution	affects not only workers but also the rest of our society and other societies (e.g., lead pollution locally; sulphur dioxide and other pollution problems in Norway which are created in the U.K.)
ii. Conditions primarily related to the *level* of production	pressures leading to damaging rapidity in the production process	increased risks of accidents, e.g., diving accidents; "executive stress" leading to cigarette diseases, road accidents, alcoholism, and overeating (obesity)
	pressures related to frenzied and damaging marketing	results in various conditions as indicated in "executive stress" above and where domestic life significantly disrupted, to increased risk of mental illness
	pressures to utilize new and inadequately tested forms of energy inputs	nuclear power radiation hazards and deaths
	pressures to adopt damaging levels of goods transport and labor mobility	road traffic accidents affecting not only trucks but involving cars, coaches, and buses; disrupted domestic life as above

CONSUMPTION i. Conditions primarily related to the *nature* or *organization* of consumption	the consumption of disease- and accident-linked products	cigarette diseases; dental caries and the sweets and chocolate (and other sugar) diseases—including obesity and some diabetes; road accidents secondary to alcohol, hypnotic, or tranquilizer consumption; poisoning from weed-killers and pesticides; aerosol sprays
	the consumption of nutritionally deficient products	refined flour and sugar, i.e., fiber-deficient carbohydrates leading to diverticulitis, some cancer of the colon, etc.
ii. Conditions primarily related to the *level* of consumption	pressures to consume more in an absolute sense, e.g., advertising of the form "eat more, drink more"	advertising that contributes to overeating and therefore to our major nutritional problem, obesity and associated diseases, e.g., heart disease
	pressures to replace/update consumer durables and other products at an ever-increasing pace (including "planned obsolescence")	anxiety states and depression that arise from financial and other pressures to "keep up with the Joneses"
DISTRIBUTION i. Conditions primarily related to the maldistribution of economic opportunities or resources	chronic persistence of shortages and inadequacies in housing and basic amenities despite ever-increasing levels of productive output and energy consumption	hypothermia; respiratory and gastrointestinal conditions that arise from grossly inadequate housing and sanitation, overcrowding, and homelessness; accidents to children from the lack of safe and attractive play facilities, e.g., the special problems of high-rise apartments
	chronic problems of unemployment and poverty among specific subgroups of the population	many single-parent families; immigrants living in overcrowded and decaying urban areas with high unemployment; middle-aged and older unskilled workers whose physical fitness has been lost; agricultural workers who have little or no land of their own for vegetables, chickens, etc. Generally, poverty and unemployment effects such as malnutrition or subnutrition, anxiety and depression and associated cigarette diseases, methyl-alcohol drinking, etc.

Fig. 9

	Convention or Assumption	Principle	Example
Indexes of economic value	1. In general, the economic value of a good or service produced in the economy is equated with the price it fetches in the market place.	If one good or service is sold for x and a second is sold for 2x, then the contribution of the second to society's economic "welfare" is conventionally taken to be twice that of the first.	In terms of national accounts, if £1m. is spent on anti-smoking educational measures and £83m. is spent on advertising and promoting tobacco, then tobacco advertising is viewed as over eighty times as valuable—in economic terms—as are the educational measures.
Indexes of economic progress	2. In general, the health of the economy is seen to depend in part on increases in the aggregate sales value of the goods and services that exchange in the market place.	Measures such as GNP or national income measure the *level* of economic activity; that is, they increase (or decrease) as the total (price-corrected) sales value of goods and services increases (or decreases).	£1,000 spent on frozen vegetables (and their packaging, retailing, etc.) has a positive effect on GNP: £1,000 "worth" of home-grown vegetables that are consumed (or informally exchanged) by the growers do not enter national accounts (and therefore GNP).
Indexes of welfare	3. In general, many unintended side effects of market activities (e.g., noise, pollution) are omitted from measures of economic welfare. Moreover, measures of economic welfare are often confused with measures of social or general welfare.	The production and consumption of goods and services are conventionally viewed as the *primary* activities of the economy: the unintended "production" of, for example, air or water pollution or occupational "stress" or accidents are viewed as "external effects," the costs of which are rarely reflected in measures such as GNP. Indeed, they sometimes are viewed as benefits.	"Defensive" expenditures such as those incurred in cleaning up the air or water, or in "repairing" people following preventable accidents, are added to rather than subtracted from measures such as GNP. Many "external" outputs of the economy for which no "defensive" arrangements exist, do not enter national accounts at all. Measures of economic welfare can increase therefore even in situations where general welfare may be in decline.

Distinctions between productive and nonproductive	**4.** In general, the production of many "public" goods and services is viewed as a "drain" on the wealth-producing (marketed) sector of the economy.	The economic value of many goods and services that are central to the quality of life but that are not, in general, marketed (e.g., health, education, etc.) is regarded as proportional to the market value of the productive "inputs" they consume; they are therefore viewed as "nonproductive" uses of productive (i.e., marketed) resources.	Marketing health, education, etc. renders them productive in terms of national accounts. In the extreme, putting health care "on the market" can be seen as a measure that might well improve "economic welfare" despite the fact that, again, general welfare may decline.
Distinctions between economic and noneconomic	**5.** Many essential productive activities that are important not only to the economy but to social welfare are not counted as contributions to society's economic welfare.	Much work that is undertaken without monetary remuneration has no economic value.	Child rearing, housework, voluntary activities (e.g., blood donations) and any other charitable or benevolent activities, the outcomes of which do not enter into national accounts, have no economic value and therefore do not contribute to economic welfare.

There are a variety of assumptions as well as accounting and linguistic conventions that are commonly adopted in discussions of policy when attempts are made to try to understand the "state of the economy," or, for example, the nation's rate of economic progress. Although such conventions and assumptions have been frequently criticized by economists and others, they are still used as primary "benchmarks" in the formulation and implementation of national economic policy and, as such, severely restrict the range of policy options considered and seriously discussed. Among other objections, these conventions—or perhaps more accurately, habits of thought—serve to obscure the antithetical relationship between many of the goals of conventional economic policy and the desire to promote human health and well-being. The table summarizes some of these conventions and gives examples of the ways in which they serve to conceal many of the health and wealth conflicts mentioned in the text.

Fig. 10

ceptualization. For example, the British National Health Service must concern itself with social costs that are inflicted, in the form of ill health, by the private sector. Thus public-health planning must consider not only environmental pollution as a growing cost of health care but also advertising that generates overconsumption of sugar and refined foods, and the microprocessor revolution, which generates fears of job loss and psychological stress and family pathology due to increased levels of unemployment, all of which burden the public-health system. Thus Peter Draper, an innovative health-policy analyst, has developed methodological approaches to studying these impacts (Figs. 9 and 10). Draper is still critical of the lag in developing these impact studies and shifting the focus of health care to prevention. He congratulates the British Royal Commission on the National Health Service for fearlessly stating that "there is no doubt that television and radio, certainly in their commercial forms, do a great deal of harm by promoting excessive consumption of alcohol, tobacco, and sweets, for example" but then zeroes in on the same evasion of responsibility we see in U.S. regulation of pollutants. The Commission lamely buys into the economists' free-market model (which avoids dealing with power) and simply calls for more tax-payer-funded "counteradvertising" to warn people of the dangers, rather than regulating commercial advertising that promotes consumption pathologies in the first place. This kind of "add-on" social cost can, of course, be labeled as "creeping bureaucracy," but in reality it is one of the inevitable systemic causes of inflation and "declining productivity" now pervading all industrial societies, as the logic of maximizing industrial production efficiency bogs down in its own inconsistencies.[10]

As mentioned, this last act of the drama of industrialism is what I have summarized as the "entropy state," in which more effort is spent in this type of "add-on," ameliorative activity and attempts to coordinate all the conflicts, legal battles over damage and liability, caring for the casualties and dropouts, coping with structural unemployment, cleaning up the mess and pollution, and managing the impacts of careless technologies (some reaching crisis levels, as with the Love Canal chemical dump, which has engendered claims of $2.6 billion, in Niagara, New York, and the release of dioxin in Seveso, Italy, which poisoned a large area) than is spent in producing useful goods and services. No one has studied this "social entropy" syndrome of late-stage industrialism, which manifests itself in systemic inflation, more closely than Belgian information theorist Jean Voge. He concurs with my own view that the much-vaunted "postindustrial society" envisioned by Daniel Bell, in which activity will shift to the knowledge industry and the services sector, when viewed less optimistically is merely the growth of the social-costs sector and the burgeoning of paper-pushers and bureaucracy attempting unsuccessfully to deal with the market system's failures, using linear methods of regulation. Voge, in his paper "Information and Information Technologies in Growth and the Economic Crisis," in *Technological Forecasting and Social Change,* Volume 14, Number 1 (June 1979), cites evidence confirming the phenomenon described by Bell as the "postindustrial society," including economist Fritz Machlup's study *The Production and Distribution of Knowledge* (Princeton University Press, 1962) and his 1975 paper "Workers Who Produce Knowledge: a Steady Increase 1900–1970," *Weltwertschaftel Arch.* 111, 4, pp. 752–579 (with T. Kronwinkler), showing that the "information ratio" (percentage of information producers and distributors in the work force) had grown from 10 percent in 1910 to 40 percent in 1970 in the United States. Voge also cites Marc Porat's thesis *The Information Economy,* a doctoral dissertation (Stanford University, August 1976) showing that the trend continues but tends to reach a saturation point at approximately 50 percent. Voge cites similar trends toward information-based activity in Britain, France, and West Germany, documented in 1977 by the OECD. The point is: are we to consider this as the "good news" that Daniel Bell sees, or the "bad news" that I describe as the entropy state of growing social and transaction costs leading to the explosion of bureaucracy? Voge takes the evidence collected by Machlup, Porat, and the OECD and plots the growing information sector, showing rather startling similarities to the local entropy effect thermodynamicists demonstrate in the performance of an engine: i.e., one can push the efficiency in converting energy to useful work only so far—up to approximately 50 percent—and the rest is then lost in waste heat and friction. This is a specific *local* example of the working of the second law of thermodynamics, the entropy law, which states that this effect on a *universal* scale is a gradual winding down of the available potential energy from differential states of matter and energy, from ordered states toward general disorder. In Chapter 5 of *Creating Alternative Futures,*

"The Entropy State," I related these thermodynamic concepts of the performance of a physical system (e.g., an engine's Carnot cycle) to the performance in increasing the production of a society. I asserted that industrial societies dedicated to narrow goals of maximizing production would reach a stage of diminishing returns and become swamped with the problems of coordination of the ever-increasing complexity they created in this drive for production. A systemic trade-off would be reached where the efficiencies achieved in production by increasing specialization, division of labor, and capital intensity would be offset by the rising "transaction costs" of coordination and maintaining social and environmental "overhead." Voge shows that the concept of a mature entropy state in industrial societies fits the thermodynamic equations of a Carnot-cycle engine, i.e., that there is an upper limit to maximizing material production per worker at approximately 50 percent, when the information workload (i.e., bureaucracy and managerial "overhead") grows faster than production. Voge states that this 50 percent level, beyond which information requirements swamp additional production, corresponds to a maximum level of "maturity" for economic growth (as traditionally defined) and that productivity increases among material workers due to better equipment and know-how can no longer compensate for the steady loss of workers to bureaucracy. I would emphasize an important point: that bureaucracy, in my view of the mature industrial society approaching the entropy state, not only refers to government but also to the growing corporate bureaucracies, the layers of management of pyramided, diverse conglomerates, in which the coordination effort multiplies in just the same fashion as in government; I discuss this in Chapter 5. Thus Voge's reconceptualization permits us to see the systemic problems inherent in pursuing a specific course with a specific goal and logic, seeking to maximize single variables, as industrialism has attempted in maximizing production—which I see as the evolutionary riddle: nothing fails like success. We must now rethink the logic that has guided us in industrial societies: that if something is good, then more of it is better. Growth of production is a good thing up to a point—then it begins to create its own problems.

Thus economists' continually insisting on seeing our dilemma of stagflation as a "problem" of "declining productivity" causes policy makers to throw good money after bad by trying to hype capital formation and investment and stoke up the engines of production beyond their 50 percent thermodynamic potential, where these new funds simply "boil off" in more waste heat. Similarly, in resource extraction, there is a trade-off between speed and thermodynamic efficiency.[11] Only changing the configuration of the society and redesigning technologies with better "thermodynamic potential" built in can produce results, as, for example, the vastly increased efficiency of new electric motors, such as the Wanless and Exxon ACS designs (*Soft Energy Notes,* April 1980). Yet, as we have seen in the example of the war between nuclear technology and its solar successors, shifting these technological directions is agonizingly slow because of the entrenched corporations, which cannot change themselves but still have political power to prevent the social system from adapting. The same limits are being encountered in the Soviet Union, in spite of its huge petroleum and mineral wealth, as it encounters the thermodynamic problems of declining *net* production (*Fortune,* July 28, 1980) and as its entrenched, inefficient bureaucracies stifle its growth, as acknowledged in its 1980 five-year plan. Voge sees the situation now approaching industrial stagnation as having two basic theoretical solutions in similar terms to my own: either the design and implementing of a "perfect" information control system (what I term the Orwellian view of the computerized Leviathan state) or a shift in direction toward decentralization and modularization of the society, with greater localized control both economically and politically. Yet, there is a third way: examine the structure using much sharper tools than the economists' heroic averaging of data into abstractions of supply, demand, rates and levels of investment, "productivity," etc., and look more closely at where the rubber hits the road, so as to redesign real technologies with better real thermodynamic performance, to fit real ecosystems and the real needs of people. Here again, either/or thinking would lead us to throw out the real gains in productivity and the knowledge accumulated during the industrialism phase, rather than conserve them and only decentralize decision bottlenecks and simplify technology where we have most obviously overshot the mark. Or, as E. F. Schumacher's writings show so clearly, we need to restore a balance that has been lost, but to do that we now need new criteria and new, appropriate methodology as well as appropriate technology.[12]

Another crucial aspect of the "declining productivity" flap that we now address is the subtheme of the same refrain: the U.S.A.'s "de-

cline in technological innovation." This is also a novel dilemma that
indicates the exhaustion of the economic logic much more than any
real concern that we no longer know how to innovate. It is my contention, spelled out throughout this book, that the United States is
still the most innovative society in the world and that, in fact, there is
a veritable avalanche of new inventions in the enormous field of
renewable resources, solar, and the biological and ecological sciences[13] stored behind the dam of political inertia of existing corporations unwilling to write off their old technologies whether nuclear
power, gas-guzzling cars, automated junk-food vending machines,
high-technology, curative surgical and medical intervention to manage
industrial diseases, petroleum-subsidized agriculture, or all the other
now unsustainable industrial enterprises. In the growing countereconomy, where ingenuity, improvisation, and entrepreneurial
spirit are still the rule and must, for the time being, substitute for
capital investments, the experiences of young inventors are pervasive: they cannot get government research and development grants
(because they are too small and bureaucrats are nervous about their
unconventional approaches), and they cannot find venture capital
(because it is retreating into short-term treasury bills or looking for
"safe" investments, such as electric utility bonds to build nuclear
power plants). Similarly, innovators' need for funds is small and specialized, and understanding of their innovations requires a knowledge of the new science of renewable resource management, and this
new awareness is only slowly filtering into the minds of Wall Street
analysts, bankers, and money managers, who are among those most
enmeshed in economics. Glimmers of the new paradigm occasionally get through; for example, *The Wall Street Journal,* whose
editorials are firmly anchored in the Golden Goose model, recently
ran an article on differences between economists' and thermodynamicists' concepts of looking at a "net-energy" balance sheet,
rather than one denominated in dollars. The article, "Energy-costly
Energy Is Wasting Resources, Some Analysts Worry" (May 3,
1979), describes some of the conclusions of energy analysts, including Earl T. Hayes, former chief scientist at the U. S. Bureau of
Mines, that the average U.S. net energy is currently 80 percent of the
gross (meaning that 20 percent of all energy is consumed in mining
and delivering the energy) but that by the year 2000, due to declining quality of resources for exploitation, the net energy will be re-

duced to 75 percent, and that in the 1990s energy growth would come to a halt. This was the thesis of my own paper in the *Financial Analysts Journal* (May 1973). My conclusion was that energy growth was *not* a precondition for economic growth and that therefore we would need to redirect our investments toward capitalizing a renewable-resource production system and develop all the innovations in solar and conservation technologies and postpetroleum agriculture that were already on the shelf. Jimmy Carter's Agriculture Secretary, Bob Bergland, was among the first to switch his research-and-development policies, promoting only research that did not increase farm automation and energy use.[14]

However, such arguments concerning real thermodynamic productivity and efficiencies and the importance of bioproductivity can all too easily be ignored if they fall outside the scope of the dominant economic paradigm, since they can be treated as "exogenous" factors, such as Arab sheiks, weather, and sun spots. Some progress has been made in trying to turn them into language that economists can hear, such as the concept of "life-cycle costing," which solar-energy companies use to show that on a lifetime-use basis, solar systems are actually competitive with traditional energy, because although the initial investment is higher, the "fuel" (sun, wind, falling water) is free. It is possible to really develop a dialogue with economists on these technological choices (which are no longer guided by the invisible hand but must be selected consciously by consumers, investors, and policy makers) only when one can enter their "head space" and translate the new concepts into terms analogous to their economic[15] models. Thus, in an economists' model, technology is a *coordinate,* rather than a *variable* (as it is in reality), because in the Panglossian world of economic theory, the consumers in the free market decide by their purchases which technologies will be developed, and this produces the best of all possible worlds. Thus a real breakthrough occurs when someone trained as an economist transcends the discipline and makes the necessary leap to a more systemic model and then takes the trouble to communicate the new model to economists in their own terms. Such a breakthrough has now been made by Italian economist Orio Giarini and French economist Henri Louberge in their book *Diminishing Returns to Technology* (1979). Their analysis does not rely on exogenous arguments as to the decline of quality of the nonrenewable resource base or pollution but, like the entropy-state thesis, it is *intra*systemic and *grows out* of the over-shoot in the logical progression of the existing institutions, technologies, and social goals. Karl Marx, building on Hegel, called these "dialectical processes," as discussed earlier, and since they represent a much more systemic and dynamic model of evolutionary human developmental processes, they explain the continued power of the Marxian method, i.e., to *look for* such contradictory processes within a system; or what the Oriental cultures call the complementary rhythms of yin and yang, and ecologists call the natural cyclic, balancing processes of energy and materials in an ecosystem.

In *Diminishing Returns to Technology,* Giarini and Louberge demonstrate the intrasystemic problem of diminishing returns to the particular type of technological development path that industrialized societies chose: that of maximizing material productivity per capita (or labor productivity), which is then equated in public policy with "productivity" in general. This was also my thesis in *Creating Alternative Futures;* maximizing this type of technological productivity would push its innovation path toward greater capital-intensity and resource-intensity and create more and more devastating impacts on the social and ecological systems. However, the novel approach taken by Giarini and Louberge demonstrates that even before such an innovation trajectory bogs down in the impacts it creates in the larger system, it will bog down for inherent reasons. The concept of diminishing returns goes back to the great classical economists David Ricardo and Thomas Malthus who related it to the limits of increasing the fertility of land, as discussed in Chapter 8. But, as we saw, it was applied subsequently in very arbitrary ways, even though it could have been one of the sharpest tools in the economics kit bag. The concept of diminishing returns could have been used to predict the saturation stage of consumerism and the rise of a countertrend toward "voluntary simplicity." Similarly the concept is closely matched in the physical and biological sciences by that of the familiar S curves found in nature; for example, in the multiplication of a population of fruit flies in a fixed environment until their increase runs into limiting factors. Thus the general thesis of Giarini and Louberge is that the historical growth of industrial societies has relied on continually sustained rates of technological innovation. Up to a point, there are all manner of opportunities to exploit basic scientific breakthroughs and employ and commercialize technologies that already exist in the society. But finally a point is reached when the growth rate of these societies begins to *require* steady rates of technological innovation on a technological base that is already very large and complex, and exploitable margins of incremental innovations to it do not yield much payoff. Here we see the optimistic assumptions made by Edward Denison, based on traditional models of production functions, for example, that of Robert Solow, that includes in the factors of production, together with land, labor, capital, and managerial skill the additional increment of "know-how" and "technological progress" as a constant factor! (See Solow, "Technical Change and the Aggregate Production Function," *Review of Economics and Statistics,* August 1957, pp. 312–20.) Thus the fact that industrial societies have become addicted to technological innovation and now cannot do without a steady stream of new technologies *maintained at historical rates* reveals another new vulnerability. As Giarini and Louberge point out, "The flow of innovations cannot be speeded up in the same way as an investment flow, by diverting production factors away from the consumer goods industries and toward the capital goods sector. In the case of *technology,* a larger share of available resources of research and development does not mean that there will be more innovations, if the invention *rate* is simultaneously falling and the average *time lag* between an invention and the corresponding innovation is increasing" (italics added). I would add that this falloff in the marginal returns applies only to the existing technological configuration and trajectory; i.e., it is this specific set of S curves that is saturating. New potential innovation curves with plenty of steam in them are waiting, *if* we change the direction of technology and redesign it with new criteria fitted to the new resource situation: i.e., begin reaping the inventions in the new, unexploited areas of renewable-resource, sustained-yield ecosciences of the dawning solar age.

Giarini and Louberge take case studies of specific industries based on historical technological innovation, such as the textile industry, and show that this industry is stagnating because the chemistry of textile innovation, based on the realization that cellulose and natural fibers could be replaced by petroleum, has now also reached maturity, limited by the basic science of modifying matter. Similar declines in the frequency of innovation in the chemical, aircraft, and automobile industries are discussed. Yet, in industrial economies, which now rely on constant rates of technological innovation for maintaining economic growth rates, each company must "stay competitive" with other companies by trying to constantly innovate, if necessary by plowing more and more money back into research and development. Here yet another problem arises: in addition to the intrinsic diminishing returns to these private research-and-development investments, one must view the individual firms as "actors" in game-theory terms. As their individual research-and-development investments increase in competition with each other, the resulting innovations (however trivial) begin to crowd each other in the marketplace, leading to earlier and earlier obsolescence of each one. Thus a new "tragedy of the commons" is in the making: the percentage of total sales of an industry invested in research and development correlates with the rate of obsolescence that this same investment causes in this industry! Thus Giarini and Louberge show that a new stage of the industrial process has been reached, in which 1) the cost of research and development in certain industries exceeds that of the other factors of production, while 2) the obsolescence of the product (in years) is *inversely proportional to the amount invested in research!* In other words, as the research-and-development investment grows, the payback period for each succeeding product innovated becomes shorter as the products are superseded more rapidly by newer products and driven from the marketplace before they are amortized. Thus whole sectors of industrial economies are experiencing this kind of diminishing returns, and it is for *this very sensible reason* that many firms reduce their levels of research-and-development investment. Thus we now see the misguided attempts of economists like Denison et al. to persuade policy makers that this "declining productivity and innovation" phenomenon can be reversed with even *more* government subsidies, faster write-offs, and tax credits for further capital investments in the very industries that are experiencing its diminishing returns firsthand. This course is tantamount to throwing tax dollars and capital investments out the window, rather than disaggregating the view of "investments" and examining precisely *which* of these industries can no longer usefully employ capital and which new areas, in the renewable-resources sciences, are still exploitable and waiting to be capitalized. Policy makers must now face the fact that in many sectors that have been the mainstay of industrial economies in the past, as Giarini and Louberge warn, "the mainspring of growth is broken" (p. 59).

Nowhere is this policy bombshell more devastating than in its implications even in a sector of the economy based on technologies still in a rich stage of exploitation: computers and microprocessors. Here the inherent technological innovations flowing from electronic and materials sciences are undiminished, and the rate of innovation is astonishingly rapid. However, in this case, a limiting factor looms, still insufficiently examined: the industrial purchasers of these microprocessors have only so much capital to *absorb* the innovations, since buying the latest generation of microprocessor technology inevitably means writing off the existing capital stock of those in use. At some point, even the largest and richest companies will have to cry halt, since a company cannot afford to junk its existing capital equipment every couple of years. A case in point may be the American Telephone and Telegraph Company, a huge user of microprocessors and already the largest consumer of capital in the U.S.A. Thus we may run into internal limits to the microprocessor revolution well before we run into the host of intractable social problems of displacement, deskilling of jobs, and rising unemployment levels discussed in the previous chapter.

Let us now turn to examining some further evidence that increasing government subsidies to the research-and-development process will *not* increase the rate of technological innovation, as is so widely believed. First, as Devandra Sahal shows in *Technological Forecasting and Social Change*, Number 16, 1980: the stagnation of growth of a system in a certain dimension (e.g., in the inputs employed) does not necessarily prevent its growth in some other dimension (e.g., productivity). He cites the well-known cases of the steel works in Horndal, Sweden, and the textile factory in Lowell, Massachusetts, which, with no additional inputs of capital over fifteen and twenty years, continuously increased productivity at an average of 2 percent annually. This effect is due to the experience gained by workers and managers on the job. Furthermore, a team of researchers at the National Science Foundation reported in *Spectrum*, the journal of the Institute of Electrical and Electronics Engineers (October 1978), that the three basic policy assumptions are unfounded: 1) that U.S. growth and productivity improvement have been slowed compared with other industrial nations, 2) that U.S. research-and-development

expenditures first declined and then leveled off during the past decade
in comparison with these other countries, and 3) that compared with
these same countries, U.S. research-and-development investment is
heavily concentrated in defense and space-related activities. When the
group compared the extent of *enterprise*-funded research and development in twelve industrial sectors for the period between 1963 and
1973, they found that although there was a decline as an *aggregated*
public and private level of funding as a ratio of research-and-development investment to GNP, that the *enterprise*-funding levels had remained constant. Since 1973, the total ratio to GNP has stabilized
in the United States, while a number of other countries have *reduced*
their share of GNP allocated to research and development. Admitting
that the data were still hard to interpret, the NSF group then explored
the question of whether, in fact, government funding of research and
development made all that much difference in increasing the rate of
productivity and innovation. Their conclusion was that "arguments
for Federal actions to stimulate industrial research and development
cannot be based soundly on the three observations described earlier."
Further, the group found there was for manufacturing, "no clear relationship between international differences in research-and-development intensities and economic growth." They also discovered that,
in any case, "in the aggregate the impact of government research-
and-development spending tends to have only a short-term effect."
Thus it appears that the demands of U.S. industry to increase levels
of government subsidies to research and development and to hype
capital investments generally have no scientific basis, even though
they have a good deal of logic in terms of enabling companies to
"externalize" more of the costs of research and development and reduce their capital investment costs, thus improving their balance
sheets.

The NSF researchers underline the fact that heroic abstractions
such as aggregated "investment" and "research" conceal the most important question: investment and research on *what*? That is, the
direction of research, development, or investment is the prime determinant of whether or not it will be fruitful.[16] Without knowledge and
a systemic understanding of the industrial society and its new situation at this historical stage of its evolution, investments can be so
misdirected as to be essentially useless and wasteful. This brings us
to realize how carefully we must define such sloppy terms as "inno-

vation." As mentioned, if we define "innovation" in average terms, then we give equivalence to the actual value of very disparate *types* of innovation: weighting innovations of the forty-seventh new brand of protein-enriched dog shampoo, patent headache pill, or cosmetic equally with the advances in microprocessor technology. Yet, incredible as it may seem, economic theory does not differentiate them, and research-and-development expenditures in all companies are compared simply in the aggregate coefficient of dollar amounts, however stupidly they have been applied, as summed up in *Business Week*'s inane "R&D Scoreboards" cheery headline "More Speed Behind R&D Spending" (July 7, 1980). This mental trap in economics by now will be familiar to the reader; as encoded in the assumptions that consumer preferences as revealed in the free market are assumed in welfare theory to be all equivalent in "value" (subjectively defined), thus yielding an innovation path that is, by definition, unquestionable and optimal.[17]

Having discovered that, in fact, research and development may *not* be declining, and that, in any case, there may be little or no connection between "productivity" and R&D funding, and that government research and development may not have much impact anyway; let us turn our attention back to the macro level of the general evolutionary transition that industrial societies are now undergoing and examine some systemic theories of the process of technological innovation. The longest-term and most comprehensive of these are based on the biological evolution of life on this planet and are best expressed by Kenneth Boulding, who has long since transcended the economic method, using it as an appropriate tool to examine micro areas, rather than as a belief system. Boulding's view of productivity is biological and evolutionary, pointing out that there are many productivity concepts, all involving a ratio of some kind of output to some kind of input, and that there are as many "productivity" concepts as there are processes that transform inputs to outputs. Boulding cautions that "general productivity concepts . . . are more difficult, since virtually all processes have many inputs and outputs . . . so that to measure general productivity we must have a set of "shadow prices" or value-coefficients. . . . The commonest measure, though not always the most significant, is the monetary unit." Herein, of course, lies the mental trap that economics has woven for us all. Boulding zeroes in on the crucial issue in productivity: knowledge, which produces evolutionary innovation (as opposed to misguided, ignorant investment, which produces nothing fundamentally new even though it may be a profitable new product). "Production," Boulding states, "originates from know-how . . . similar to the processes of evolution based on genetic information. The chicken egg never produces a hippopotamus. It doesn't have the know-how. It only knows how to make a chicken. . . . The ongoing processes of evolution of which economic development is merely a recent example, are limited by the evolutionary potential of the system. The development of evolutionary potential . . . never produces exponential growth. It produces a *pattern* of growth, maturity, stability and eventually death. These considerations may seem rather remote from the problem of U.S. productivity, but in fact they underlie that problem" (*Spectrum* IEEE, October 1978, p. 42).

Another systemic view of technological innovation and productivity is the "long-wave" theory of Russian economist Nikolai Kondratieff, dismissed for decades by both Western and Marxist economists. As discussed in Chapter 4, Kondratieff claimed that there were distinguishable long cycles of economic growth and decline that generally embraced epochs of about fifty years, and that these waves of economic growth and decline were due to phases of scientific and technological innovation and the specific industrial complexes to which they gave rise. Kondratieff's view was one of continual *dis*equilibrium associated with such specific technological innovations as railroads, as well as wars, revolutions, and the bringing into the world's monetized economy of new nations, and thus he was naturally unpopular with economists, who were more comfortable with their reversible, equilibrium models of economic processes as analogous to locomotion. An example that Kondratieff might have used of a technologically based long wave, one that is particularly germane today, is that of the innovation and commercialization of the automobile in this century and its proliferation to the point that the industrial configuration to which it gave rise—factories, dealers, suppliers, credit companies, repair shops, as well as the public-sector infrastructure (the national highway system, traffic courts, laws, police administration, air pollution)—today account for one out of every six jobs in the U.S. economy. Yet this technological wave is now clearly exhausting itself, since it has also been based on mining, in a very brief historical period, the planet's store of petroleum. In fact, the U. S. Office

Plate 32

YOUR TAX DOLLARS COULD BUILD PEACE

$YOUR TAX DOLLARS$

TODAY 53% OF YOUR TAX $ PAY FOR WAR. THE CURRENT MILITARY BUDGET REQUEST IS $104 BILLION – LARGEST IN U. S. HISTORY – AND ALMOST EQUAL EXPECTED REVENUES FROM ALL U. S. WORKER INCOME TAXES – ESTIMATED AT $106 BILLION

SUPPORT THE WORLD PEACE TAX FUND ACT H.R. 4897

Proposed by Congressman Ron Dellums and Don Edwards, Pete Stark, John Burton, George Brown, Augustus Hawkins, of California; Walter Fauntroy of D. C.; Parren Mitchell of Maryland; Joe Moakley, Michael Harrington, of Massachusetts; Charles Diggs, John Conyers, Bob Carr, of Michigan; Henry Helstoski of New Jersey; Bella Abzug, Ben Rosenthal, Edward Pattison, of New York; and Robert Kastenmeier of Wisconsin

WHAT THE WORLD PEACE TAX FUND WOULD DO:

- Give taxpayers who oppose war a LEGAL alternative to paying taxes for military purposes.

- Establish a World Peace Tax Fund to use these taxes for peace-related projects.

- Provide that World Peace Tax Fund must be in addition to regular appropriations for domestic and United Nations programs.

- Model the World Peace Tax Fund after existing Federal trust funds.

- Help to build a peaceful society by supporting research and other efforts to foster non-violent methods of resolving international disputes.

THIS BILL NEEDS

YOUR SUPPORT TO BECOME LAW

For more information write:

World Peace Tax Fund Box C
2111 Florida Avenue, N. W.
Washington, D. C. 20008

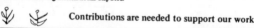

Contributions are needed to support our work

Plate 33

AND A LITTLE CHILD SHALL LEAD THEM.

Desegregating New Orleans' William Frantz Public School, November 14, 1960. Protected by U.S. Marshals, this little girl, now 20, led America's black children toward equal education.

On May 17, 1954, the Chief Justice of the United States handed down the Court's opinion:

"We conclude," said Chief Justice Earl Warren, "that in the field of public education the doctrine of 'separate but equal' has no place. Separate educational facilities are inherently unequal."

With these now familiar words, the Supreme Court not only outlawed racial segregation in public schools, but signaled the move to dismantle statutory segregation in all walks of life.

The *Brown* decision, which ended the legality and asserted morality of segregation, did not just happen. It resulted from years of carefully planned litigation by the Legal Defense Fund. And it happened because black families were willing to challenge white supremacy despite economic and physical reprisals.

Nor was the law self-executing. It had to be tried and tested in thousands of instances. Above all, it had to protect and defend the rights of little children who were trying to obtain the kind of education the Constitution guaranteed.

Remarking on the massive resistance and violence in Little Rock during 1957 and New Orleans in 1960, one eminent jurist bitterly recalls, "Here were grown men and women furiously confronting their enemy: two, three, a half-dozen scrubbed, starched, scared and incredibly brave colored children. The moral bankruptcy, the shame of the thing was evident."

Yes, it was the Justices of the High Tribunal — and the courage of little children in communities throughout the South — who gave heart to black families everywhere, who sparked the civil rights movement of the 1960's. And it was a nation shocked by the bombings of black homes and churches, by segregationists with pick handles and attack dogs, which influenced Congressional enactment of a series of monumental Civil Rights Acts.

The tragic and heroic events of yesterday are now history. They were part of a new America, awakened to the evils of segregation. Millions of citizens in that era responded to the quest for equality and full participation in American life.

· · · · ·

During the past 20 years, Legal Defense Fund lawyers have argued thousands of cases which give new content and meaning to *Brown* and other rights flowing from that opinion.

• Between 1955 and 1960, LDF lawyers chipped away at the "all deliberate speed" doctrine, helped put down Governor Faubus's mini-revolution, and pushed desegregation in the South on almost a student-by-student basis.

• In the early 1960's LDF lawyers defended tens of thousands of demonstrators in the sit-ins and freedom rides. We took 45 of these cases to the Supreme Court, winning virtually all of them. As a result, we were able to establish major legal precedents — and free practically every demonstrator.

• Since passage of the Civil Rights Act of 1964 — and subsequent enactment of other major civil rights legislation — LDF lawyers have been upholding, interpreting and enforcing these laws.

Will you — in 1975 — follow the lead of the little children?

Every generation has to fight its own battles to preserve and expand the democratic process. The *Brown* decision gives us — in 1975 — the leverage we need to turn the clock ahead . . . to make the ideals expressed in our Constitution work for millions of families.

Education is the great equalizer: LDF is engaged in 150 court actions, South and North, to help improve educational opportunities for all.

Poverty breeds poverty: One-third of all black families exist *below the poverty line*—20 years after *Brown.* The simplest way to break this vicious circle is to put dollars—earned dollars—into the pockets of the poor. LDF is fighting more than 200 employment discrimination cases to gain equal job opportunities for black workers.

Fair housing for all: Where a man lives determines, to a large degree, where and how he will work — and the kind of school his children will attend. LDF is fighting 100 lawsuits to overcome housing discrimination.

Help America close the grap between the cherished ideal and the bleak reality that confronts millions of black Americans — every living day of their lives. Put your beliefs and convictions to work. Give to the Legal Defense Fund.

LEGAL DEFENSE FUND

Plate 34

Women's International League for Peace and Freedom

it will be a great day
when
our schools
get all the money
they need
and the air force
has to hold
a bake sale
to buy a
bomber

WILPF provides opportunity to work for a demilitarized society; to stop government repression and surveillance; to end racism and sexism; and for an economy that puts people before profit.

WILPF was founded in 1915. It has 140 branches in the United States and national sections in 25 countries. WILPF also has consultative status at the United Nations. Its program is implemented through lobbying, publications, study groups, and non-violent action. Its journal, PEACE & FREEDOM, is published 9 times yearly, and contains timely articles and suggestions for action.

JOIN US! THE WORLD NEEDS
WILPF AND WE NEED YOU.

Plate 35

of Technology Assessment reported, in March 1979, that overemphasis on auto transportation had become a threat to the entire U.S. economy (New York *Times,* March 10, 1979). Kondratieff's theory was advanced in the 1920s. The theory predicted that capitalism would experience the downswing of the 1929–30s period but that it would not constitute the long-awaited end of capitalism predicted by Marx, but that another rise would occur as a new technological wave developed. Thus Kondratieff was also excoriated in the U.S.S.R., and as nearly as can be established, was put in one of Stalin's prison camps for his unorthodoxy.

Interest in Kondratieff has been revived by Walt W. Rostow, mentioned earlier in this book, and systems dynamicist Jay Forrester, pioneering author of *World Dynamics* (Wright-Allen Press, 1971). Based on a national model of U.S. economic dynamics as they have evolved for two hundred years, and constructed by the Systems Dynamics Group, at Massachusetts Institute of Technology, Forrester's model analyzes various cyclic behavior patterns in the economy, superimposes their differing periodicities on one another, and tries to identify their causes and relationships and to fit them to the more traditional data of economists. Needless to say, an approach that sees economic data as a subset of a much larger set of systemic interactions in a society aroused wrath among economists, who had earlier attacked Forrester's *World Dynamics*. Basically, the economists' criticisms were couched in the belief that the price system would eventually work to adequately ration scarce resources and would lead to substitution and innovation, as their market theories prescribed, and thus achieve equilibrium. The facts that prices tell only about the past, not about the future, and that the problem of absolute scarcity is considered theoretically impossible, were usually overlooked.

Forrester's national model was able to simulate three distinct cycles in the U.S. economy:

1) A 5.5-year cycle, normally referred to as the business cycle (however, of great significance for our puzzle) this cycle is *less* related to capital investment than to *employment* vis-à-vis *inventories.* Thus, any legislation to subsidize capital investment to counteract such short-term inventory fluctuations would be wasted.

2) A 16.6-year cycle, normally referred to by economists as the Kuznets cycle (after Simon Kuznets, the inventor of the GNP). This longer cycle receives very little attention from economists, who focus

on the oscillations of the shorter, so-called business, cycle. The Kuznets cycle is related, accurately, according to Forrester, to investment, but more specifically to lags inherent in procuring and depreciating investment capital. Forrester agrees with economists that this cycle is not of primary importance.

3) The Kondratieff cycle, already described, which Forrester sees as the most important cycle for the determination of the economy's behavior, comprising in its fifty years a peak of economic activity followed by a ten-year plateau, then a drop into a depression period for about a decade, and a long climb over the next thirty years to the next peak. Thus, Forrester concludes, "We are coming to believe that there is little effect on business cycles from interest rates, credit availability and investment tax incentives" (*Changing Economic Patterns,* paper before the M.I.T. Club of Chicago, May 4, 1978). In fact, Forrester asserts that "Reaction of the economic system to a change in money policy occurs smoothly over as much as two decades, not suddenly within a year or less as is assumed" (*Planning Review,* November 1980).

So we find that the predictions flowing from the thermodynamic view of economic evolution coincide with both the system-dynamics view and the intrasystemic view (Giarini and Louberge), as well as the biological, evolutionary view of Boulding, elaborated mathematically into models of morphogenesis and "catastrophe theory" and the complementary, nonequilibrium, thermodynamic models of living-cell chemistry of Ilya Prigogine, which we explore in Chapters 11 and 13.[18]

The underlying themes of all these larger models of the transformation processes that industrial societies are undergoing are structural change (not inferable from any of the system's existing states or variables) and increasing uncertainty and indeterminacy in research and experimental methods (described by Alfred North Whitehead as "the fallacy of misplaced concreteness"). The implications are that the researcher's *hypotheses* and theories or models, paradigms, world views (the lexicon increases with growing awareness of the dilemma) are now the obvious key to the "results" and the "facts." Thus the debate moves toward appropriate methodology, appropriate epistomology, and inevitably from there to what E. F. Schumacher called the pressing task of "metaphysical reconstruction."

On the macro level of the dilemma of how to deal with the "declining productivity and innovation" issue, which economists are determined to keep as the focus of our attention, the issue must be stated as: what is the appropriate new *direction* for industrial societies in the face of totally new conditions? This implies the reexamination of goals and values and the need for conscious choices, a terrible responsibility that can no longer be avoided. The only question is will the new goals, directions, and choices be made democratically, by engaging in a great debate over the next decade, a politics of reconceptualization, or will the choices be forced on us by an Energy Mobilization Board? As Forrester said, "So we face a major policy dichotomy: Does the government support those who are becoming unemployed in the capital sector . . . or do what is done in wartime and redirect the capital-producing capability into an area that we are going to need in the future? I am no more sure of the shape of the next technological wave than other people. I expect that energy will move toward renewable and more decentralized sources, not only because of the nature of the new energy sources, but because of changes in our social system. If our society goes to still bigger and more centralized energy sources . . . we will produce a more and more vulnerable social system." Much as I agree with Forrester about the general shape of the next technological wave, based, I believe, on the ecosciences, I do not share Forrester's view that only an increase in the money supply beyond the rate of increase in real output can produce sustained inflation. This smacks of illusory, money-based definitions. Nor do I share his reliance on the price system, i.e., continually increasing energy prices alone, as sufficient to push the social system toward the new wave of renewable-resource innovation. As shown in Fig. 5, there are too many subsidies to continued overinvestment in capital vis-à-vis labor that will tend to overwhelm the effect of energy price increases, thus skewing the system further toward the evolutionary dead end of greater capital, energy, and resource intensity. Thus we can expect a further period of redoubling efforts to address the nonexistent problem of "declining productivity," until we reconceptualize the real dilemma of institutional stagnation and then begin our new journey on the third way: toward the renewable-resource-based economies of the dawning solar age.

Finally, it is sad to see that the economic model of "efficiency," a carryover from the industrial age, is likely to cast its long shadow over the design of the renewable-resources societies and technologies, which will distort their most ecologically fitted performance for long-term sustainability. For example, by 1980, *Business Week,* as well as *Fortune* and other business journals, were reporting the run-up of stocks in these fields, as speculators caught glimmers of the new possibilities.[19] But the bias toward the needs of the *existing* corporate structure and institutional configuration is alarming, since the huge overhead of these dinosaur corporations will overburden the new technologies with overhead costs and the bias toward short-term, profit-maximizing, capital-intensive, and overengineered approaches. To capitalize the new wave with the greatest thermodynamic rather than economic efficiency will require a new generation of managers and new companies more specifically designed for the task as well as attuned to the moral and social issues posed by manipulating biological systems and our genetic heritage, as pointed out by Ted Howard and Jeremy Rifkin in *Who Shall Play God?* (1977). As Forrester notes, "A unique technological infrastructure goes with each succeeding wave. For example, the infrastructure that supported railroads was incompatible with the one that grew up around airlines. Historically, the down-slope of a Kondratieff wave has been a period in which the economy extracts itself from the old technology by using up the old capital plant, while the up-slope has been a period of rebuilding along new technological lines" (*Fortune,* January 16, 1978). Thus, as mentioned earlier, it would have been better not to subsidize the Chrysler Corporation but, rather, to subsidize the workers so they could redeploy themselves in the growing edge of the renewable-resource economy. And the entropy law, as we shall see in Chapter 13, is not the whole story.

NOTES – CHAPTER 10

[1] Amitai Etzioni, "Reindustrialization: View from the Source," New York *Times,* June 29, 1980.

[2] As the energy/inflation-driven stagnation affected all industrial economies in the winter of 1979–80, variations of these approaches to "restoring productivity" were the order of the day. By mid-1980, the U.S. jobless rate had increased to 7.5 percent. The British Government stoically took a bitter steel strike when workers refused to accept an 8 percent pay increase in the face of 17 percent inflation. Sir Arthur Knight, Mrs. Thatcher's appointee to the National Enterprise Board (NEB), which was set up to bail out lame-duck corporations, faced

up to the bankruptcy of that policy, and reverted to Adam Smith. Coming full circle, Sir Arthur outlined a "new" policy for the NEB, saying that he wanted "market forces to rule and that NEB should only become involved when market forces produced unacceptable consequences." Again one must ask, "Unacceptable to whom?" In Denmark, the 1979 election pitted workers against rugged individualists, and West Germany, Japan, France, Australia, and the Scandinavian countries all tried the old formulas for curing their "sagging productivity and growth" (*Business Week*, "World Economic Outlook. The Industrial Nations," February 4, 1980). Only oil-rich Mexico and China would "forge ahead on economic growth," while the troubled Soviet and East European economies would have to cope with the same "productivity" problems and declines in economic growth of two to three percentage points. The Soviets were making the same kinds of supply-oriented energy mistakes as the United States, force-feeding nuclear, oil-exploration, and coal industries beyond their thermodynamic limits of production while shortchanging renewable resources ("Soviet Energy Choices," *Plowshare Press*, January–February 1980, p. 6, from the Mid-Peninsula Conversion Project, 867 W. Dana, No. 203, Mountain View, CA 94041, annual subscription $6).

[3] While most of the vocal business community presented a united front on these diagnoses, some, such as William Sneath, chairman of Union Carbide Corporation, whose own mental paradigms were obviously in transition, were more ecumenical: "A system determined to have efficiency prevail over other human values will one day find itself without support. At the same time, a society with an underpowered economy will fail to achieve either social or economic goals." However, Sneath then called for more tax breaks for business and investors and all the rest (speech at the Woodlands Conference on Growth Policy, Woodlands, Texas, October 31, 1979).

By contrast, Simon Ramo, former chairman of TRW Corporation, noted that "The overall problem in the U.S. is poor *management* [italics added] of America's ability to apply science and technology to its problems" (*The Christian Science Monitor* interview, January 15, 1980).

[4] Meanwhile, the Corporate Accountability Research Group released a study showing that health and safety regulations protecting workers, consumers, and the environment provided Americans with more than $35 billion in *benefits* in 1979. The study refuted Murray Weidenbaum's estimate for the American Enterprise Institute that all federal regulations had cost $66 billion in 1976. The Corporate Accountability Research Group noted also that Dr. Weidenbaum's Center for the Study of American Business received $700,000 in one year from business donors.

[5] *Business Week*, "Why Managers Are No Longer Entrepreneurs," June 30, 1980, p. 72.

[6] See, for example, *Fortune*, "The New Down-to-Earth Economics," December 31, 1978, p. 72. A memorial volume to Shackle, *The Information Revolution*, ed. Lamberton, appeared in 1974 (*The Annals* of the American Academy of Political and Social Science, Philadelphia).

[7] For example, in 1977 Japan's level of per-capita productivity was only 68 percent of that of the U.S.A., although the Japanese *rate* of increase was much more rapid. Source: University of Pennsylvania, Summers, Kravis and Heston, in *U.S. News & World Report*, April 26, 1980.

[8] Proceedings of this symposium are available from the U. S. Government

Printing Office, Washington, D.C. Ninety-sixth U.S. Congress, Sub-Committee on Science, Research, and Technology, May 1980.

9 *The Christian Science Monitor,* September 27, 1979.

10 In democratic societies it also leads to citizen action, as well as that of the American Cancer Society's belated drive aimed at ending government tobacco subsidies and to ban all advertising of cigarettes except those containing 50 percent less of harmful ingredients than the year before. "We believe this program will save seventy thousand lives a year," said Allan K. Jonas, Chairman of the Society's Task Force on tobacco and health (New York *Times,* January 14, 1979).

11 The latest, most dramatic example of this thermodynamic "boil-off" effect of attempting too rapid gear-ups of production is that which the United States experienced in early 1980 as it tried vainly to gear up its military arsenals. It became clear that however much *money* was thrown into the process, the inevitable encounter with the second law of thermodynamics led to the usual delays, personnel shortages, unavailability of key components such as large forgings and castings, bearings, machine-tool capacity, semiconductors, etc., as well as the general friction that policy makers persist in not associating with the effects of the entropy law. ("Why the U.S. Can't Rearm Fast," *Business Week,* February 4, 1980, explained all the fragmented particulars of the situation —but lacked a model of the entropy law that would have explained the phenomenon in scientific terms, rather than by means of the opinions of economists.)

12 No better example of the economists' methodological mumbo jumbo exists than the Central Intelligence Agency's economists' estimates of the supposed "build-up" of Soviet military strength, which hawks used in early 1980 budget debates to protect military budget increases at the expense of cutting social services (over 90 percent of which is Social Security, Medicare, and Medicaid, in any case). James J. Trieres, chief economist for the Center for Defense Information, in Washington, revealed the CIA's methodology thus: "What the CIA measures is not what the Soviets actually spend, but what they *would* have spent if they had paid their servicemen at U.S. rates, bought their weapons at U.S. prices and spent the same amount as the U.S. to keep each unit in operation —none of which they actually do." He adds that since no credible data on Soviet military spending are available, anyone can play the game of estimating Soviet military spending. He adds that the CIA's use of U.S. market-based prices have little relation to Soviet prices, which are set by government policy and are hardly translatable into U.S. economic equivalents ("The CIA's Fairyland Estimates of Moscow's Build-up," *The Christian Science Monitor,* February 1, 1980).

18 See, for example, *Energy Strategies: Toward a Solar Future,* eds. Kendall and Nadis (Ballinger, 1980); Denis Hayes, *Rays of Hope,* (Norton, 1977); Lovins; Commoner, and others mentioned previously. The Center for Renewable Resources, of Washington, D.C.; the Solar Energy Research Institute, of Golden, Colorado; *Solar Age Magazine;* the International Solar Energy Society; and the Wind Industries Association are some of the best sources of further information.

14 "Trouble for Tomato-picking Machines," *The Christian Science Monitor,* December 24, 1979.

15 For example, at the October 1979 UN Environment Program (UNEP) Conference on Cost-Benefit Evaluation of Environmental Protection Measures, the chairman noted at the outset, "cost-benefit analyses could not be considered

in isolation, as a unique methodology. They were part and parcel of a whole range of analytical tools that must be utilized to improve judgmental decision-making." He added, "It is necessary to begin with environmental impact assessments" (Participants Communication, UNEP/IG. 17/1, Restricted Distribution, June 14, 1979).

16 For example, what will it avail our innovators to develop products that may, through lack of coordination and price lags, compete with each other for rapidly dwindling raw materials, such as those the United States must also import? The Soviet Union, South Africa, and Zimbabwe (formerly Rhodesia) control more than two thirds of six minerals that are critical to current U.S. needs: 96.5 percent of chromium, 90.5 percent of manganese (without which the U.S. steel industry would be out of business), 99.7 percent of platinum, 74.6 percent of tungsten, 69.4 percent of nickel, and 69 percent of cobalt, according to a 1980 study by Amos Jordan and Robert Kilmarx, of Georgetown University's Center for Strategic and International Studies (*The Christian Science Monitor,* January 11, 1980).

17 Before long, the old paradigm of industrial "technology transfer" to the Third World and debates about whether we in the United States should let the Soviets and the Chinese have the benefit of our technology, may be reversed. The first World Congress on Social Prospects of the World Association for Social Prospects Study (AMPS), in Dakar, Senegal, which I attended, produced a Declaration of Dakar, one of whose articles related to the expunging of the term "technology transfer" from the UN language as a holdover of colonialism. Another example is that of China, from whom we may have to learn the technology of deploying the 7 million methane digesters that reprocess animal and human wastes into farm fertilizer. The digesters have improved sanitation in the farm areas by the use of technologies of anaerobic microbiological conversion.

In 1980, the Chinese had some 13 million more digesters and planned to have 70 million units by 1985, which would imply that two thirds of all China's rural households would use this biogas for heating and lighting by 1985 (*Soft Energy Notes,* December 1979, p. 77).

18 A major controversy between thermodynamics and economics will become a confrontation in the 1980s as the realities of the entropy law become more apparent. Already, the Department of Energy (which, more correctly, should be called the Department of Entropy) convened in August 1979 a conference on the second law of thermodynamics. However, Nicholas Georgescu-Roegen is still the scourge of the economics profession, and his papers are too scientific for most economists (the Editorial Committee of the *American Economic Association Journal* recently turned down a paper by Georgescu-Roegen with the effrontery of contending that it was "not rigorous"! in fact, his papers are too rigorous for economists to comprehend). An important treatment of the entropy law as an emerging corrective world view for industrial societies is *Entropy: a New Worldview* (Viking Press, 1980), by Jeremy Rifkin with Ted Howard. However, the entropy law should not become a new cosmology.

19 See, for example, "Where Genetic Engineering Will Change Industry," *Business Week,* October 22, 1979, p. 160.

CHAPTER 11

<hr>

The Future of Risk, Insurance, and Uncertainty

The socioeconomic transition now underway in all mature industrial societies is characterized by growing complexity, saturation of their traditional growth in rising social costs, and a "plateauing" of their historical trajectories of technological development. This transition process is still so little understood that despite its ubiquitous recognition in most futures research, it is still conceived, in rearview-mirror imagery, as an emerging "postindustrial" era. Daniel Bell, Herman Kahn, and others use this term to imply a shift of the industrial system to a service- and knowledge-based, tertiary economy.[1]

Alternative scenarios of the future of industrial societies include decentralization of economic power and activities and the spontaneous devolution of overgrown bureaucracies and centralized, capital-intensive technologies as well as disintermediation of overextended trading and exchange systems. They explore the concomitant shifts in the loci of control of societal functions both downward to local levels for more precise implementation, and upward to global and regional levels for conceptual realignment of planetary coordination of functions beyond the management of nations individually; e.g., environmental affairs, oceans, and space policies.

The first view of "postindustrial" life, typified by Daniel Bell, is essentially an extrapolative one: a relatively smooth transition based

This chapter is reprinted from *Best's Review*, May 1978, and from *Risk Management*, May 1978, with permission.

on and arising from the assumed increasing efficiency of existing agricultural, resource-extraction, and primary-production systems, i.e., continued growth along current pathways leading to a flowering of a tertiary, services economy. The second view involves major discontinuities, in which nothing less than a shift of basic paradigm (i.e., world view) is necessary to reconceptualize the situation. The historical, fossil-fueled growth pattern creates technological complexity that is unmodelable and therefore unmanageable. Unanticipated social impacts, environmental depletion, and proliferating bureaucratic attempts to regulate and coordinate the situation add to the general transaction costs, until the social costs of the system begin to exceed actual production. Since the social costs are added to the gross national product instead of subtracted, the GNP goes up, while inflation begins to mask the declining situation.

Alvin Toffler describes, in *Future Shock,* the social effects of a superindustrial future, and in *The Third Wave* (1980) he vividly portrays the new global turbulences it can unleash. But, as Michael Marien points out, in his *Societal Directions and Alternatives* (1976), other major interpretations of the "postindustrial" era, for example Belloc's *The Servile State* (1913), Arthur Penty's *Old Words for New; a Study of the Post-Industrial State* (1917), and *Prosperity and Security* (1938) by Ralph Borsodi, have predicted a collapse of industrialized societies due to the vulnerabilities of excessive interlinkage and other factors. Other thinkers have sounded a continual note of caution during the intervening years, including Karl Polanyi in *The Great Transformation* (1944), K. W. Kapp in *The Social Costs of Private Enterprise* (1953), the late E. F. Schumacher in *Small Is Beautiful* (1973), and more recently, Amory Lovins in his *Soft Energy Paths* (1977) and myself in *Creating Alternative Futures* (1978).

More important, the two-hundred-year development of industrialism constitutes an irreversible evolutionary process, and the notion of many economists that we can turn the clock back merely by such means as deregulation is a profound misunderstanding. The studies by Edward Denison, of the Brookings Institution, cited in Chapter 10, are a case in point. He correctly assessed the rising social costs of such increasing societal complexity and quantified these costs quite credibly as approximately 2 percent of GNP.[2] But Mr. Denison then draws an incorrect conclusion from his data: i.e., that we should try to focus on reducing regulations and their costs, rather than deal with the much more structural problem of identifying and modifying the particular types of socially and ecologically disruptive technologies that inevitably require regulation, and where necessary, replacing them with more socially and ecologically compatible technologies, with fewer impacts.

These transition processes, which late-stage industrial societies are experiencing, are analogous to a class of systemic change processes familiar to general-systems theorists as morphogenesis (i.e., structural transformations not inferable from the existing state of a system or its properties or variables). Morphogenesis can be studied in biology (e.g., the process by which a chrysalis turns into a butterfly) and in engineering, hydrology, and human and animal behavior. As Magoroh Maruyama has shown, there are two basic types of cybernetic systems: 1) steady-state systems, which maintain their structure over time by negative-feedback loops and deviation-damping, mutual-causal processes, and 2) morphogenetic systems, undergoing structural evolution and governed by positive-feedback loops and deviation-amplifying mutual-causal processes (see Fig. 11).[3]

What does all this mean to insurance companies and risk managers in general? We are aware of the turbulent, intensified risk environment in which insurers and risk managers operate today. For some years the industry has focused on such underlying causes as inflation and its effects on capital investment and all business decisions.

Another problem is what many refer to as "social inflation"[4]: the mushrooming of tort-liability problems, speculative claims, entrepreneurial lawyers, ever-more-pervasive third-party interpretations of product liability, medical malpractice, as well as the increasing "psychology of entitlement," which has led to huge awards for wholly new classes of risks, such as ski accidents and the now famous Connie Francis $2.5 million rape award against Howard Johnson's motel chain.[5] All this is familiar and has led to many predictable responses from insurers, including higher deductibles, higher premiums, encouragement of self-insurance, broader risk-management strategies, and the development and growth of captive insurance companies—as well as the increasing outright avoidance of much high-risk insurance and the wise refusal to underwrite more than a fraction of massive un-

STABLE, EQUILIBRIATING SYSTEM (morphostatic) (structurally stable), e.g., thermostat-controlled mechanical system, early agrarian or small-scale production economies (as conceived in market equilibrium supply-demand theories), reversible components and decisions

UNSTABLE, DIS-EQUILIBRIUM SYSTEM (morphogenetic) (evolving new structure), e.g., living, biological systems, human societies, large-scale sociotechnical economic systems, rapid innovation and structural evolution, many irreversible components and decisions

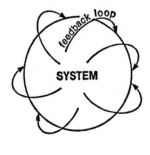

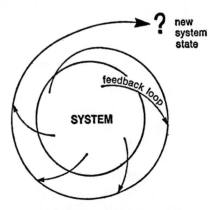

System internally dynamic, but stable structure maintained and governed by *negative* feedback loops.

System internally dynamic *and* structurally dynamic, governed by *positive* feedback loops, which can amplify small initial deviations, which sometimes break through thresholds and push the system to a new structural state.

Fig. 11 Simple Diagrams of Two Major Types of Cybernetic Systems

tried programs such as the ill-fated government swine-flu-vaccine program.

Such responses have done much to stabilize the situation. But they have, in turn, produced counterresponses, such as mandatory state auto-insurance pooling and municipal property-insurance pools such as the FAIR plans. These plans are now under broad attack from consumers and government officials for their self-rating and high premiums and what many see as a new form of "redlining" of many urban areas and their residents.[6] The socializing of risks in the health-care sector (now a staggering 10.5 percent of GNP) has led to widespread abuse and an inflation-ridden quagmire. The insurance mess is fueling citizen protest organizations such as Massachusetts Fair Share, which not only fights high premiums but challenges the underlying concept of private insurance: the investment of premiums as the source of income to pay future claims.[7] No wonder, in the light of today's situation, that some insightful insurance executives are beginning to see the exhaustion of the logic of some of our current industrial assumptions. Roy R. Anderson, vice-president of Allstate, noted in a speech on "The Future of the Insurance Business in a Changing Society": "We have broken loose from Western industrial society—our institutions are now changing—and a new society will emerge, probably within the next couple of decades. It is this concept of changing civilization that explains why some of the problems we have encountered in our systems of insurance seem to defy solution, that is, liability insurance, health insurance, and Social Security . . . in each of them I see evidence of a systemic breakdown, of the possibility of an impending traumatic change."[8] (Fig. 11 provides a context for interpreting this change.)

Under this growing pressure, it is fashionable for insurers and risk managers to emphasize their problems with tort liability and mandatory assignments and what they often see as the increasingly unrealistic behavior of the public and its elected officials in demanding ever wider, "pie-in-the-sky" protection from not only risk but "pain and suffering" to the point of denial of the daily uncertainties of life.[9] I believe that while there is truth in this interpretation, it blinds the industry to the need to recognize and address a set of much deeper, more-structural problems: those involving systemic transition of industrial societies and their particular, capital-intensive, socially and ecologically disruptive technologies and the whole class of novel vulnerabilities they generate. These range from the systemic inflation and "stagflation" impasse and the unmanageable complexity I termed "the entropy state" to the unsolved theoretical problems of modeling the new class of "ticking time bomb" risk probabilities, i.e., those ever more widely displaced in space and extensively delayed in time, for example current research in recombinant DNA (dioxyribonucleic acid) and the resulting commercial development of new organisms that, according to Dr. L. Cavalieri, of the Sloan-Kettering Institute, might unleash dangerous mutant viruses upon human populations with no immunity.[10] Companies already engaged in this research include Miles Laboratories, Eli Lilly & Company, Hoffmann-La Roche, and the Upjohn Company. There is, of course, increasing pressure to underwrite such enterprise, as well as many other new technologies with the inherent potential for producing major and sometimes, as is the potential with DNA recombinance, irreversible catastrophes.

Today's most ubiquitous "ticking time bomb" risks have been sneaking up on us for almost fifty years in the proliferation of technologies based on petrochemicals, from vinyl chloride and polychlorinated biphenyls to ever-more-exotic food, drug, and cosmetic compounds, polymers, plastics, pesticides, and aerosols and the resulting increase in the burden of carcinogens in the air, the water, and the food chain.[11] Federal agencies charged with regulating toxic substances—the Food and Drug Administration, the Environmental Protection Agency, and the Occupational Safety and Health Administration—are attempting to coordinate their authority to regulate the some thirty substances already known to cause cancer in humans and to launch a program of confirmatory identification of an additional 2,156 suspected carcinogenic agents.[12] The federal government, already under heavy fire for its bias toward funding research to "find a cure" for cancer, has been forced to shift to a preventive strategy and now admits that between 60 and 90 percent of the cancers that kill a thousand Americans a day are caused by substances in the environment.[13] However, the writing on the wall is clear for insurers of both life and casualty: the proliferation of carcinogens in workplaces, food, water, and air, together with tobacco smoke, increased exposure to radiation (which alone accounts for twenty-two thousand cases of leukemia per year),[14] as well as the countless additional stresses of industrial life, lack of exercise, and poor diet is now in many subtle and cumulative ways compromising the health of millions of citizens of such societies, and altering toward greater unpredictability the etiology of disease and the accepted actuarial calculations of life expectancy.

The 1977 banning of the fire-retardant chemical Tris[15]; the FDA's initiation of procedures to remove from the marketplace such sedatives as Cope, Compoz, Miles Nervine, Tranquim, and Quiet World because they incur "risks for users with no demonstrated medical benefits"[16]; as well as the billions of dollars' worth of damage suits filed against drug companies, doctors, and hospitals by adeno-cancer victims of diethylstilbestrol (DES), a now proven totally ineffective drug thought to inhibit miscarriages[17]; and the incalculable insurance exposure of Hoffmann-La Roche Company, of

Switzerland, stemming from the dioxin-release incident in Seveso, Italy, where claims may continue emerging for decades, are just a few specific cases in point in the drug industry. The latest horror story is that of human exposure to microwave radiation from TV sets, microwave ovens, CB radios, burglar alarms, garage door openers, and other consumer products, as well as workers' exposure to such radiation leaking out of video display terminals in banks and in airline and hotel reservation desks.

The Environmental Protection Agency has found that some highrise city office buildings are constantly exposed to radiation several times greater than that the Soviets beamed at the American Embassy in Moscow.[18] Blood disorders, hearing loss, and dizziness as well as a rare form of cataract on the rear of the human eyeball—all are now being documented in operators of microwave radar equipment, including air traffic controllers, some of whom have early-stage blind spots that already impair their vision in tracking passenger planes on their radar screens. Microwave disability victims are at last being awarded benefits by the Veterans Administration, and many more claims are now being filed under workmen's compensation, according to Paul Brodeur, in *The Zapping of America* (1978). The more pervasive nature of the microwave radiation problem involves defense radar installations all over this country and the world, as well as the reported Soviet weather-modification experiments—thought by some to be affecting U.S. weather patterns—which involve microwave technology based on theories of the Yugoslavian-born inventor Nikola Tesla.[19]

The capital intensity and scale of many of today's energy technologies create catastrophic levels of risk, whether nuclear-fission power plants, liquefied-natural-gas (LNG) tankers, or oil transported in supertankers. The essential uninsurability of LNG stems from the fact that, according to the report issued by the U. S. Congress Office of Technology Assessment in October 1977, "The igniting of a spill of this extremely volatile substance is quite beyond the capabilities of any known firefighting methods to extinguish it."[20] The weak link in such a vulnerable energy supply is not so much in LNG tanker construction as in poor traffic-control methods in busy harbors and waterways. The recent collisions of oil supertankers in busy shipping lanes highlight their vulnerabilities.[21] In addition, some naval engineers have voiced the possibility that, technologically speaking, these ships have pushed too far too fast, already ominously suggested by the buckling that many of them have incurred before or after launching and on sea trials. Noted marine scientist and presidential adviser Dr. Edward Wenk, Jr., states in *The Politics of the Ocean* (1972) that "the very use of supertankers needs reevaluation."[22] Noël Mostert notes in *Supership* (1974) that for the largest class of tankers, ultra large crude carriers, insurance now accounts for 70 percent of the operating costs.[23] He contends that these ships may be fatally flawed as a species and that little prior thought was given to the fact that they also offer an entirely new and incalculable set of fire and explosive hazards. Likewise, undersea oil-drilling blowouts incur such enormous liability that in the case of the April 1977 North Sea blowout of the Phillips Petroleum drilling platform, the stock of the company dropped almost five points on a 302,800-share turnover in one day on the New York Stock Exchange, until Phillips assured its investors that insurance coverage was adequate to meet all claims.[24]

The point is, of course, as stated by Princeton University statistician Lawrence S. Mayer in the New York *Times* (March 20, 1979): "First, the risk of certain acts, such as being killed by a traffic accident, can be estimated by a simple division involving the number of deaths and the total number of passengers. The risk of other acts, such as building nuclear power plants, cannot be estimated. The claimed risks are merely the subjective opinions of people on one side of the issue or the other.

"Second, the acceptability of the risk of any act depends on the expected benefits of the act. People travel by car in spite of the risks, not in order to expose themselves to risks. It would be irrational to apply a single risk standard to all societal acts independent of their benefit.

"Third, risks are not additive. As citizens are exposed to more acts that risk their lives, the probability that they will not be killed by one of the acts decreases much faster than might be expected. If a citizen is exposed to six acts, each of which risks his life with probability .1 (i.e., one in ten), the probability of being killed by one of the acts is .5 (i.e., five in ten)."

Similarly, incalculable risks inhere in nuclear power and were recognized over a decade ago in their socialization via the Price-Anderson Act, which was extended in 1975 to cover these uninsurable risks until 1987. It is interesting to note that the extension of the Price-Anderson Act was based on the now largely discredited Nuclear Regulatory Commission report known as WASH-1400, which calculated the probabilities of a range of nuclear-power-plant failure modes using independent analyses of separate components, rather than the more likely scenarios of systemic interdependence in actual operating conditions.[25] WASH-1400 used fault-tree analysis methodology, which the National Aeronautical and Space Administration had discarded ten years before, after it was found to have underestimated the failure rates of the ill-fated Apollo engines by over forty to one. The earlier, Brookhaven study, WASH-740, in 1965, was more forthright. It stated, "There is no objective quantitative means of assessing that all possible paths leading to catastrophe have been recognized and safeguarded, or that safeguards will in every case function as intended when needed."[26]

WASH-740 was withheld from the public until a suit under the Freedom of Information Act forced the Atomic Energy Commission to release it in 1973. In the WASH-1400 study, which persuaded Congress to extend the federal underwriting of nuclear power plant risks, the probability of reactor meltdown was calculated at one in twenty thousand per reactor year. But, as pointed out in the review by the Union of Concerned Scientists, this was based on a mistaken application of statistical theory (the use of the median in log-normal distributions), which introduced a factor-of-2.5 understatement.[27] Correction of this error alone increases the probability to one in eight thousand per reactor year, in addition to the understatement errors of the fault-tree analysis and other problems caused by overoptimistic assumptions. Thus, private insurers sensibly avoided shouldering incalculable risks now borne by an unsuspecting public.

Similarly, taxpayers now shoulder an increasing burden of investment risk in nuclear construction through tax credits to utilities and new subsidies such as passing forward to utility customers the construction work in progress (CWIP) charges on their bills. These construction investments are now at a standstill, not primarily due to the opposition of their environmentally-aware investor/consumers but due to rising capital costs, engineering problems, and unforeseen snags such as sites chosen at earthquake fault lines or those where droughts have reduced cooling water available to below operating requirements.[28] Such was the case with the $5-billion reactor now at a standstill in California's San Joaquin Valley, which, if it is ever built, will compete bitterly with irrigation farmers for its required 20 billion gallons of water annually.[29]

The point of elaborating on such cases is merely to illustrate that in today's mature industrial societies, whole classes of their capital-intensive technological trajectories are now encountering systemic diseconomies of scale and social costs pushing boundary conditions rooted in the basic laws of thermodynamics and natural systems.

If all these types of literally uninsurable risks are viewed as stemming from a specific class of technologies now encountering visibly diminishing returns, we see a deeper structural pattern affecting risk management. These technologies have recently and rapidly evolved from the much slower development rate of industrial production methods over the past two hundred years. As Dr. Orio Giarini, secretary-general of the International Association for Risk and Insurance Economics Research, in Geneva, has so well shown,[30] the discipline of economics has neither captured this increasingly dynamic technological development in its equilibrium models of supply and demand nor has it noticed that these rates of innovation along current pathways are meeting diminishing returns.

Let us now refer to Figure 11 which shows the two basic types of cybernetic systems which help us to interpret the behavior of the late-stage industrial societies with which we are dealing. Economics is still undergirded by simple models of mechanical locomotion, such as the equilibriating model of supply and demand, however baroque the input-output and econometric models spun from such concepts appear to the uncritical eye. The equilibrium model of supply and demand undergirding the free-market, laissez-faire philosophy is now a too atomistic and linear concept, lacking the systemic consideration of a host of new variables its practitioners still consider "exogenous." As Nicholas Georgescu-Roegen showed in *The Entropy Law and the Economic Process* (1971), economics still uses arithmetical and reversible models of locomotion to portray many components of economic systems. Inasmuch as these components can be studied separately, such microeconomic views may be adequate. However, when one views economic processes in extended time/space dimensions, one sees that they are not separate and reversible processes but evolve symbiotically and structurally, as they have for the past two hundred years of the industrial revolution. This misapplication of micro theory to the interacting macroeconomic system introduces order-of-magnitude errors now accounting for the disintegration of macroeconomic management in all mature industrial countries.[31] Economists still conceive of this rapidly evolving, disequilibrium macro economy as if it were still an equilibrium system that could be managed with the uncomplicated hydraulics of aggregate supply and demand, as discussed in Part One.

Thus, while economics has begun to learn from general systems theory that industrial societies are nonlinear, cybernetic systems governed by thousands of feedback loops, they are naturally still more at home dealing with equilibrium, "thermostat"-type systems that maintain their structure by negative-feedback loops (Type 1) than with the now emerging understanding of disequilibrium systems governed by positive-feedback loops (Type 2), which are in the process of irreversibly transforming themselves into improbable new structures. If, as I contend, late-stage industrial societies are of the second, morphogenetic type, in constant disequilibrium and evolving toward, *by definition*, unpredictable new states, this would require a completely new theoretical basis for risk management.

One of the conventional assumptions underlying risk-management probability calculations (including that of the nuclear safety study WASH-1400) is that the greater the catastrophe the less likely its occurrence. This may be warranted in static or slowly moving, equilibrium systems, in which deviations are *damped* by negative feedbacks. However, as Maruyama has shown, in morphogenetic systems, deviation-*amplifying*, positive-feedback, mutual-causal processes govern. In such systems, the normal laws of causality must be revised to state that similar conditions may result in dissimilar outcomes. Maruyama adds, "It is important to note that this revision is made without the introduction of indeterminism and probabilism. Deviation-amplifying mutual-causal processes are possible even within a deterministic universe, and make for the revision of the laws of causality even within the determinism. Furthermore, when the deviation-amplifying mutual-causal process is combined with indeterminism, here again the revision of the basic law becomes necessary. The revision states, '*A small initial deviation, which is within the range of high probability, may develop into a deviation which is very improbable within the framework of probabilistic unidirectional causality.*' "[32]

It is this set of deviation-amplifying probabilities that set apart the class of capital-intensive, excessively interlinked, high-risk technologies I have mentioned. Under such conditions, it might well be that the greater the accident the greater the chance of its occurrence! The point is that we do not know, and the theoretical foundations of statistical probability modeling are now inadequate. *Until they are reformulated, insurance in this sphere has become indistinguishable from gambling.* Research in this area is proceeding in conditional probability modeling, cross-impact analysis, technology assessment, as well as time-dependent conditional-probability modeling. But time sequence is the key to determining absolute, conditional, and joint probabilities, and there is still a need, as pointed out by Roy Amara, president of the Institute for the Future, to develop a calculus for such time-sequence interactions.[33] As M. McLean, a technology forecaster, noted in *Futures* (1976): "By concentrating attention on the manipulation and refinement of probability estimates, the cross-impact approach lost sight of the fact that such estimates can only be temporary substitutes for an understanding of the *causal structure* of socioeconomic processes" (italics added). Since the proliferation of cross-impacts characterizes such socioeconomic processes, we must note the caveat of Professor Murray Turoff, of the New Jersey Institute of Technology, in *Technological Forecasting and Social Change* (1972), that "traditional probability relationships are irrelevant to cross-impact studies."[34]

A key to the time-sequence problem is the fact that in complex, nonlinear systems, distinguishing between "cause" and "effect" becomes almost impossible—the now familiar modeling and decision problem of proliferating "chicken-and-egg"-type situations. For example, as mentioned, risk managers view inflation as a "cause" of their problems, but inflation is also an "effect" of deeper structural "causes," which in turn are "effects" of yet other "causes," and so on. This illustrates the inadequacy of the simple trade-off view of inflation in economic theory and the conceptual confusion of believing that we understand a phenomenon by naming it! Economics is not a substitute for thought, computation cannot substitute for conceptualization, nor do correlations imply causality.

The most spectacular work on modeling morphogenetic systems is growing out of the qualitative mathematics of René Thom's *Structural Stability and Morphogenesis* (1972, English translation 1975) and his catastrophe theory. This has led to an explosion of interest, and later work in this field by others is now available in *Catastrophe Theory*, edited by E. C. Zeeman.[35] Other "qualitative modeling" based on biological and ecological theories, rather than on inert, physical theory, is that of such systemic theorists as C. S. Holling, Roy A. Rappaport, H. H. Pattee, Gregory Bateson, Erich Jantsch, Ingemar Falkehag, and the late Conrad Waddington, as well as the brilliant theories of "order through fluctuation" in dissipative structures of physicist Ilya Prigogine.[36]

Time and its flow is also the key problem in modeling morphogenetic systems. As Maruyama points out, the error in current laws of causality involves the assumption of the unidirectional, orderly flow of time, giving us the underlying "unidirectional causal paradigm" and the "random process" paradigms of Western science and culture. Mr. Maruyama describes the unidirectional causal paradigm as "that which has become fashionable since the discovery of indeterminism and informational indeterminabilism in quantum mechanics . . . and remains fashionable in the philosophy of science and sociology, even though physical and biological sciences have already moved out of it. According to this paradigm, there is a one-way flow of influence from cause to effect, but occurring with some probability rather than with certainty." Effect can be predicted from cause with some probability, as can cause be likewise inferred from effect.

Mr. Maruyama notes that the "scientific method" consists in discovering the probability distribution and in establishing the limits of accuracy of observation. He adds, "Multi-variate statistical analysis, correlation analysis, regression analysis, etc. can be attempted for phenomena not completely amenable to laboratory experimentation, e.g., weather or tropospheric scattering of electromagnetic waves. If statistical relations between two variables are found, this may be due to one of the following unidirectional causal relations: (a) one causes the other with some probability, either directly or through other, intermediate variables, or (b) both are influenced by some common cause, with some probability. However, the causal direction cannot be known from statistics alone and must be determined by logical considerations" (i.e., a priori).[37] Mr. Maruyama describes the second paradigm as the "stochastic paradigm," or the random-process paradigm, thus: "This paradigm is due to the development of thermodynamics in the nineteenth century, based on the logic of coin tossing, where each toss is considered to be independent from other tosses. Thus the outcome of the first toss should not influence the outcomes of the subsequent tosses. This paradigm is similar to the distribution of temperature in thermodynamics and the second law, of increasing entropy: the higher the degree of homogeneity in temperature the higher the entropy. There is some degree of continuity in the sense that the state of the system at a given time is related to the state of the system at a previous time—related with a certain probability distribution, i.e., this type of change is termed 'stochastic process.' "[88]

Mr. Maruyama, a Japanese, can clearly point out such contrasting, but not necessarily mutually exclusive, paradigms of time and causality in Western vis-à-vis other cultures. Similar insightful contrasting of paradigms is expounded by Erich Jantsch in *The Self-Organizing Universe* (Pergamon Press, 1980). In today's situation it would appear that risk managers must first be aware of the extent to which paradigms of thought are the conceptual tools underlying all our theoretical models, and secondly, in dealing with various probability problems, determine how to select fitting paradigms for developing alternative theoretical approaches. In Figure 12, I have adapted Maruyama's Simplified Table of Three "Pure" Paradigms from his paper in *Cybernetica* (1974) entitled appropriately "Paradigmatology and its Applications."[89]

Traditional Western logic is still most often grounded in the first two paradigms described above. However, the needed research for prudent risk management during the transition decades ahead for late-stage industrial societies will probably involve much deeper explorations of the behavior of cybernetic and morphogenetic systems. They are best viewed from the vantage point of Mr. Maruyama's Paradigm 3, the "mutual-causal paradigm" portrayed in column three of Figure 12. This newer paradigm is derived from Norbert Wiener's 1949 formulations of cybernetic processes and Stanislaw Ulam's mathematical formulation for morphogenetic systems in 1960, which showed that complex patterns can be generated by means of simple rules of interaction. The ongoing research on morphogenetic systems mentioned earlier breaks further ground within this same paradigm.

Meanwhile, in the real world, risks continue to proliferate, despite the lack of needed theory. Pragmatically, as more and more classes of risk fall into the category of the theoretically uninsurable, pressure

SIMPLIFIED TABLE OF THREE "PURE" PARADIGMS,
after M. Maruyama.

	1. Unidirectional Causal Paradigm	*2.* Random Process Paradigm	*3.* Mutual Causal Paradigm
Science:	traditional "cause" and "effect" model	thermodynamics; Shannon information theory	post-Shannon information theory
Information:	past and future inferable form	information decays and gets lost; blueprint must contain more information than finished product	information can be generated; nonredundant complexity can be generated without preestablished blueprint
Cosmology:	predetermined universe	decaying universe	self-generating and self-organizing universe
Logic:	deductive, axiomatic	inductive, empirical	complementary
Perception:	categorical	atomistic	contextual
Knowledge:	belief in one truth, if people are informed they will agree	why bother to learn beyond one's own interest?	polyocular: must learn various views and take them into consideration
Methodology:	classificational, taxonomic	statistical	relational, contextual analysis, network analysis
Research hypothesis and research strategy:	dissimilar results must have been caused by dis-similar conditions; differences must be traced to conditions producing them	there is a probability distribution: find out probability distribution	dissimilar results may come from similar conditions due to mutually amplifying network; Network analysis instead of tracing of the difference back to initial conditions in such cases
Analysis:	preset categories used for all situations	limited categories for individual's own use	changeable categories depending on situation
Assessment:	"impact" analysis	what does it do to *me?*	look for feedback loops for self-cancellation or self-reinforcement

Fig. 12

mounts to socialize them at higher levels in the society, from municipal property insurance pools to auto insurance assignment at the
state level, and finally to the proliferating demands for the federal
government to act as "insurer of last resort." We see a new "tragedy
of the commons"[40] scenario unfolding with these demands coming
from a widening array of sectors of our national life; from the
forerunners in nuclear power liability, federally insured bank deposits, workmen's compensation, and Social Security, to the new
pressures for socializing comprehensive health care and medical malpractice, compensation of crime victims, underwriting municipal debt
as in the case of New York City, bailing out faltering corporations
and real estate investment trust speculators, to the many enormous
capital investment schemes for energy development from synthetic
fuels to nuclear fusion. This does not even include the costs of moth-
balling aborted energy projects, such as the Clinch River Breeder Reactor, for which $80 million was recently appropriated to close it
down, or the horrendous costs of decommissioning some of the nuclear fission light water reactors now becoming obsolete.[41] Recent reports, on actual decommissioning, in *Electrical World* suggest that
such costs in some cases can exceed the initial investment cost of the
plants themselves.[42]

Whenever risks are socialized at successively higher levels in society, some time is bought to work out solutions, but at the trade-off of
greater systemic vulnerability to cascading breakdowns due to the
greater interlinkage, as well as the eventual result of pathological
diffusion of responsibility. At the private, local, and individual level
there is stability in the widely dispersed, large universe of small,
decoupled risks. Each time such risks are socialized at a higher level
of societal management, there is a greater aggregation of risks into
fewer, larger units; i.e., each universe of risks grows smaller as the
agglomerations of risks grow more unwieldy. Finally, at the federal
level, there are few checks on the simultaneous socializing of diverse
risks, and few conceptual overviews are available to legislators as to
the total federal exposure, other than that inferable from monitoring
total debt, projected federal budget deficits, and general inflation
rates. From such ominous statistics, we might calculate what proportion of these statistics might be accounted for separately as a "socialized risk fraction" or indicator. However, until government statisticians produce such a new index, we might err on the side of caution

and assume that we have already mortgaged a large portion of our foreseeable "social risk future." Such a view, of course, is reflected finally at the international level, where the tribulations of many mature industrial countries undergoing similar systemic transitions are reflected in the instabilities of the international monetary system itself. The "risk aggregation buck" stops eventually at the level of the national currency. Today, with our energy habit still uncurbed and our grandiose technological plans blinded by lack of theory to looming future scenarios of rapidly eroding returns, one might say in a very real sense that part of the declining value of the U.S. dollar is attributable to the excessive socialization of risk. We may have reached the point where we are blinded to this fact, since, to paraphrase an old adage: everybody's risk has become nobody's risk. At least, the danger is that nobody is making it his or her business to assess it systematically.

Lastly, if any of the foregoing analysis seems relevant, what might risk managers do about it? I will dare to offer some modest prescriptions:

1. Hire some thermodynamicists to check your economists' models (particularly regarding their assumptions as to future capital productivity, investment decisions, and inflation causes beyond the Phillips Curve trade-off interpretation).
2. Fund through professional and trade associations further research on the behavior of disequilibrium and morphogenetic systems.
3. Examine innovation options in inherently lower-risk, decentralized technological modes: e.g. solar, wind, and geothermal energy, conversion of electric power to co-generation with district heating, as well as recycling systems, bioconversion of wastes, and investments in energy conservation. One insurance company has taken this to heart. New York Life Insurance Company's new office building is warmed with the waste heat from its computers (New York *Times,* January 27, 1979).
4. Continue refusing to underwrite unnecessarily hazardous technologies, whether nuclear power, LNG tankers, or recombinant DNA research, or ill-advised government crash schemes such as the Project Independence energy scheme (later proved infeasible because of unrealistic capital requirements) or the swine-flu fiasco. Forcefully bring to the attention of the public and their leg-

There are half a million men and women in prisons around the world for the simple crime of disagreeing with their governments.

From South Africa to the Soviet Union, from Brazil to Korea, authoritarian regimes persist in the barbarian practice of jailing, often torturing, their citizens not for anything they've done, but for what they believe.

These prisoners of conscience have only one hope — that someone outside will care about what is happening to them.

Amnesty International has helped free over 14,000 political prisoners by marshaling world public opinion through international letter-writing campaigns.

Your pen can become a powerful weapon against repression, injustice and inhumanity.

Join with us today in this important effort.

Because if we do not help today's victims, who will help us if we become tomorrow's?

This powerful weapon can help free prisoners of conscience all over the world.

Prepared by Public Media Center, San Francisco.

Plate 36

We Don't Need Food For Thought We Need Facts For Action

If we really want a world without hunger we need facts, not fancy. Unfortunately, there are a lot of myths about hunger which actually inflate the problem, instead of helping us solve it. Here are some myths you've probably heard—and the facts to match:

MYTH #1: People are hungry because there's too little food and land to go around.

The fact is, there's enough grain to feed everybody on the planet 3000 calories a day. And no country lacks ample food-producing resources of its own—even so-called "basketcases" like Bangladesh. Scarcity is an illusion fostered by the concentration of control over food and land in the hands of a few. Large landholders (often the least productive) grow cash crops for export instead of planting food first; hunger is the result.

MYTH #2: We can eliminate hunger by redistributing food.

The fact is, food distribution reflects the distribution of control over resources like land and credit inside a society. The poor go hungry in India for the same reason they go hungry in America: they're cut out of the economy. Too few people control the land, what it grows, and where it goes. Redistributing food solves nothing. Only by redistributing control over food-producing resources can we build the basis of food security.

MYTH #3: Global interdependence is the ultimate answer.

The fact is, exporting cash crops from the Third World doesn't benefit the hungry in the least. Workers on Phillipine banana plantations receive less than 2¢ out of every dollar spent by Japanese banana consumers. Multinational corporations and domestic elites are the only winners in the world agricultural trade. And they take their profits in cash, not in food for the hungry poor.

MYTH #4: Hunger is a contest between the "rich world" and the "poor world."

The fact is, talking about "rich" countries and "poor" countries obscures the truth that every country is rich at the top and poor at the bottom. The hungry aren't our enemies; they're our allies in the struggle to democratize control over food resources at home and abroad. In America, the top 50 food corporations reap more than 90 percent of the profits in the entire food industry. The same companies are taking control of land and food in the Third World.

MYTH #5: Hungry people are too weak to help themselves.

The fact is, this is the most destructive myth of all. 40 percent of the Third World has freed itself from famine and hunger in our lifetimes. The poor aren't passive or resigned; they're blocked by political and economic structures which have-frozen the status quo of hunger. And right now, US government and corporate policies are hurting the poor, not helping.

So don't skip lunch.

You need all the strength you can get to help stop devastating US political, military, economic and corporate interventions which shore up regimes at war with their own people's fight for food. And to support worker-managed food alternatives battling the handful of corporations taking control of our land and food here at home.

Get more facts for action. Begin by reading **FOOD FIRST: Beyond the Myth of Scarcity** (Ballantine 1979) by Frances Moore Lappe and Joseph Collins with Cary Fowler.

And become a Friend of the Institute for Food and Development Policy. The Institute is doing hard research on world hunger, unsupported by government. We don't have to defend mistaken policies based on myths. Your tax-deductible contribution of $25 or more entitles you to all Institute publications at half-price. Contribution of $100 or more brings you the same publications absolutely free.

Our address: **Institute for Food and Development Policy, 2588 Mission Street, San Francisco, CA 94110.**

Plate 39

STOP TOURISM

make where you *ARE* paradise

Plate 40

islators the reasons for refusing to privately insure such risks, and explain that neither should the government underwrite them, in spite of the railroading and lobbying by special interest groups and their "research."

Finally, it is encouraging to note the self-adjusting, healing mechanisms already at work in late-stage industrial societies. They include the trend toward decentralization of technologies, urban areas, and their populations typified by the "small is beautiful" movement. There is a new surge of interest in greater personal self-reliance, autonomy, and responsibility, as documented in this book and elsewhere. And as Ronald Inglehart has shown in *The Silent Revolution* (1977), similar movements exist in most mature industrialized countries, where the diffusion of "postmaterialist values" is matched by the growing political skill of these movements. Scenarios of spontaneous devolution, in which formerly delegated authority is recalled to more functional levels and relocalized, should be welcomed rather than feared. Societies with smaller-scale enterprises may prove more systemically efficient, and the recent trend toward self-employment (10 percent of the increase in employment during the fiscal year 1977–78 was due to this factor), as well as to neighborhood economic revival, the mushrooming of cooperatives (50 million Americans now belong to some form of co-op), and the growth of "self-insurance" implicit in the phenomenal popularity of the health, fitness, and better nutrition movements, may help in the transition of today's crisis- and conflict-ridden industrial societies into saner, healthier, "postindustrial" patterns for the future.

NOTES – CHAPTER 11

Since the paper from which this chapter was derived was first presented to the American Risk Management Association's meeting in New Orleans, April 1977, the whole question of risk analyses, and what are and are not "acceptable risks" has taken on new urgency. Lloyd's of London began feeling the results of past acceptance of too many improperly assessed risks, such as the debacle over their insuring of the U.S. computer-leasing companies without taking into account the rapidity of technological obsolescence of each generation of computers and their liabilities to the leasing companies who had to replace them with new models. In addition, Lloyd's always prided itself on taking on the largest commercial risks, from the old days of insuring shipping to its recent overcommitment to some of the new, incalculable risks I described, involving "mega-risk" technologies, whether jumbo jets, liquefied-natural-gas tankers, large crude-oil carriers, nuclear power plants, or, one that I did not mention, which Lloyd's took on for the RCA Corporation: its $77-million communications satellite, which disappeared off Cape Canaveral, Florida, in December 1979. Lloyd's new troubles and internal bickering within its syndicates of insurers over its huge losses and who was to blame, led to the stepping down of Chairman Ian Findlay after only two years, and to official recommendations for government regulation of Lloyd's operations (*The Christian Science Monitor*, January 30, 1980).

Also since I wrote this paper, the nuclear-power insurance industry has felt firsthand the consequences of the faulty probability calculations on which their liabilities and those borne by taxpayers rested. Nuclear Insurers, of Farmington, Connecticut, the largest of the U.S.A.'s three nuclear-insurance pools, has experienced substantial rate increases and new inspection procedures in the wake of Three Mile Island. Thus, insurers who had looked to high-risk underwriting of what they assumed were progressively improbable accidents turned from the profitable operation they expected, to a new world of increasingly unmodelable uncertainty (New York *Times*, February 24, 1980). The politics of energy risk analysis was typified by the furor over the Canadian Atomic Energy Agency's fraudulent Inhaber report (see Chapter 6 and *Soft Energy Notes*, December 1979).

The debate became acrimonious as the assumptions under risk-taking were revealed, particularly that there *was* a difference between known risks (historical data) and future uncertainties that did not fit the traditional probability theories, as mentioned. The respected British science magazine *Nature* was moved to editorialize in its November 30, 1978, issue: "Known risks, such as car accidents —where risk is simply calculated from past events [are fundamentally distinct from] unknown risks—such as terrorists taking over a fast breeder [nuclear reactor]—which are matters of estimating the future." Economic theory labors under similar delusions, since all its data are historical and it cannot model expectations.

Finally, just as there is excess liquidity in the banking system and not enough good investment risks, so in insurance there is excess capacity and not enough insurable risks to go around (*The Guardian*, February 21, 1980).

[1] Daniel Bell, *The Coming of Post-Industrial Society*, Basic Books, 1973.

[2] *Forbes*, March 6, 1978.

[3] *American Scientist*, Volume 54, No. 2, June 1963.

[4] *Business Week*, "The Overload on the Nation's Insurance System," September 6, 1976, p. 46.

[5] *Forbes*, "Who Is Raping Whom?", September 1, 1976, p. 61.

[6] New York *Times*, "High-Risk Urban Insurance Under Attack in State," April 24, 1977.

[7] *Just Economics*, Volume 5, No. 4, "Fair Share Stops FAIR Insurance Increase," May 1977.

[8] Roy R. Anderson, "The Future of the Insurance Business in a Changing Society," speech before Alex Brown Conference, February 23, 1977.

[9] *Business Week*, "The Way to Ease Soaring Product Liability Costs," January 17, 1977, p. 62.

[10] Ted Howard and Jeremy Rifkin, *Who Should Play God?* Dell Publishing Co., 1977.

11 New York *Times,* "Surveillance Widens for Cancer Research," May 11, 1977.

12 New York *Times,* "Agency Plan to Identify and Control Toxic Chemicals," March 5, 1978.

13 Ibid.

14 New York *Times,* "Panel Criticises U. U. on Radiation Hazard" (General Accounting Office Report to the Senate Governmental Affairs Commission), December 23, 1977.

15 New York *Times,* "Court Orders Chemical Concerns and Stores to Share TRIS Burden," May 4, 1977.

16 H. E. W. News Release, June 21, 1977. U. S. Dept. of Health, Education and Welfare.

17 *Time,* "Taking DES To Court," May 9, 1977, p. 44.

18 *New Times,* "The Air Pollution You Can't See," May 6, 1978, p. 64.

19 *The Application of Tesla's Technology in Today's World.* Investment research report by Lafferty, Harwood & Partners, Ltd. Members Montreal Stock Exchange, January 27, 1978.

20 U. S. Congress Office of Technology Assessment, OTA-O-53, September 1977.

21 New York *Times,* "Supertankers Collide Off S. Africa and Catch Fire," December 17, 1977.

22 Edward Wenk, Jr., *The Politics of the Ocean,* University of Washington Press, 1972, p. 419.

23 Noël Mostert, *Supership,* Warner Books, 1975, p. 168.

24 New York *Times,* April 26, 1977.

25 Nuclear Regulatory Commission, WASH-1400, Reactor Safety Study, October 1975.

26 *The Risks of Nuclear Power Reactors,* Study Dir. Dr. Henry W. Kendall, Union of Concerned Scientists, Cambridge, Mass., August 1977.

27 Ibid.

28 See, for example, *Power Plant Performance,* Council on Economic Priorities, New York, 1976.

29 New York *Times,* "Western Drought Imperils Building of World's Biggest Nuclear Plant," May 9, 1977.

30 Orio Giarini, "Economics, Vulnerability and the Diminishing Returns of Technology," *The Geneva Papers on Risk and Insurance,* No. 6, October 1977.

31 Hazel Henderson, *Creating Alternative Futures,* Part 1, "The End of Economics," G. P. Putnam's Perigee Books, 1978.

32 Magoroh Maruyama, "Paradigmatology and Its Application," *Cybernetica* No. 2, 1974.

33 Technological Forecasting and Social Change, quoted in "Subjective Conditional Probability Modelling," Mitchell and Tydman, Volume 11, No. 2, 1978, p. 134.

34 Technological Forecasting and Social Change, An Alternative Approach to Cross-Impact Analysis, M. Turoff, No. 3, 1972, p. 311.

35 E. C. Zeeman, *Catastrophe Theory 1972–1977,* Addison-Wesley Publishing Co., 1978.

36 See, for example, contributions to *Evolution and Consciousness: Human Systems in Transition,* eds. Erich Jantsch and Conrad Waddington, Addison-Wesley, 1976.

37 Maruyama, op. cit., p. 145.

[38] Maruyama, op. cit., p. 146.

[39] Maruyama, op. cit., p. 143.

[40] The tragedy of the commons refers to the parable of the use of a common grazing pasture in feudal England, by all the peasants for grazing their sheep. It was not long before some peasants discovered that it was possible to increase the number of their own animals grazing the commons. However, once all the rest began to follow suit, the commons became overgrazed and was soon destroyed for everyone. Garrett Hardin cast this in rigorous terms in his much-footnoted article in *Science*, December 13, 1968, p. 1243, entitled "The Tragedy of The Commons."

[41] New York *Times*, "Carter Signs Appropriations Bill: Will Close Clinch River Project," March 8, 1978.

[42] Gordon D. Friedlander, "De-Commissioning Nuclear Reactors," *Electrical World*, February 15, 1978, pp. 44–48.

CHAPTER 12

Science and Technology:
The Revolution from Hardware
to Software

Today all mature industrial societies have already reached their conceptual limits to growth, long before the actual exhaustion of their physical resources. We face, first and foremost, a metaphysical impasse that now impedes our efforts to create alternative technological futures. It is in this sense that our research emphasis will need to be on developing "software" rather than more "hardware." When societies or individuals face rapidly changing conditions, the two most likely responses are 1) to rigidify and redouble their efforts at maintaining their present course, and 2) to reconceptualize their situation and redefine their problems. Our task now involves the latter course, which, in turn, requires a broad definition of technology: human knowledge applied to human problem solving—i.e., *both hardware and software.* The science and technology policy agenda for the next decade and beyond must be seen in a profoundly changed context. This emerging agenda in all mature industrial societies must now take into account the new paradoxes that the trajectory of the industrial innovation process itself is now visibly generating. This two-hundred-year-old process of technological innovation, which we also

This chapter is derived from invited testimony before the Joint Congressional Hearings of the U. S. Senate Commerce Committee, Sub-Committee on Science, Technology, and Space, and the U. S. House of Representatives Committee on Science and Technology, Sub-Committee on Science, Research, and Technology, February 14, 1978, Washington, D.C.

know as the industrial revolution, has been based on premises and logic that are now exhausted: the maximization of material production measured by narrow criteria of "efficiency" (which has suboptimized social and ecological efficiency).

Thus we have overlooked the looming crises in distribution of the fruits of productivity, as well as the increasing strain on the natural resource base of our capital/energy/materials-intensive forms of technological innovation. Worse, the lion's share of funds for technological research and development go to the military. From 1975 to 1977 the military share of U.S. and Western European research and development budgets rose from 38.9 percent to 39.2 percent, and total world military spending was $425 billion, according to the U.S. Arms Control and Disarmament Agency. This sum far exceeds funds for social, health and civilian research, agricultural and ecological innovation, which are now top-priority areas for research, if we are to preserve our domestic security. Justification for stepping up this military effort has been predicated on the supposed existence of a worsening balance with the Soviet Union. This is called into question by the errors in *The Military Balance, 1980–1981,* by the International Institute for Strategic Studies, of Great Britain, which underestimated U.S. arsenals and contained over one hundred inaccuracies, according to the U. S. Center for Defense Information, Washington, D.C. (*The Christian Science Monitor,* December 16, 1980)

The entire industrial innovation process and the current statistical apparatus used to measure its progress must undergo redesign before a corrected course for technological innovation can be pursued. Our current stage of technological "plateau" in the evolution of industrial societies constitutes a saturation of their particular growth curve, and unanticipated social impacts and costs have begun to exceed the real productivity of the economy. A society-wide trade-off has been reached between 1) specialization, centralization, and division of labor, and 2) the social and transaction costs incurred. We need to remember that each order of magnitude of technological complexity and managerial scale inevitably calls forth an equivalent order of magnitude of government effort at coordination and control. As in a physical system, the society winds down of its own weight to a state of maximum "entropy," at which no further useful work is produced. Short of a complete reconceptualization of the situation, macroeconomic management breaks down, and inflation begins managing the system. Similar analyses of the inevitable fate of large sociotechnical systems are now gaining acceptance, for example in the Stanford Research Institute study by Elgin and Bushnell, described in Chapter 5. An example of the growing list of "double binds" in today's macroeconomic management is evident in the U. S. Congress Joint Economic Committee *Mid-Year Review of the Economy* (August 9, 1979). The report bemoans the problem of uncertainty created by secretive Federal Reserve Board policy making, since this makes planning more difficult, yet it acknowledges that the Fed must keep its action a surprise, or investors, firms, and policy makers will try to compensate by shifting their own strategies to discount the Fed—thus rendering it ineffective. This is simply one more example of the problem of applying linear, either/or logic to nonlinear systems, being discussed, if not addressed, by the economists of the "rational expectations" school, mentioned earlier.

Clearly, a reconceptualization of the underlying premises of the industrialization process as viewed in economic theory is one of the most urgent items on our national research agenda. Indeed, Chairman Richard Bolling, of the Joint Economic Committee, in releasing its study entitled *U.S. Long-term Economic Growth Prospects: Entering a New Era* (1978), stated that the lower growth scenarios it portrayed "would challenge economic policy as never before." The study emphasized that "social limits to growth . . . may be more important than the earth's physical limits in curbing the economy's development over the next quarter century." This new focus on "software" is welcome. The essence of the issue is that whereas we are all familiar with the typical S curves associated with diminishing returns and the plateau stage of many *specific* technologies, (e.g., the progression from radio and the vacuum tube, to the transistor, to integrated circuits and microprocessors), we are less familiar with the idea of a scenario of diminishing returns and the plateauing of *an entire constellation of technologies underpinning an entire type of society:* industrialism itself. We must now examine industrialism as a particular type of society, with its own world view and set of beliefs as to the nature of reality, with its own self-referential logic, paradigms, values, and goals, and buttressed with its own intellectual paraphernalia, science, and validation system. It is now necessary, in Thomas Kuhn's term,[1] to restructure the belief system within which knowledge acquisition takes place.

The major belief systems of industrialism—continual economic expansion, technological determinism, and the linear logic of left-brain-hemisphere dominance, of narrow, Cartesian reductionism[2]—must now give way to a more balanced, transdisciplinary, holistic world view and the reintegrating of the capabilities of the right brain hemisphere. The linear, reductionist logic inherited from Aristotle and Descartes has been brilliantly successful in its own terms: the focus on maximizing specific variables. This "tunnel vision" has also led to the now familiar explosion of negative feedback from the global ecological system: climatic variability, increasing desertification, and worldwide air and marine pollution, as well as diminishing availability of resources, notably petroleum. For these and other reasons, global interdependence is now a fact of life.

The new occidental awareness of Islamic culture, forced by overdependence on petroleum, may be useful in effecting needed paradigm shifts. From the sense of social embeddedness and reverence for the Creator of the Muslim, we may better see our perceptual fragmentation and reductionism, as described by Pakistani physicist Ziauddin Sardar.[3] Sardar describes the Islamic use of variable ranges of methodologies and acceptance of the limits of each, together with the reverent approach to science that Occidentals have lost in the past three hundred years.

Therefore, the future context for our science and technology agenda must also be planetary. The transition from Cartesian, linear logic to new paradigms integrating the global, systemic nature of our situation into our systems of knowledge must now be recognized more fully in all our research-and-development activities. We have created with our globe-girdling technologies of communications, transportation, military, and space, the "hardware" of global interdependence. The greatest task before us is to now write the new programs of "software" needed to manage this global system: the monetary agreements, the conflict-resolution and peace-keeping mechanisms, the systems of law to manage our common property resources, and the maps of its various cultures' value systems, which show where they converge.[4]

It is encouraging to see already the subtle shifts in our paradigms from the knee-jerk reliance on "hardware" and the "technological fix" and "supply-side" approaches to our problems toward the "software" approach, in which we are beginning to look at ourselves and our social and instrumental frameworks as the targets of modification. Nowhere is this shift more visible than in energy systems, in which the limitations of the old paradigms have become painfully obvious as they encounter the dismal laws of thermodynamics. The economists claim that "there is no such thing as a free lunch," but today "each lunch costs more than the last"! There are trade-offs in speed of exploitation versus efficiency and other limiting factors such as capital availability, process water and other inputs, and institutional barriers. Our energy policy has been forced to address the software aspects of the problem: the demand side, the institutional design, the behavioral, attitudinal, sociological, and political questions. An invisible hand can no longer be relied on; in fact the short-term adjustment of the no-longer-free market-price system is now actually preventing us from dealing with the long-term structural adjustments (e.g., the temporary oil gluts due to recessions in industrial countries can lower prices in the short-run and foster a "back to normalcy" complaisance).

This shift to the software focus implies a reexamination of the composition of all our scientific boards and advisory committees to correct for the current overrepresentation of hardware-oriented, hard sciences and engineering personnel and to add more social and behavioral scientists. The shift to software is also an inevitable aspect of the broader paradigm shift now underway, from material-based, empirical, objective, instrumental rationality to more subjective, value-oriented cultures.[5] This shift is also visible in the sciences, for example in the imaginative new hypotheses of physicists Wheeler, Everett, Bohm, Wigner, and others endeavoring to "write the observer back into the equation," and in such new epistemology as that of Capra.

Similarly, neuroscientists and psychologists are converging on exploring the uncharted reaches of inner space: the powers of the human mind. This subjective, software orientation implies a major redirection of scientific resources so that we can repattern our knowledge as an appropriate basis for our next technological trajectory. The entire emphasis on hard sciences and reductionist research will need to yield to the much more difficult transdisciplinary research, using models that capture dynamic, qualitative change processes, rather than simpler, Newtonian, mechanistic models.

Before our society's gears can reengage, major goals and values need to be clarified and reformulated, and new contexts need to be mapped, a process already underway in our political system (e.g., the debate over the bankrupt logic of a medical system costing 10.5 percent of GNP, predicated on ever more technology and research to "cure" diseases, rather than a preventive health-maintenance approach to reducing stress and hazards of industrial culture that lead to "disease"). Similarly, before our science-and-technology enterprise can embark on a new and more fruitful course, it must address this task of metaphysical reconstruction. Thus our first agenda item, over all, must be research in epistemology. We simply cannot proceed without designing more explanatory models of where we are, of causation, of nonlinear systems of interacting variables and the deeper structure of our sociotechnical systems, and of ourselves. Our more inclusive new research methods—technology assessment, general systems research, environmental impact statements, and futures studies—all present our decision makers with even *greater* uncertainty. Symptomatically, our decision makers become more uncertain (if they are honest). They admit that they do not know what to do and cannot master the avalanche of data in thousands of unrelated studies of interacting issues. Political issues are fought with intellectual mercenaries, marshaled into producing ever-more-prestigious reports buttressing opposing positions and interest groups. The political arena has become an information war, fought with data and symbols and often decided by the computer "firepower" and research "foot soldiers" that each group can afford to mobilize. But today much of the data we are drowning in are poor data, inappropriately collected, based on the obsolete paradigms of the past. There is a hierarchy of information quality, as represented by Figure 13. The raw, unpatterned data in which we are all drowning are meaningfully patterned by the use of models; the models are driven by assumptions, concepts, and a world view, which in turn are controlled by goals and purposes, all of which are driven by values. Thus values drive entire information systems, knowledge constructs, and the economic and technological systems of any culture.

The most obvious signals of the need for metaphysical reconstruction of the foundations of our knowledge is the proliferation of paradoxes. Paradoxes indicate only that the boundary of a particular system of logic has been reached. Viewed from higher system levels and broader perspectives, paradoxes are complementarities. Today, such "paradoxes" abound in physics, mathematics, psychology, and

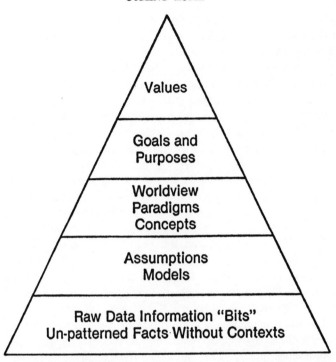

Fig. 13 Information Quality Scale (i.e., meaning of information as relevant to human purposes)

most evidently in economics, in which errors in analysis of factors of production and in productivity measures have long skewed the technological innovation process toward excessive capital-intensity and toward today's producer-driven, rather than consumer-responsive, technology. As the old consensus on what constitutes "proof" breaks down, we see science and technology politicized, as they must be. For example, the U. S. Congress Office of Technology Assessment now accepts the role of "honest broker" between clashing technical views, a stance I espoused during my six-year term (1974–80) on its Advisory Council, as more realistic than the now-threadbare claims of "scientific objectivity."

The overarching paradoxes of our industrial societies are that the current trajectory of technological innovation creates interdependencies that destroy conditions for free markets to allocate resources efficiently. Yet, at the same time that this market-choice system is failing, we do not have an adequate social-choice system, and neither do we know how to plan these complex societies. Furthermore, this same capital-intensive technological trajectory leads to larger-scale, more vulnerable, risky configurations, which require social investment and risk underwriting at the same time that their very complexity and centralization disenfranchise taxpayers and consumers from democratic participation in these technological decisions. Indeed, many of these technologies (e.g., nuclear power) are *inherently* totalitarian, therefore, by definition, *unconstitutional*. The growing awareness that whole classes of the scientific and technological enterprise may simply be incompatible with democratic forms of government is seen in the rising public opposition to nuclear power (our first socialist technology), genetic manipulation via recombinant-DNA research, and the pervasive, "creeping Big Brotherism" of electronic data-processing and funds-transfer systems.

The public's healthy reaction to such technologies, justified by unscientific sloganeering about their supposed "productivity" and greater "efficiency" is skeptical: "Efficient for whom"—the consumer or the producers, the corporation or the society? Asking intelligent questions is the most vital role played by citizens in science-and-technology policy. They provide priceless rigor and sometimes necessary negative feedback to the mindless maximizing of subsystem efficiency when this threatens other values or the society as a whole. If citizen participation had not become the rallying cry of mature industrial societies, they would have had to invent it! These technological issues are so pervasive and their impacts so widespread that they have become, per se, value issues and pose not only political questions but epistemological questions that can be dealt with only by reformulating our research agenda and rethinking our current technological commitments.

What is required for our scientific and technological enterprise is not so much micro-rigor and more data collection but paradigmatic rigor, in which we unravel the models and examine their deeply embedded assumptions. This also requires that we expose intellectual, as well as financial, investments; full public disclosure of both is essential. Such paradigmatic rigor will require sustained, well-funded research into the epistemological bases of economics (the most urgent, because this discipline has preempted the debate over resource allocation), physics, and computer modeling, as well as those underlying our hardware-oriented scientific enterprise. We also need to explore new as well as unfashionable older approaches and compare them with other, quite different systems of logic and knowledge, such as those of Eastern cultures. Examples of such paradigmatic research include neuroscientist Karl Pribram and physicist David Bohm's joint exploration[6] of holographic models of both the function of the human brain and the nature of the universe, and the works of Pitirim Sorokin, Margaret Mead, Ian Mitroff, Murray Turoff, Ida Hoos, Marilyn Ferguson, Erich Jantsch, Jean Houston, Magoroh Maruyama, John Platt, Fritjof Capra, Willis Harman, and others.

One of the key areas of software innovation will continue to be that of shifting our conceptual basis from that of traditional equilibrium systems, such as those underlying economics, to portraying dynamic, disequilibrium systems undergoing irreversible, qualitative change and structural transformations (i.e., morphogenetic systems). This seemingly abstract field of morphogenetic modeling is, however, immediately relevant to pragmatic issues involving whole sectors of our existing economy. For example, as discussed in Chapter 11, the quiet crisis in our insurance industry involves the paradox of literally incalculable risks/in our unstable sociotechnical systems in constant disequilibrium, while the basic models for calculating probabilities are still largely predicated on equilibrium assumptions and what Maruyama describes as the probabilistic, unidirectional, causal paradigm. Under such a drastic shift of conditions toward greater uncertainty, private-enterprise-based insurance becomes indistinguishable from gambling. Demands appear on all sides to socialize such risks at local and state levels (e.g., in automobile-insurance pooling), and at the federal levels (for medical malpractice, pensions, and so on, or in the socializing of nuclear-energy risks in the Price-Anderson Act). Pushing such uninsurable risks up to the higher levels of society merely buys some time—but at the expense of greater systemic vulnerability and cascading breakdowns.

Another key area of needed software research is global modeling in order to better grasp the global interactions that human activities have set in motion that are now modifying our planetary habitat. Anthony Fedanzo, Jr., explores the epistemological bases and assumptions of the current generation of global models[7] as to whether they postulate, for example, cyclical theories of sociopolitical change

(Forrester, Lévy-Pascal), theories of hierarchical, multilevel systems (Mesarović, Pestel), or self-organizing systems based on biological processes. The larger paradox, Fedanzo notes, that might be inferred from global modeling may be the sad conclusion that although the surface phenomena may be captured and modeled by various quantitative methodologies, the problem is that statistical correlations do not explain deeper, causal relationships. Although we have no choice but to continue this crucial research on global interactions, Fedanzo likens their stage of development to that of trying to practice psychology by studying only reflex behaviors of the autonomic nervous system! Examples of the misuse of econometric models in the policy process are legion in which these models, prepared by private consulting firms that often have large corporate clients whose interests are at stake in the decisions, will crank out "results" for gullible lawmakers, "proving" that some piece of special-interest legislation is "in the public interest." Such use of econometrics was discussed with refreshing candor in *Fortune*.[8]

As mentioned earlier, the most fruitful models and analogies for portraying the extraordinary, complex sociotechnical systems we have created will be those based on biological, organic models, rather than Newtonian, mechanistic processes. As Georgescu-Roegen has pointed out, this is the fundamental flaw at the basis of economics and its compounded epistemological errors of econometrics and of input-output and arithmetical modeling. Economic processes are not equilibrating and reversible, but irreversible, and involve qualitative energy/material transformations, usually associated with rising levels of entropy. Such processes cannot be captured using reversible arithmetic models of locomotion. Economics is not a science, but a value-based set of assumptions too often paraded as science.

I have drawn attention to this long agenda of needed software innovation because I fear that in our objectified, hardware-oriented world, it will be the aspect most likely to be overlooked. Rethinking our situation, rather than committing our resources to hasty and potentially irreversible or disastrous investments along the old, capital-intensive technological path, is the wisest course, since our store of capital is "fat" accumulated in easier times and represents our last store of cheap flexibility for future adaptions. We are face to face with the oldest riddle of evolution: "nothing fails like success."

Growth requires structure ... and structure gradually chokes growth. New trajectories for technological innovation have not yet been charted. But we know some of the parameters of the emerging regenerative, resource-based societies within which our technological innovations—both hardware and software—will unfold: greater global equity in access to resources within ecological tolerances and within the range of human psychological and social adaptability. For each specific technology, we might ask whether it is labor-intensive, rather than capital- and energy-intensive, and how much capital is required to create each workplace. Does it dislocate settled communities and cultural patterns, and if so, at what social cost? Is it based on renewable or exhaustible resource utilization? Does it increase or decrease societal flexibility? Is it centralizing or decentralizing? Does it increase human liberty and widen the distribution of power, knowledge, and wealth in societies or concentrate them? Does it embody multidisciplinary thinking and global interactions or is it parochial and one-dimensional? Does it favor self-reliance or create further dependency on large institutions? Does it make maximum use of existing infrastructure, or will it entail costly or duplicative infrastructure? Are its cost, benefits, and risks equally borne by all groups in society, and if not, who will be the winners and who the losers? What risks does it pose to workers, consumers, society at large, and future generations, and can they be assessed by current probability calculations? If it is irreversible and poses massive inter-generational transfers of risk (e.g., breeder-reactor technology), it should be assumed socially unacceptable until proved otherwise. The very shifting of burdens of proof to the producers of technological hardware in itself constitutes an important paradigm shift toward greater human maturity and responsibility for future generations.[9]

NOTES—CHAPTER 12

Since I delivered this testimony, the paradigm of globalism is taking root, however slowly. The interactive nature of all the problems faced by humanity makes this unavoidable. Another aspect of this paradigm shift is the view we in the northern hemisphere can obtain of ourselves by tuning in to the observations of our gyrations by the peoples of the southern hemisphere. They see all the crises of the northern hemisphere in the growing crises of industrialism and westernized science and technology: the monetary crises, the overdependence on energy and raw materials, and the increasing entropy, waste, and pollution. Many now talk more of how to specifically decouple from these crises not of their causing, which industrial countries are now trying to "export" to them. The clearest example to the Third World of the monolithic nature of the northern hemisphere and its industrialized countries, which invalidates simple communism-socialism-capitalism debates, was the fact that the U.S.A. and the U.S.S.R. voted together against the interests of Third World countries in most cases at the 1979 UN Conference in Vienna on Science and Technology for Development. Even the Eastern bloc's strong vocal support for the Third World was belied by their voting record (see " 'Heads I Win Tails You Lose,' a Retrospective on UNCSTD," by Ziauddin Sardar, *Impact International*, December 28, 1979 to January 10, 1980).

Finally, in a debate with Samuel C. Florman at Texas A & M University's Twenty-Fifth National Student Conference, in February 1980, it became clearer than ever to me that the "macho" and "big-bang" approaches to technology were failing. I found it necessary to talk of the choices we must now make between the technologies of fear—fear of death and the desire to control and manipulate the world (the technologies of Thanatos)—and the technologies of love grounded in that sense of belonging and embeddedness in nature that leads to the gentle descriptive sciences pursued in the desire to commune with the Creation—the technologies of Eros. These humble technologies stem from our matriarchal past and are often seen by men as threatening to their autonomy. Thus, backlash to this "small is beautiful" approach is growing, as typified by *Paper Heroes*, by Witold Rybczynski (1979).

An all-encompassing approach to science and technology must now emerge. The Swiss Chapter of the Society for International Development pointed out in its North-South Roundtable, in Colombo, in August 1979, "Technology is everywhere out of control, in the sense of being beyond national or international management especially in relation to its power to create, destroy and redistribute employment and income. . . . Individuals and states in the North and South share an interest in gaining better control of technology, its impacts on employment and societies, and the distribution of its benefits. There is also a clear common interest in the promotion of indigenous Third World technological capabilities, as appropriate, to meet priority needs or take up market opportunities." In the technology debate, too, the politics of reconceptualization has begun.

[1] Thomas Kuhn, *The Structure of Scientific Revolutions*, University of Chicago Press, 1962.

[2] Hazel Henderson, "Re-Directing the Goals of Knowledge," *Publ. Admin. Rev.*, January 1975.

[3] Z. Sardar, "Science in the Muslim World," *Nature*, November 1979, p. 354.

[4] E. Laszlo, *Goals for Mankind*, E. P. Dutton, 1976.

[5] Pitirim Sorokin, *Sensate Culture*, in *Social and Cultural Dynamics*, 1937–41.

[6] Karl Pribram and David Bohm, *Brain Mind Bull.*, July 1977.

[7] *Technological Forecasting. Social Change*, Volume 11, No. 2, 1978.

[8] *Fortune*, November 20, 1978, "The Economic Modelers Vie for Washington's Ear."

[9] The growing debate about the "appropriateness" of various technologies in the 1980s evidenced this maturing, based on such ground-breaking studies as Lewis Mumford's *The Myth of the Machine* (1964) and Jacques Ellul's *The Technological Society* (1964).

CHAPTER 13

Thinking Globally, Acting Locally:
Ethics for the Dawning Solar Age

To sum up: we see aging industrial societies undergoing a profound transition, actually a confluence of at least six historic transitions of differing periodicities (see Fig. 1):

1. The transition from the petroleum age to the now emerging solar age (a very rapid cycle, most of which is confined to this century).
2. The transition from the fossil-fuel age (coal, gas, and oil), which began in the early seventeen hundreds in England and will peak sometime around 2100 and be exhausted around 2300, according to geologist M. King Hubbert's no longer controversial estimates.

   This transition from human societies living on the earth's stored fossil-fuel "capital" to its daily "income," i.e., solar-driven energy, will mean an economic transition for all societies. This transition is already underway, from economies that have maximized material production, mass consumption, and planned obsolescence based on nonrenewable resources and energy, to economies that minimize waste by recycling, reusing, and maintainence based on renewable resources and energy managed for sustained-yield, long-term productivity.

3. The transition of *industrialism* itself, as it matures and makes this painful resource-base shift, whether in Britain (where it began), Western Europe, North America, Japan (where the process was

vastly accelerated), or the Soviet Union, whose younger industrial economy also shows the same signs of "plateauing" as it runs into the same inexorable energy crunch and the same sort of social bottlenecks in managing the complexity that is one of the most characteristic features of industrialism. The situation the Soviet economy faces today was outlined in *Fortune* (January 29, 1979, pp. 90–95), and their recent resource crunch is examined in *Fortune* (July 28, 1980, pp. 43–44). And even though, theoretically, socialism is supposed to preclude environmental costs and pollution, in practice, of course, ecological ignorance on the part of commissars and central planning committees can be just as environmentally devastating as that perpetrated by ecologically ignorant corporate executives and their economists.

4. The socioeconomic transition will be accompanied by a *conceptual transition* as the three-hundred-year-old logic undergirding industrialism's rise also reaches exhaustion. The logic stemming from Galileo, Bacon, and Descartes and continuing with Newton, Leibniz, and the Enlightenment philosophers—reductionism, materialism, technological determinism, and instrumental rationality—will fail us. Even the fierce ideological battles of the nineteenth century between capitalism, socialism, and communism, which continue today, will realign, since it is no longer only a matter of *who owns* the means of production but also the need to address the ecological, social, and spiritual dilemmas posed by the *means of production themselves.*

5. We are also undergoing a *cultural transition.* Will it be breakdown or breakthrough? Stress is one of evolution's tools. Human social systems and individuals, like those of all other species, *need* to be stressed in order to change. So I don't share the despairing counsel of policy makers who say, "You can't change human nature." First, we don't really know what constitutes "human nature," as feminist research has so well documented. Second, value shifts among humans are common and evident in all cultures. In fact, value changes and shifting cultural styles and metaphysical versus materialistic philosophies are the stuff of all human history.[1]

DECLARATION OF THE RIGHTS OF THE CHILD

PREAMBLE

Whereas the peoples of the United Nations have, in the Charter, reaffirmed their faith in fundamental human rights, and in the dignity and worth of the human person, and have determined to promote social progress and better standards of life in larger freedom.

Whereas the United Nations has, in the Universal Declaration of Human Rights, proclaimed that everyone is entitled to all the rights and freedoms set forth therein, without distinction of any kind, such as race, colour, sex, language, religion, political or other opinion, national or social origin, property, birth or other status.

Whereas the child, by reason of his physical and mental immaturity, needs special safeguards and care, including appropriate legal protection, before as well as after birth.

Whereas the need for such special safeguards has been stated in the Geneva Declaration of the Rights of the Child of 1924, and recognized in the Universal Declaration of Human Rights and in the statutes of specialized agencies and international organizations concerned with the welfare of children.

Whereas mankind owes to the child the best it has to give.

Now therefore,
The General Assembly
Proclaims this Declaration of the Rights of the Child to the end that he may have a happy childhood and enjoy for his own good and for the good of society the rights and freedoms herein set forth, and calls upon parents, upon men and women as individuals and upon voluntary organizations, local authorities and national Governments to recognize these rights and strive for their observance by legislative and other measures progressively taken in accordance with the following principles:

PRINCIPLE 1

The child shall enjoy all the rights set forth in this Declaration. All children, without any exception whatsoever, shall be entitled to these rights, without distinction or discrimination on account of race, colour, sex, language, religion, political or other opinion, national or social origin, property, birth or other status, whether of himself or of his family.

PRINCIPLE 2

The child shall enjoy special protection, and shall be given opportunities and facilities, by law and by other means, to enable him to develop physically, mentally, morally, spiritually and socially in a healthy and normal manner and in conditions of freedom and dignity. In the enactment of laws for this purpose the best interests of the child shall be the paramount consideration.

PRINCIPLE 3

The child shall be entitled from his birth to a name and a nationality.

PRINCIPLE 4

The child shall enjoy the benefits of social security. He shall be entitled to grow and develop in health; to this end special care and protection shall be provided both to him and to his mother, including adequate pre-natal and post-natal care. The child shall have the right to adequate nutrition, housing, recreation and medical services.

PRINCIPLE 5

The child who is physically, mentally or socially handicapped shall be given the special treatment, education and care required by his particular condition.

PRINCIPLE 6

The child, for the full and harmonious development of his personality, needs love and understanding. He shall, wherever possible, grow up in the care and under the responsibility of his parents, and in any case in an atmosphere of affection and of moral and material security; a child of tender years shall not, save in exceptional circumstances, be separated from his mother. Society and the public authorities shall have the duty to extend particular care to children without a family and to those without adequate means of support. Payment of State and other assistance towards the maintenance of children of large families is desirable.

PRINCIPLE 7

The child is entitled to receive education, which shall be free and compulsory, at least in the elementary stages. He shall be given an education which will promote his general culture, and enable him on a basis of equal opportunity to develop his abilities, his individual judgement, and his sense of moral and social responsibility, and to become a useful member of society.

The best interests of the child shall be the guiding principle of those responsible for his education and guidance; that responsibility lies in the first place with his parents.

The child shall have full opportunity for play and recreation, which should be directed to the same purposes as education; society and the public authorities shall endeavour to promote the enjoyment of this right.

PRINCIPLE 8

The child shall in all circumstances be among the first to receive protection and relief.

PRINCIPLE 9

The child shall be protected against all forms of neglect, cruelty and exploitation. He shall not be the subject of traffic, in any form.

The child shall not be admitted to employment before an appropriate minimum age; he shall in no case be caused or permitted to engage in any occupation or employment which would prejudice his health or education, or interfere with his physical, mental or moral development.

PRINCIPLE 10

The child shall be protected from practices which may foster racial, religious and any other form of discrimination. He shall be brought up in a spirit of understanding, tolerance, friendship among peoples, peace and universal brotherhood and in full consciousness that his energy and talents should be devoted to the service of his fellow men.

Publicity to be given to the Declaration of the Rights of the Child
The General Assembly,

Considering that the Declaration of the Rights of the Child calls upon parents, upon men and women as individuals, and upon voluntary organizations, local authorities and national Governments to recognize the rights set forth therein and strive for their observance.

1. *Recommends* Governments of Member States, the specialized agencies concerned and the appropriate non-governmental organizations to publicize as widely as possible the text of this Declaration;

2. *Requests* the Secretary-General to have this Declaration widely disseminated and, to that end, to use every means at his disposal to publish and distribute texts in all languages possible.

Fig. 14

We are also learning fast in mass-consumerist societies that overemphasis on material things diverts us from achieving fuller human maturity. Here it is interesting to contrast the views of human needs held by economists and psychologists. While economists see human needs in material terms, not only food, clothing, and shelter but actually postulating that material needs and desires are, *in principle,* insatiable; psychologists, on the contrary, see most human needs as *non*material: acceptance by others, self-esteem, loving interpersonal relationships, the challenge of useful and interesting work, the desire for meaning, purpose and harmony, and the urge to pattern and make sense of our experience, as described by Abraham Maslow in *Toward a Psychology of Being.* An expanding calculus of self-interest, increasingly coterminous with group and species "self-interest," culminating in the holistic perception of oneness with the global ecosystem, is characteristic of this progression toward maturity. This evolution of the human personality forms the theoretical foundation of humanistic psychology. It also provides the formula whereby the either/or contradictions of the Newtonian world view and rugged-individualist economics can be overcome in the growing awareness of the realities of global interdependence (see Plate 30). We see these new perceptions of reality emerging in the spate of recent manifestoes and declarations of principles derived from them. They include, for example, the Declaration on the Human Environment, the Declaration of the Rights of the Child, the Charter of Economic Rights and Duties of States, and the many other United Nations initiatives discussed in *Building the Infrastructure of World Order: a Survey of Global Policy Development from 1945–1977*, by Robert H. Manley.[2] We are beginning to see the subtle function of the United Nations in nudging each member state to consider not only those policies on cooperative, global, ecosystem management and peaceful uses of oceans and space but also in forcing such issues as basic human needs for food, shelter, employment, education, and greater social equity onto the domestic agendas of these member states.[3] Similar statements of principles arising out of new acceptance of the reality of an interdependent planet range from The Common Heritage Principles growing out of the efforts to create a law of the seas by such pioneers as Arvid Pardo, of Malta, and Elizabeth Mann Borghese (see Fig. 15), the Cooperative Principles for economic enterprise developed by the 1966 Congress of the International Cooperative Alliance, based in London, England (see Fig. 16), to the more personal behavioral principles of the Shakertown Pledge, drawn up by a small group of religious-retreat directors concerned with the inextricably linked problems of poverty, overconsumption, and ecological exploitation (see Fig. 17). This personal awareness and conscientization process has even led to a Declaration of Principles for the Conscientized International Expert (see Plate 31) and its excruciating sensitivity to the anomalies of jet-set academic elites in first-class hotels at meetings purporting to address hunger and poverty.

The pragmatic implications of global human interdependence on a crowded planet are also producing new domestic scenarios, such as those produced by the Swedish Secretariat for Future Studies on changing to more frugal life-styles (*How Much Is Enough? Sweden in a New International Economic Order*[4]) and the Science Council of Canada's series of studies (*The Conserver Society: a Blueprint for Canada's Future?*). We see the increasing use by "hard-nosed" government agencies in many countries of scenarios, rather than trend projections, so that "contingencies" (inevitabilities, of course!) such as raw-material and energy scarcities and the now obvious loss of control of domestic policy making due to global-interactive events (whether in Iran, in Afghanistan, or due to random terrorism) can be addressed in more palatable ways.

Increasingly, we see the denouement of the nation-state as a viable unit of governance; becoming too big for the small problems of its own local populations and at the same time too small for the big problems of global relations and ecosystems as documented by the 1980 report of the Brandt Commission.[5] This evidence was summed up in an editorial by James Reston in the New York *Times* (January 12, 1980) entitled "The World's Hidden Leaders." He quoted French President Giscard d'Estaing: "We are living through one of those periods when world balance depends on the level-headedness of a few men," and added that we do not even know who those men are. Reston cited the Iranian-American confrontation over the hostages taken at the American embassy, in which it was never quite clear to anyone who was in charge of negotiations, as well as the Soviet movement into Afghanistan, in which intelligence sources speculated that President Carter was deceived by the domestic maneuvering of "post-Brezhnev" factions operating at variance to the official statements of Brezhnev himself. Neither do we really know who is in charge in Peking, adds Reston, and indeed, Washington policies vacillate likewise as the tug-of-war between the President and Congress continues. What Reston does not question, however, is the basic paradigm: that it is indeed *possible* for one man to be "in charge" of any of these complex multidimensional societies and ever-more-interactive global situations. Our common sense tells us that the whole concept is untenable and an illusion. Yet the illusion is embraced beyond reason by most "leaders" and "decision makers": the neat orderly world of the geopolitical strategists, the war gamers with their erroneous concept that the world is operated upon by "rational actors" as in the appallingly simplistic views of top officials of the U. S. Department of Defense.[6]

THE COMMON HERITAGE OF HUMANITY

The penetration of ocean space by the technological/industrial revolution has undermined the traditional law of the sea. Neither coastal State sovereignty nor the freedom of the seas—the two principles on which traditional law of the sea was based—can solve the problems created by the intensifying exploitation and diversified use of ocean space made possible by modern technology.

Ocean space can no longer remain a global commons. Exercise of recognized authority is a necessary condition of intensive ocean space development, to protect investments, conserve living resources, control marine pollution, reconcile competing uses and, most importantly, to facilitate the participation, as equals, of poor and technologically-less-advanced countries in the coming era of ocean development.

At the same time, the excessive extension of insufficiently constrained coastal State sovereignty over ocean space exacerbates inequalities between States and could hamper vital transnational uses of the marine environment, from navigation to scientific research and pollution abatement.

In 1967 the Government of Malta proposed that the traditional freedom of the high seas should be replaced by the principle that ocean space and its resources beyond national jurisdiction are the Common Heritage of Mankind; initially applied to the seabed, this proposal was expanded in 1971 to embrace the high seas. This concept had five major implications:

(a) the Common Heritage cannot be appropriated: it can be used but not owned;

(b) the Common Heritage requires a system of management in which all users share;

(c) the use of the Common Heritage requires a sharing both of financial benefits and of benefits derived from shared management and technologies, on a basis yet to be specified (the latter two implications—shared management and benefit sharing—change the structural relationship between rich and poor countries and the traditional concept of development aid);

(d) the Common Heritage must be used for peaceful purposes only (disarmament implications); and

(e) the Common Heritage must be preserved for future generations (environmental implications).

From ocean space, the concept of the Common Heritage of Mankind may be expanded to other areas. Some legal experts, for instance, now consider outer space and the resources of the moon and other celestial bodies to be a Common Heritage of Mankind. More broadly, many proponents of a New International Economic Order have stressed that the NIEO as a whole must be based on the Common Heritage principle.

Fig. 15

THE COOPERATIVE PRINCIPLES

The 1966 Congress of the International Cooperative Alliance has approved these wordings of six Cooperative Principles.

1 Membership of a cooperative society should be voluntary and available without artificial restriction or any social, political, racial or religious discrimination, to all persons who can make use of its services and are willing to accept the responsibilities of membership.

2 Cooperative societies are democratic organisations. Their affairs should be administered by persons elected or appointed in a manner agreed by the members and accountable to them. Members of primary societies should enjoy equal rights of voting (one member, one vote) and participation in decisions affecting their societies. In other than primary societies the administration should be conducted on a democratic basis in a suitable form.

3 Share capital should only receive a strictly limited rate of interest.

4 The economic results arising out of the operations of a society belong to the members of that society and should be distributed in such a manner as would avoid one member gaining at the expense of others. This may be done by decision of the members as follows: (a) by provision for development of the business of the cooperative; (b) by provision of common services; or, (c) by distribution among the members in proportion to their transactions with the society.

5 All cooperative societies should make provision for the education of their members, officers, and employees and of the general public in the principles and techniques of cooperation, both economic and democratic.

6 All cooperative organisations, in order to best serve the interest of their members and their communities, should actively cooperate in every practical way with other cooperatives at local, national, and international levels.

International Cooperative Alliance, 11 Upper Grosvenor Street, London, England W1X 9PA

Fig. 16

Acceptance of the reality that humans do not "manage" the planet calls for new models of organization, governance, and decentralization, and wherever possible, the localizing of production, consumption, and participation, together with the democratic formulation of planetary agreements, declarations of principles, and the rights and responsibilities of all people. This "thinking globally, acting locally" formula, the theme of the first Global Conference on the Future, in Toronto, July 1980, can inform local action with the understanding of the requirements of planetary interdependence and limits. It can also fuse thought and action, a prerequisite for integrated personalities (i.e., whole people) functioning in integrated, whole systems, recognized by psychologists, anthropologists, revolutionaries, and theologians, as Renee Marie Croose Parry reminds us in "Human Needs and the New Society," and espoused by Margaret Mead, Kurt Lewin, Mao Tse-tung and Teilhard de Chardin.[7]

6. Today's transition is also fundamentally marked by the *decline of systems of patriarchy* that have predominated in most of the world's nation-states for some three thousand years as the earlier matrilineal societies and matriarchal religions were superseded. The nation-state, like all patriarchal systems, is hierarchically structured; it is based on rigid divisions of labor (as well as polarization of sex roles); manipulative technology; instrumentalist, reductionist philosophies; the control of information; and on competition, both internally as well as between nations. Unlike the earlier, smaller city-states and fiefdoms, these nation-states have proved, as Toynbee showed, to be highly unstable, perhaps somewhat like large, unstable macromolecules. Indeed, nation-states are quintessential expressions of patriarchal dominance systems; from the family to the workplace, community, academia, the church, and all levels of government. They are characterized by extreme polarizations of conceptual, bureaucratic, academic, intellectual work in centralized, urbanized, metropolitanized complexes. They are rendered operational by the now failing macroeconomic management and centralized political decisions based on the large statistical aggregates of the formal, monetized GNP economy. At the other end of the scale are the undervalued manual tasks, rural and agricultural life, and the unpaid work of the nonmonetized, "informal" economy of household production,

gardening, canning, home repairs, nurturing and parenting, volunteer community service, and all the cooperative activities that permit the overrewarded competitive activities to appear "successful."

Thus, patriarchal modes are also reaching logical limits. Hierarchies become bottlenecks; excessively conceptual governance becomes divorced from reality whether in Washington, Brussels, or Moscow, where bureaucrats try to govern by manipulating statistical illusions, using highly aggregated, averaged data that do not fit one single real-world case or situation. Corporate executives make momentous technological and economic decisions using highly selective marketing studies, isolating "effective demand" from real-world need as well as social and ecological impacts. Similarly, patriarchal academic hierarchies in science and technology are now overspecialized and abstract, perhaps because they have systematically excluded women, as well as minorities, with challenging, alternative views.

Most of all, it is obvious in the nuclear arms race that competitive patriarchal nations buy not more "security" but national bankruptcy and ruin with their horrendous military machines. Far from the Defense Department and Ronald Reagan's demands for ever higher military budgets, the Washington-based Center for Defense Information found the United States and its allies already superior to the Soviet Union in all elements of national power.[8] As Lester Brown points out in *Redefining National Security* (Worldwatch Institute, 1977), these nations cannibalize both social and ecological systems. The latest example of this military insanity is the MX Missile's requirement of $30 billion, 22,330 square miles of land, 90 billion gallons of water, 22 million gallons of petroleum a year, 10,000 miles of roads, and 2,200 miles of rails.[9] The MX has sparked widespread opposition in Utah and Nevada in the Sagebrush Rebellion against federal encroachment on all local land use.

Patriarchal domination has been discussed in a continuous stream of social criticism since Charlotte Perkins Gilman's *Women and Economics* (1898), culminating in the burgeoning literature of feminist scholars of today. However painful it is for the "men in charge" of most nations, corporations, academia, and other institutions, there is much evidence associating patriarchal societies with oppression, violence, and militarism. The great British economist John Stuart Mill pointed to the "general unfitness of men for power" in his essay *The*

Subjection of Women (1869), but it is, rather, the overemphasis and glorification of qualities that have come to be equated with "masculinity." At the same time, such patriarchal value systems breed contempt for cooperation, humility, nurturing, and acceptance of natural rhythms of life and nature—all have been decried as "womanish." Both sexes are quite capable of the full range of these ways of being and behaving, and it is, rather, the institutionalization of only the "masculine" value system in the social, institutional, and political spheres and the ghettoizing of the nurturing, cooperative value system within the family and the female roles that are now causing disastrous imbalances. In fact, this insane specialization and division of labor, in which the males are supposed to do the thinking, acting, and competing, and the females are supposed to do all the feeling and cooperating, has produced severe personal problems for both men and women.

Thus it is the *institutionalizing* of these so-called "masculine" value systems in patriarchal societies with which we are concerned. Indeed, patriarchal societies are also characterized by the dominance by the old men and elites in their ubiquitous "old-boy networks," which continue to oppress their younger, subordinated males, for example by sending them off to fight the wars. Friedrich Engels, coauthor with Karl Marx of the *Communist Manifesto,* wrote of this domination in *"The Origin of the Family, Private Property, and the State,* in 1884 (*The Essential Feminist Writings,* edited by Miriam Schneir, Vintage, 1972). Engels viewed the first instance of class oppression as that of the female sex by the male and pointed out that the Latin root of family, *familia,* meant originally all the slaves belonging to one man, over whom he had the power of life and death. James Robertson elaborates, in *Power, Money, and Sex* (Marion Boyars, London, 1976, pp. 89–90), the links between extreme nationalism and male chauvinism, quoting Nazi Germany's Goebbels: "The National Socialist Movement is in its nature a masculine movement"; and Adolf Hitler (who is even more revealing of Nazism's extreme imbalance between male and female roles): "We do not find it right when woman presses into the world of men. To one belongs the power of feeling, the power of the soul . . . to the other belongs the strength of vision, the strength of hardness."

In a similar vein, political scientist Ali A. Mazrui, of Kenya, observes in "The Warrior Tradition and the Masculinity of War"

(*Journal of Asian and African Studies*, XII, 1–4, pp. 70–81), "In cultures which are otherwise vastly different, the role of warrior has been reserved for men . . . crimes of violence have been disproportionately committed by men. The jails of the world bear solemn testimony to the basic masculinity of violent crime." Mazrui contrasts this orientation with that of Mohandas Gandhi's militant nonviolence and his androgynous qualities, as well as his striving toward motherly, nurturing modes of living, closely related to the household, the village, and the land. John Lennon, in the last few years of his brief life, took up a nurturing, household role as an important learning experience. Economist Thorstein Veblen, in his *Theory of the Leisure Class* (1899), saw a future in which all people would display a more generic, less differentiated expression of human nature, and asserted that ". . . the average, dispassionate sense of men says that the ideal character is a character which makes for peace, goodwill and economic efficiency, rather than for a life of self-seeking force, fraud and mastery." And Frederick Douglass, born a slave, wrote in his autobiography, explaining his staunch support of women's rights, "When a true history of the anti-slavery cause is written, women will occupy a large space in its pages, for the cause of the slave has been peculiarly a women's cause."[10] In the past century, women from many countries spearheaded the world peace movement and founded a large number of the nongovernmental organizations functioning for global humanitarian causes today.[11] These voluntary, citizen-based associations have provided a wealth of alternative, innovative organizational forms: cooperative, information-sharing networks; political action models based on decentralized "cells"; consciousness-raising and heterarchical (as opposed to hierarchical), participatory organizations, and the strategies of nonviolent protest and conscientizing the leaders of existing power structures. For example, even in the Soviet Union, the feminist magazine *The Woman and Russia* calls male comrades to account for shirking their fair share of work, and has urged men not to fight in Afghanistan.[12] Even in male-dominated Japan, the Asian Women's Congress to Fight Discrimination and Aggression campaigns against economic aggression and exploitation in Asia by Japanese businessmen, while the European-based Internationale Féministe's slogan is "No feminism without the liberation of all the oppressed. No liberation of the oppressed without feminism."

The drive for economic growth can now be seen as a crucial vehicle for oppression, domination, and exploitation, not only of women by men by assigning them the unpaid, low-paid, or low-status work of society,[13] but its accompanying economic reasoning, which abets the subordination of minorities in all countries by designating vital but nonspecialized, nonacademic, and bureaucratic work as "less productive." Similarly the economically powerful countries can dominate and downgrade the role in the world economy of so-called "less developed" countries and justify the role of multinational corporations and capital investment as well as the world trade game they have designed under the guise of the "free-market system." As Professor Joseph Huber, of the Free University of Berlin, puts it, "Up till now, the development of the market (monetized) economy and the industrial system has not only led to increasing exploitation of natural resources and increasing turnover of goods and services. It has also prompted a drop in the local, subsistence economy, i.e., the self-supporting economy, both collective and individual." Huber calls for this "dual economy" to be rebalanced, with the *monetized, institutionalized sector* strictly limited, so that it is prevented from parasitically destroying the fragile social ecology of the *informal sector,* which he defines as nonmarket work: housework, neighborhood cooperation, and nonpaid pursuits and "leisure jobs" not appearing in the GNP, all held together by the social life of its members, their community life, values, and norms.

Contrary to economists' beliefs, the informal sectors of the world's economies, in total, are predominant, and the institutionalized, monetized sectors grow out of them and rest upon them, rather than the reverse. Even in the industrialized nations, this submerged and surprising reality can be documented—although the bias of economic statistics virtually precludes this type of analysis. In France, for example, a 1975 study calculated that while 43 percent of the total working hours of the French population were devoted to formal employment, 57 percent of the working hours were in the informal sector (Adret, *Travailler deux heures par jour,* Éditions du Seuil, Paris, 1977). While it is clearly necessary for any society to have both an institutional and an informal sector in its total economy, the danger has arisen in industrialism's focus on economic efficiency (measured by money), and goals and values derived thereby, that the overgrowth of the institutionalized sector has created huge imbalances that now threaten to destroy the informal sector, which is the bedrock of all societies. This cannibalizing of the informal sectors, where the monetized sectors' social costs have been buried, is most visible, of course, in the "most advanced" industrial societies, and luckily is providing an important object lesson for other countries who want to avoid the same trap. Thus the task is that of rebalancing societies so that informal-sector values and functions are revived and restored, while the institutionalized economy and its money values are limited and put back in their place. Huber sees possible *negative* and *positive* scenarios of the dual economy in the future. In the *negative* scenario, the two sectors would become increasingly separated, with two societies or two classes of human beings existing in parallel, one class employed in the institutionalized sector, enjoying the benefits of secure positions and a certain degree of influence and respectability, and the members of the other class more or less "unemployed" and leading a marginal existence in the informal sector, without social insurance, as day laborers and charity beneficiaries. In the *positive* scenario of the dual economy, all members of the society have a professional job in the institutionalized sector and, at the same time, work part time, or in alternating phases, in the informal sector, and all people participate in both sides of the dual economy and share the same social security. Huber adds, "The *negative* scenario is patriarchic. Hiring practices are preferential to men, whereas housework in the informal sector is performed mostly by women as 'housewives.' In the *positive* scenario, the sexes are on a socially equal standing. Men and women share the paid positions, as well as the household tasks, child rearing, and other social activities. The model is simple: instead of a male earner in a small family, with a forty-hour week and a monthly income of [fifteen hundred dollars], two persons work a twenty-hour week and receive [seven hundred fifty dollars]. Instead of an unpaid, fulltime 'housewife,' there are two part-time, unpaid household workers. In the *negative* scenario, there are very unequal wage levels, high tax burdens and high demand on state bureaucracy. In the *positive* scenario, there is a balanced income distribution, low taxes, and a reduced need for state intervention, since people perform for themselves community services which today are performed by government bureaucracies with tax monies."[14] Ironically, failure to attend to the requirements of the informal economy only leads to the growth of yet a third sector of quasi-public and voluntary organizations to ameliorate these social costs and bridge the gulf between the so-called "public" and "private" sectors. In the United States this "third sector" now employs 5 million people at a payroll of $36 billion and outlays some $80 billion annually.[15]

A rebalancing of the dual economies in societies is really the only way to reduce centralism, Big Brotherism, bureaucracies, mindless hierarchies, and bottlenecks, as well as the accumulations of power and wealth they always create—which in turn have always led to expansionism, institutional aggrandizement, military adventures, technological mischief, fantasies of omnipotence and control, and the inevitable exploitation of subordinated groups and the environment. Such clearly achievable restructuring of the patriarchal ordering of societies that could defuse these dangerous imbalances by such means is described by Huber, James Robertson, Scott Burns, Amory Lovins, E. F. Schumacher, Leopold Kohr, and the following representatives of a long decentralist, libertarian tradition stretching from the utopian theorists of the nineteenth century: Proudhon, Kropotkin, Bakunin, Robert Owen, Saint-Simon, and others in Europe, to the American tradition expressed by Jefferson, Ezra Heywood, William B. Greene, J. K. Ingalls, Henry George, Josiah Warren, Benjamin Tucker, and contemporary libertarian reformers Ralph Borsodi, Mildred Loomis, Stuart Chase, Helen and Scott Nearing, and Agnes Inglis, curator of the unique decentralist library housed at the University of Michigan, Ann Arbor.

Mildred Loomis, the venerable grandmother of the decentralist movement in the United States, is publisher of *The Green Revolution,* its journal, founded in 1946.[16] Loomis authored *Decentralism: Where It Came From. Where Is It Going?* School of Living Press 1980, Box 3233, York, PA 17402, a definitive history of the decentralist movement. The libertarian-anarchist-utopian tradition contains much wisdom and, although its premises were based, almost without exception, on maintaining the unpaid subjection of women to all household, child-rearing, and support services,[17] there is still much to be derived from this literature. The incorporation of much of this tradition into feminist theory has produced a rich synthesis infused with much earlier anthropological and archeological studies of matrilineal, matriarchal, and polyandrous traditions, and producing innovative concepts of social organizations characterized by androgynizing human behavior and social roles and liberating both sexes from their current roles as prisoners of gender.[18]

The fascinating history of decentralist thought and action provides indispensable grounding for today's futurists and new-age activists in America. It is an important clarification of the philosophies of anarchism and mutual aid and their views on property. We are shown how the perversion of private-property rights for the *individual* led to the monstrous inequities that allowed multinational corporations to masquerade under this protection as "individual persons" under law. This has led to today's confusion over property rights, which no longer distinguishes between the necessary inviolability of individual property rights needed to assure personal autonomy and self-reliance, self-respect, and self-motivation, and the endless accumulation of property by corporations and institutions to the point where they have the power to oppress and disenfranchise individuals and smaller groups. Mildred Loomis, in *Decentralism,* helps us see today's resurgence of co-ops, neighborhood revival, community economic reconstruction, and land trusts in the context of past efforts. By so doing, it helps demonstrate the irrelevance of old political labels, whether they be Republican or Democrat, Liberal or Conservative, capitalist or socialist. Decentralism is one of the keys to understanding the new politics of our time and how some of its contemporary figures, such as Jerry Brown, can be interpreted. For those who wish to interpret the new politics of reconceptualization, earlier experiments and theories of decentralism must be examined to see how they were overwhelmed by the rising tide of industrialism. Its faulty logic was concealed by the cornucopia of resources that earlier, smaller populations could exploit for two centuries before the social bills began coming due. The early decentralists, with their more profound ecological and social logic, battled the tide and left us their precious legacy—ready for today's new-age decentralists to apply successfully in the receding backwash of the now exhausted industrial era. We need to reassess their hard-won experience in *today's* situation of social breakdown, environmental devastation, and resource depletion, in order to see that their ideas of humanistic and ecologically harmonious, social productivity can be brought to fruition. We can see how the new politics of "small is beautiful" springs from the oldest traditions in our national history, and we can thus interpret today's convergence of apparent polar opposites and new groups of strange bedfellows: including many Libertarians and tax revolters, appropriate-technology innovators, small-business people, ecology activists, holistic-health-care promoters, advocates of state's rights and consumer protection, together with new labor unions for farm and household workers, and advocates of worker ownership of businesses and of neighborhood economic development (see Fig. 2).

We can even see how the extreme laissez-faire tradition of "free market" economics, typified by Friedrich von Hayek and the Austrian school, mentioned in Chapter 8, has common roots with the ideas of libertarian-anarchist economic proposals to "privatize" money, i.e., create a free market in which many competing monetary systems exist and no institution, particularly the state, is permitted a monopoly on this crucial function. This permits an appropriately *delimited* extension of the barter system, without today's totalitarian overemphasis on the monetized, institutionalized Big Brother economic sectors.[19]

Yet it is obvious that the decentralist, libertarian, "matri-anarchist" restructuring of patriarchal institutions, from the nation-state to the traditional division of labor within the family, are profoundly threatening to all those whose identity is entwined and who benefit economically and politically from the existing structures. They all see their traditional positions, however relatively low on the totem pole they may be, eroding through no particular fault of their own, as, for example, the outrage of medical student Alan Bakke at having lost out in the more equitable admissions system at the University of California due to the number of similarly qualified minority and women students competing for places. In the sense that these social processes are *inter*generational and reflect the continual need to redress imbalances and inequities of the past, for example the historic discrimination against blacks documented by Ray Marshall, Secretary of Labor in the Carter administration, in "Black Employment in the South,"[20] children always seem to pay for the sins of their forefathers. This almost karmic element in social systems seems unavoidable, from the redressing of the oppressions and grievances of slavery and witch-hunting, in which millions of women were murdered, to the billions of dollars' worth of reparations that the current generation of Germans has paid to the Jewish victims of the Nazis' holocaust. Similarly, we see the same principle at work in the social bills now coming due for the past exploitation of the environment, the unconscionable exhausting by four generations of industrial citizens of the planet's 60-million-year endowment of petroleum, not to mention the grim legacy of radioactive wastes. It is our children and grandchildren that will pay these bills, and suffer the social, physical, and genetic consequences. Here, the new politics of reconceptualization is typified by the slogan of the Values Party, in New Zealand: "We do not inherit the world from our parents, we *borrow* it from our children."

The deeper dilemma in shifting our unsustainably imbalanced patriarchal societies lies in their very long traditions, and that they have operated as positive-feedback, self-reinforcing systems, interplaying the extensions of male experience and identity to the point where this "masculinity" aspect of all human beings has been overamplified in institutions, technologies, and dominance-submission interactions and reinforced by "false-positive" feedback and acculturation. This whole set of "masculinized" values is now deeply associated with male identity, and thus any attempt to dig deeper, to this more fundamental level of social analysis, is extremely threatening personally and is usually energetically resisted, denied, reversed, or repressed, with all the classic defense mechanisms described by psychologists. For example, Philip Slater's discussion in *Earthwalk* of the roots of individualism; infantile fantasies of push-button technological gratification; obsession with control; as well as the relationship of autism, autarchy, and authority *as various aspects of men's fears of dying and their loss of awareness of connectedness*—are extremely unpalatable to current rationalization and political-legitimacy theories. Slater warns that "The result of men's fear that dying is the ultimate loneliness is an increased likelihood that all humanity will die together. . . . The technological impulse is strongly influenced by the need to deny human mortality. . . . The notion of 'push-button control' appeals to fantasies of infantile narcissism . . . the delusion that pleasure can be obtained through mastery . . . has a built-in contradiction: . . . control and pleasure destroy each other. . . . This is how push-button control leads to fantasies of pushing the nuclear button."[21] Erich Fromm, in *The Sane Society* (Fawcett Books, 1955), describes patriarchal society as characterized by respect for man-made law, rational thought, and sustained efforts to control and change the natural world; whereas matriarchal society is characterized by the importance of blood ties, close links with the land, and acceptance of human dependence on nature, while valuing love, unity, and universal harmony.

Suffice it to say that an enormous body of literature has emerged and been rediscovered in the past decade as to the division of labor, specialization, and social roles between the sexes, and the extent to which these are biologically and culturally based and reinforced. The parallel with economic theory is obvious, since most of the economic theorizing since the industrial revolution and Adam Smith has been based on concepts of various types of "efficiency" *based on specialization and division of labor,* and most theories have been spun from various assertions concerning "human nature" and the propensities and fitness of certain types of human beings for various work roles—and even extended (in the theory of comparative advantage) to the idea that some countries were more "fitted" for specialized roles in world production, while others would simply export their raw materials. It is precisely these paternalist formulas Third World countries reject.[22]

Such reexaminations of all this economic theorizing range from James Robertson's view, in *Power, Money, and Sex,* that "man may be caught in a trap which, down the millennia and the centuries, he has been making for himself. The trap is his own nature. The built-in program, inherited biologically and culturally, which now governs the desires and the behavior of the species, may be holding us firmly on a suicide course." The view of Lionel Tiger and Robin Fox, in *The Imperial Animal* (Paladin, London, 1974), is that "human males have all the enthusiasms of the hunting primate, but few of the circumstances in which this reality can be reflected. So they create their own realities: they make up teams, they set up businesses and political parties, they form secret societies and cabals for and against the government, they set up regiments; they make up fantasies about honor and dignity; they turn their enemies into "not-men"—into prey. They generate forms of automatic loyalty and complete dedication that can spread the Jesuitical message of the Church Militant, and also send screaming jets to a foreign country." A similar view of how males ritualized the hunt and thereby invented war and trade is contained in the stunning new synthesis of the human story *The Time Falling Bodies Take To Light* (1981), by metahistorian William Irwin Thompson.

Yet women have not been blameless. In their subjugation, most have pandered to and appeased their men, served as mirrors to glorify their "heroic" acts, fed their ambitions and lust for power and wealth, and shared in them or basked in their reflections. Here, too, a tangled, intergenerational web of causes and effects must be unraveled and the tragedy of generations of powerless, economically insecure women who have manipulated their sons through guilt and obligation and the overwhelming power of nurture must be faced. Indeed, *the most powerful human capability is that of nurture,* since it is literally, the life-giving capability and thus implies, in withdrawal, *death itself.* Furthermore, since most nurturing of infants has been by women, many psychologists and feminists see this single-parent early nurture as a source of much subsequent social pathology. Male children cannot identify with the power of this nurturing, so remain subconsciously afraid of women and with deep fears of the remembered dependency, for which they must continually overcompensate by efforts to dominate, control, and order the world around them to serve them predictably. Elizabeth Dodson-Gray explores this male cultural need in *Why the Green Nigger?* (Roundtable Press, 1979), and its expression both in the need to dominate women (wife as "mother-in-chains") and in the exploitation of natural resources and the proliferation of technologies of control and manipulation ("Mother" Nature must also be raped, subdued, and brought into service). Dodson-Gray then links this to religious cosmologies and various stories of the creation, most of which presuppose male gods who, in a classic denial and reversal of biological reality, give birth in some way to the female (for example God's making Eve from Adam's rib). Dodson-Gray argues, with many other feminist theologians, including Rosemary Reuther and Mary Daly, that hierarchical paradigms are, in the last analysis, rooted in these hierarchical cosmologies and anthropomorphic concepts of the Creator.

It is well beyond the scope of this book to explore these issues further, except to state my own conviction that we cannot unravel the current human dilemma without fearlessly examining such fundamental levels of our being. It is in this psychological area that Marxism has run aground, as has most economic and social theory.[23] French Marxist Roger Garaudy courageously explores this new terrain, and in a recent conversation, told me that the feminist critique of industrialism, socialist and capitalist, was the only fundamentally innovative analysis available. This view has much in common with the "dual-economy" analyses and their revelation of the oppressiveness of all money-based economic sectors, with their abstract, quantitative, symbolic "economism" and their tendencies to progressive domination of rural, household, subsistence sectors with their indigenous cooperative, cohesive values and ancient cultural wisdom learned in adaptation to specific environments and resources. Emerging out of such reassessments is the spate of new journals and conferences on regional approaches to a New International Economic Order and alternative development strategies.[24]

We are coming to realize that, ultimately, all social control systems operate at the level of language and symbol systems, encoding in various cultures their values (i.e., *what* is valuable and *who* is valuable). Then the valued people and activities are drawn inside the magic circle of monetization, while those devalued are left out. Thus, any economic system can appear successful, depending on where such boundaries are drawn. Thus, crucial relationships exist (often denied and rationalized) between culture and ethics and all economic/technical systems. *Value systems and ethics, far from being peripheral, are the dominant, driving variables in all economic and technological systems.* It is in this sense that Marxists assert, correctly, that all knowledge is political. Similarly, all science is value-based.

Thus, the task facing industrial societies as they enter the 1980s and 1990s and their coming "trial by entropy" will be to face up to the *unsustainability of their value systems*—rather than view their "problems" *as deficiencies of nature.* This kind of "gestalt switch" out of our infantile, anthropocentric preoccupations is now the prerequisite for the survival of our species. This reconceptualization and redesign of human symbolic systems is our chief means for escaping (always temporarily!) the entropy trap we designed for ourselves during the fossil-fueled industrial era. Thus it is the same kind of source of negentropy described by Ilya Prigogine as that displayed by living systems. Prof. Bartek Kaminski, of the University of Warsaw, has summarized these issues and the challenges they pose for economic theory in "Entropy and Economics," *Oeconomica Polona,* 1980, Number 1. Polish Academy of Sciences, Warsaw. At the same time, we must internalize the view of the indivisible human family that we *are* biologically and genetically: *one species,* with no scientific bases for the superstitious, parochial divisiveness based on minor differentiations, whether of skin color, ethnic stock, sex, or social functions. This next evolutionary leap in our expanding consciousness, imagination, and empathy is the only potential we possess that can hope to counter the increasing entropy we are creating on a planetary scale. We know that life forms on this planet will all be extinguished in due time with the death of the sun, our mother star, some billions of years hence. But the aeons available to us between now and that event give us all the leeway we need to put ourselves and our planetary house in order, since there is no reason to believe that we are a species irrevocably programmed to "self-destruct."

This concerted effort involves examining our value systems as clinically as we can from psychological, anthropological, biological, thermodynamic, and ecological viewpoints and then modeling the types of behavior outputs and "hardware" these packages of "cultural software" produce. Some cultures, such as the Balinese, the African Bushmen, and the Native American nations produce very little "hardware" but have extremely elaborate, ingenious, behavior-regulating "software" tuned to the requirements of their ecological niches. These cultures, of course, leave very little trace on the world, geared as they are to natural cycles of entropy and regeneration. So, in the arrogance of westernized scholarship, we assume for this reason that they are "primitive." On the other end of the hardware-software scale are the Americans, who use almost twice as much energy per capita as even Europeans, and four times the world average consumption of minerals (Office of Technology Assessment, *Technical Options for Conservation of Metals*, U. S. Government Printing Office, Washington, D.C., 1979). The U.S. value system has produced more hardware and created more entropy in a shorter period than any other culture in the history of the planet. Military machinery and expenditures take 43 cents out of every dollar spent by the U. S. Government, while the Pentagon manages to minimize their bite on taxpayers with fudged accounting, as analyzed in *Scientific American*.[25] The military budget for 1980 was $138 billion; new corporate arms sales were $13 billion in 1979 (with a backlog of $48 billion of orders), and five million citizens owed their livelihood to the Pentagon.[26] Militarism is the most energy-intensive, entropic activity of humans, since it converts stored energy and materials directly into waste and destruction without any useful intervening fulfillment of basic human needs. Ironically, the net effect of military, as opposed to civilian, expenditures is to increase unemployment *and* inflation, as documented in *The Empty Porkbarrel* (1980) and *Bankrupting America* (1980), reports from Employment Research Associates, Lansing, Michigan. It is only the insane, abstract mediation of money that gives us the appearance of the "exchange value" produced with which military workers can buy such basic necessities. Yet here, too, the mad money game is bankrupt, since we now see that the easy assumptions that military budget appropriations must be increased in the light of the instabilities in Iran and Afghanistan rest on faulty logic.[27] There is no automatic link between spending more money on "defense" and our ability to deal with the new guerrilla and anarchist-terrorist modes of such new scenarios. They lie beyond the old, predictable "great power" and "rational actor" models of the military strategists, who are still fighting the more orderly, hierarchical wars of the past. These situations are no longer static and linear, but dynamic, nonlinear, and morphogenetic. Edmund Muskie, as Secretary of State, expounded the new understanding that aiding the development of the Third World was the United States' best strategy for national security.[28]

In order to devise workable ethics for the solar age, we will be required to begin the cooperative, global task of inventorying all the world's value systems, religious beliefs, and cultures, past and present, and assessing these behavioral outputs and the hardware configurations they produce. Some unilateral efforts have been made that can serve as models, such as the Club of Rome reports; *Goals for Mankind,* edited by Ervin Laszlo (Dutton, 1977); *No Limits To Learning,* by teams from the U.S.A., Morocco, and Romania led by James Botkin, Mahdi Elmanjdra, and Mircea Malitza (Pergamon Press, 1979); and the 1977 study of a scenario of a global economy with a shift of focus toward filling basic human needs for food, clothing, education, and housing in all countries, involving value shifts toward maximizing local self-reliance and income equalization (published by the Bariloche Foundation, of Argentina, and directed by Herrera). This inventorying task is a viable project using current computer technology, with its greatly reduced cost due to microprocessor innovation. Then it will be necessary to assess which value systems have been most interpersonally harmonious and just, as well as environmentally attuned and sustainable for the longest periods, and review all historical evidence as to the reasons for any failure. Here we must be aware of the terrible trap of all history: that it records events not only with the distortion inevitable in the perceptions of the chronicler but also of the record keepers of the culture and what was deemed "important" to record. Thus most conventional history is that of human ego, pride, power, and conquest; ignoring all but fragments of the story of the humble production and activities of ordinary people. Furthermore, what have usually been deemed the historical "failures" of some cultures because of their conquest by a more "powerful" or wealthy culture may have simply reflected the sheer luck of geographical location and rich veins of mineral and energy resources available. But these caveats should in no way prevent

---

THE SHAKERTOWN PLEDGE

Recognizing that the earth and the fulness thereof is a gift from our gracious God, and that we are called to cherish, nurture, and provide loving stewardship for the earth's resources.

And recognizing that life itself is a gift, and a call to responsibility, joy, and celebration.

I make the following declarations

1. I declare myself to be a world citizen.
2. I commit myself to lead an ecologically sound life.
3. I commit myself to lead a life of creative simplicity and to share my personal wealth with the world's poor.
4. I commit myself to join with others in reshaping institutions in order to bring about a more just global society in which each person has full access to the needed resources for their physical, emotional, intellectual, and spiritual growth.
5. I commit myself to occupational accountability, and in so doing I will seek to avoid the creation of products which cause harm to others.
6. I affirm the gift of my body, and commit myself to its proper nourishment and physical well-being.
7. I commit myself to examine continually my relations with others, and to attempt to relate honestly, morally, and lovingly to those around me.
8. I commit myself to personal renewal through prayer, meditation, and study.
9. I commit myself to responsible participation in a community of faith.

Fig. 17

us from addressing such tasks, since all human learning proceeds from such model-building, testing, and experimenting. These are indeed *appropriate* applications of computer technology, similar to the useful, *descriptive* models of past climate changes, or the construction of models of water use as it exceeds ecological tolerances, where the goal is *learning to modify and align human behavior* to natural cycles, rather than the manipulation of natural systems for short-term goals of current generations.

In devising principles, and ranges of local application of them, appropriate to the coming solar age, we must take into account the depletion of the ecosystem's natural and geological resources and devise new criteria under these new conditions of scarcity for the "success" and "failure" of cultures and value systems, i.e., whether on balance, they have been entropic or negentropic. This new "hindsight" will also be vital in devising alternative sets of value systems, principles of social interaction, laws, and the evaluation and measurement tools needed to monitor their behavioral outputs and institutional and technological configurations. These quantitative and qualitative evaluations and measurements of societies and their functioning are always important and will remain so, since they provide one type of vital feedback, together with voting, local participation, and regional and global information and representation systems. All our societal feedback mechanisms, from the face-to-face community behavior norms and sanctions to the global bodies, principles, covenants, and laws for the world's oceans, atmosphere, and space must themselves be continually monitored, improved, and changed. In computer- and systems-theory terms, the value system, principles, and rules of social interaction and resource allocations can be termed the "programs" of software, while the monitoring, measuring, and evaluative mechanisms and social feedbacks serve the "comparator" functions and permit continual adjustment, as in a thermostat-controlled, cybernetic system.

Interestingly, it is precisely in the erroneous descriptions of even such complex, nonlinear, cybernetic systems as "hierarchies" that so much misunderstanding has arisen. We are so entrapped in the hierarchical, either/or, dichotomizing logic that even our best systems theorists talk of such multidimensional, heterarchical, feedback-governed systems as embodying "hierarchies of control" and see other forms of organization only as "nonhierarchical," with fewer "levels of control." There is an implicit assumption that organization is impossible *without* hierarchy: governor and governed; levels and spans of control. This language betrays a fixed vantage point, thus a narrowed, unidirectional scan. I am grateful to have had the opportunity to clarify with Erich Jantsch before his death, in December 1980, that hierarchy is not a fundamental characteristic in nature—an impression given in his otherwise monumental work, *The Self-Organizing Universe* (1980). The term hierarchy implies that all such systems have definitive boundaries, rather than the reality: that, like all biological systems, they are open and have "leaky" boundaries. In fact, "boundaries" are often set by the intent of the observer to isolate and study *some* of a system's properties, rather than observe its actual seamless, teeming, multidimensional interactions with other systems and their environments. A good illustration of the latter, more realistic approach is that of Lewis Thomas' description in *The Lives of a Cell* (Viking, 1973) of the human organism as a complex ecosystem of symbiotic subspecies, whether our intestinal bacteria or all the other colonies of single-celled organisms that orchestrate themselves into the grand symphony we call a human being.

The confusion of hierarchical thinking as it collides with real-world, cybernetic systems is evident, for example, in the terminology of cyberneticists themselves in referring to the "thermostat function" as the "governor" controlling the "system." The ambivalence is obvious, but the failure of such taxonomy is more subtle. The concept of "control" is incorrect and could better be stated as "reciprocity" or "mutual causality" (in Maruyama's term), or in the tensor/compressor relationships in which Buckminster Fuller describes the stability of structures, as summarized in *Critical Path,* 1981. In any case, in a thermostat's functioning, the feedbacks *from* the rest of the system are "governors"—actually a *reversal* of our hierarchical concept of governance "from the center" or "from the top down"! Even Kenneth Boulding, one of the founders of general systems theory, makes these kinds of errors in his discussion of allometry (study of the optimal size of biological organisms), diseconomies of scale, and limits to the size of structures in *Ecodynamics.*[29] For in any open, nonlinear, cybernetic system, we may well ask which is the "governor": the structure, the program, the feedbacks, or the environment? The answer is *all of them,* acting in concert. As we saw in Chapter 11, the essence of these open, far-from-equilibrium, morphogenetic systems, with deviation-amplifying mutual-causal processes, is *indeterminacy.* Thus, my earlier assertion that *only the system can model the system and only the system can manage the system* may be indeterminate, messy, and frustrating to orderly minds —but, then, so is the real world.[30] These realities, when accepted, deal a death blow to our either/or, hierarchical, dichotomizing, objectified, location-specific, westernized logic, and clear the way for the more subtle, "soft-focus," intuitive, simultaneous cognition typified by the oriental world view, folk wisdom, and mystic traditions. "The *tao* that can be spoken is not the true *tao*" is a useful statement about the difference between "lens-type," focused awareness and "holographic," field awareness. A *re*statement of another oriental saying springs to mind: "He who speaks does not know—*s/he* who knows does not speak." Harmony and peace of mind are achieved by releasing the busy, compulsively differentiating, westernized mind to resolve the contradictions it generates, by transcending them and enfolding them in the intuitive, meditative contemplation of the whole: the great, mysterious oneness of creation. No wonder the teaching of meditation, relaxation-response, and other calming techniques are a new "growth industry" sweeping industrial societies and their executive suites!

The outlines of the ethics for the solar age, summed up in the edict of "thinking globally, acting locally," are visible today not only in the principles and declarations mentioned but in the actions of citizens in thousands of groups and networks with their own reports and periodicals.[31] Many are now linked in budding planetary coalitions, for example Amnesty International and its spectacularly successful campaign for human rights, and the emergence of a worldwide coalition against nuclear proliferation and for peaceful, safe, renewable energy. These new, effective organizations are also nonhierarchical models of how planetary coordination of vital functions such as environmental protection and wise stewardship of oceans, land, and mineral resources can be achieved through nonbureaucratic, democratic means and through coordination of local and regional actions based on agreed-upon global principles and uniformity of constraint. Models abound, from that of the international postal system run out of a small brownstone house in Switzerland, to global telephone linkages. If even a small fraction of what is spent on armaments were channeled into the serious study and development of such global organizational forms and functions, spectacular new successes would surely follow. And at last, serious discussion of global taxation of arms sales for such purposes is beginning, spurred by the Brandt Report, mentioned earlier. General formulas for the "thinking globally, acting locally" model would seem to imply keeping production, consumption, and institutional participation in economic and political life as close to the local level as possible. Global information and laterally designed communications systems of all kinds, based on user-control, random-access principles, and two-way representation, reciprocity, and exchange of experience, ideas, and planetary learning, together with democratic global compacts and functional bodies, could then implement and monitor planetary agreements and treaties. Much of this new global apparatus is slowly and informally emerging in the citizen-based, nongovernmental networks that often do the groundwork for the organizations that follow and ratify the new values. Many examples of these citizen networks and action/research groups and their emerging planetary conscientization are illustrated in this book and speak eloquently for themselves. Some represent concepts reemerging from earlier traditions (such as the "Call to Consciousness" statement from a 1977 UN Conference in Geneva by a group of six Native American nations), some are hybrids, and some are genuine innovations. One such innovation was the drive by citizens groups in California to pass bill A.B. 23 into law in 1979, setting up a State Commission on Crime Control and Violence Prevention mandated to investigate the possible causes of crime inherent in poor nutrition, insensitive birthing practices, tactile deprivation, repressive attitudes toward sex and the human body, sex-role stereotyping, violent television programming, poverty, prejudice, and powerlessness. Passage of the bill was achieved by a coalition of humanistic psychology groups, the New Age Caucus, a statewide political activist group, and State Assemblyman John Vasconcellos, author of *A Liberating Vision: Politics for Growing Humans* (Impact Press, 1979). It is almost ludicrous to contrast this simple, sane law with the almost paranoid compendium of facts and figures on the objective occurrence of violence that resulted from the National Commission on the Causes and Prevention of Violence set up in 1969 by President Johnson in the wake of the black protests in 1968.

The Commission skirted the issues of oppression and structural violence by traditional values, mores, and authority, noting that Americans have always been a violent people; and in a classic understatement, it concluded that "we cannot assume that modernization will bring political stability in its wake"—a truism borne out vividly in Iran.[32] Another example, in contrast to the bureaucratic approach, was the policy declaration passed by 61 percent of San Francisco's voters in November 1978: "The people of the City and County of San Francisco demand that the federal government cease spending our tax money for wasteful military purposes and instead use it to provide jobs and services that our people so desperately need, thereby creating jobs with peace by cutting the military budget."

Another concept for amplifying the opinions of citizens emerged as the major outcome of the First Global Conference on the Future, Toronto, 1980, that of the global public opinion poll, to be conducted by the new Global Futures Network (of which I am a founder). These polls hope to survey citizens' opinions in all countries on global issues, over the heads of their governments and official spokespeople, in the belief that average citizens do not benefit from arms sales, multinational corporate deals, resource exploitation, and war.

The new logic from which the ethics of the solar age will derive is a synthesis of oriental, westernized, and "folk" wisdom, which acknowledges the range of applicability and the limitations of each. This type of all-encompassing image of a holographic universe recognized by an analogously functioning holographic human awareness is expressed in a spate of exciting new scientific speculations, as mentioned in Chapter 12. It has produced a whole new genre of literature, from the early, prototypical writings of Buckminster Fuller and Marshall McLuhan, with their nonlinear, compressed, endless-dependent-clause-filled sentences; as well as the pioneering style of James Joyce's portrayal of the interpenetration of subjective and objective in the simultaneous, multidimensional, teeming flow of his perceptions. Examples of the emerging, "open systems" style include Fritjof Capra's *The Tao of Physics* (Shambala, 1975), Gary Zukav's *The Dancing Wu Li Masters* (Morrow, 1979), Kenneth Boulding's *Ecodynamics* (*supra*), Erich Jantsch's *The Self-Organizing Universe* (Pergamon, 1980), Mary Daly's *Gyn/Ecology* (Beacon, 1979), Marilyn Ferguson's *The Aquarian Conspiracy* (J. P. Tarcher/St. Martin's Press, 1980), William Irwin

Thompson's *Passages About Earth* (1973) and *Evil and World Order* (1976), both Harper & Row, and his latest and best, *The Time Falling Bodies Take To Light* (St. Martin's Press, 1981), Gregory Bateson's *Mind and Nature* (Dutton, 1979), and *Gödel, Escher, Bach* (Basic Books, 1979), a poetic integration by computer scientist Douglas Hofstadter that relates principles of music, art, mathematics, and biology in a dazzling synthesis of metalogic, illustrating the uses and limits of all logical and other symbol systems. Many of these writers even break with linear style and do not number their chapters, acknowledging that in multidimensional exposition it does not matter whether the reader begins in the "middle," "end," or "beginning" of a book, since each chapter, and sometimes each paragraph or sentence is an attempted hologram. Thus, scientific exposition moves from reductionism and limited, static exactitude to flowing, cognitive acceptance of the new, indeterminate nature of reality—moving inexorably toward poetry and the gems of simultaneous, intuitive perception of haiku and the Zen koan, and beyond words and symbols, leading the reader experientially into the full awakening discovery of the eternal Now.

The logic and ethics of the solar age will flow from an underlying principle now being rediscovered by Western science: that of *interconnectedness*. This principle has been primary in all religious traditions (the very Latin root of the word religion is *religio,* meaning to bind together) and is also emerging in personal knowledge as the "peak experience" of undifferentiated unity described by psychologist Abraham Maslow. This is synthesizing to provide a powerful new rationale behind demands and aspirations for a more just world order and the efforts to build scenarios of its functioning. Old orders collapse because their allometric dimensions are exceeded and they can no longer contain or program new forms of energy. In truth, "order" and "chaos," too, are sides of the same coin and are seen in terms of the observer. It is becoming increasingly clear in the light of recent international events that the world has changed fundamentally, and a new world order is emerging inevitably. As Marilyn Ferguson notes in *The Aquarian Conspiracy,* "when society falls apart, as it is now, it is reorganizing at a higher level." The same inference can be drawn from the work of Nobelist Ilya Prigogine in his study of biological systems he calls dissipative structures, which experience transformations through the exceeding of thresholds and the amplification of disequilibrium states. But here, too, we must avoid the trap of hierarchical thinking. I prefer to view these transformations not in terms of "higher" "levels," but as encompassing greater dimensionality and openness to the eternal, ever-present total potentiality of the universe, whether represented in Lewis Thomas' single cell, Jean Houston's body/mind continuum, James Lovelock and Lyn Margulis' single planetary organism, "Gaia," or Karl Pribram and David Bohm's holographic universe/human-awareness continuum, in which there is no "higher," "lower," "up," or "down"—just as Buckminster Fuller has been telling us for so long. Thus the rest of this century will indeed be a period of chaos, terrifying to those entrenched in the old order but exhilarating to those with few stakes in upholding it, who see new possibilities. We are seeing the disintegration of the ligaments of technocratic industrial society, watching the emergence of a new world order. The new paradigms will help us keep our balance, composure, and compassion.

Thus, in general terms, we are quite aware of the basic foundations on which the new world order must be built. Fundamentally, these principles are as follows:

- the value of all human beings,
- the right to satisfaction of basic needs (physical, psychological, and metaphysical) of all human beings,
- equality of opportunity for self-development for all human beings, as expressed, for example, in new measures of human development such as the basic human needs (BHN) measure proposed by the United Nations Environment Program; (the comprehensive) physical quality of life indicator (PQLI) (see Fig. 18) of the Overseas Development Council; the measure of economic welfare (MEW) of economists James Tobin and William Nordhaus, which somewhat improves on the GNP; the Japanese net national welfare (NNW); and the U. S. Midwest Research Institute's quality of life (QOL); (even though the three latter efforts are only incrementally better than GNP),
- recognition that these principles and goals must be achieved within ecological tolerances of lands, seas, air, forests, and the total carrying capacity of the biosphere,
- recognition that all these principles apply with equal emphasis to future generations of humans and their biospheric life-support systems, and thus include the respect for all other life forms and the earth itself.

TABLE 1

	Per Capita GNP ($)	PQLI
Low-Income Countries: (under $300 per capita GNP)	**152**	**39**
Afghanistan	110	19
Egypt	280	46
Ethiopia	100	16
India	140	41
Kerala State	110	69
Indonesia	170	50
Mali	80	15
Nigeria	290	25
Sri Lanka	130	83
Lower Middle-Income Countries: ($300–$699 per capita GNP)	**338**	**59**
Albania	530	76
Cuba	640	86
Ghana	430	31
Guyana	500	84
Honduras	340	50
Korea, Republic of	480	80
Morocco	430	40
Thailand	310	70
Tunisia	650	44
Upper Middle-Income Countries: ($700–$1,999 per capita GNP)	**1,091**	**67**
Algeria	710	42
Argentina	1,520	84
Brazil	920	68
Gabon	1,960	21
Iran	1,250	38
Iraq	1,160	46
Mexico	1,090	63
South Africa	1,210	48
Taiwan (ROC)	810	88
Yugoslavia	1,310	85
High-Income Countries: ($2,000 or more per capita GNP)	**4,361**	**95**
Kuwait	11,770	76

What the PQLI Can Show Us

Table 1 compares the performance of selected countries measured by GNP and by the Physical Quality of Life Index. The average figures for countries grouped by income show a direct relationship between the level of per capita GNP and PQLI. Yet this correlation does not hold for all countries. For example, Cuba, Guyana, and Korea (with per capita GNPs of less than $700) as well as Sri Lanka and the Indian state of Kerala (with per capita GNPs well below $200) all have PQLI rankings above the average of countries with incomes between $700 and $2,000. In fact, their PQLIs are well above those of Gabon, Iran, Kuwait, and Libya (with per capita GNPs of $1,960; $1,250; $11,770; and $4,640, respectively).

These divergences from the expected relationship suggest that significant improvements in basic quality of life levels can be attained before there is any great rise in per capita GNP; conversely, a rapid rise in per capita GNP is not in itself a guarantee of good levels of infant mortality, life expectancy, or literacy.

A major advantage of the PQLI is not only that it measures the *current* level of achievement, but that it is a fairly sensitive measure of change over time as well. Table 2 shows quality of life changes for a number of countries over the last two decades.

The index also can be used for

Libya	4,640	42
Sweden	7,240	100
United Kingdom	3,590	97
United States	6,670	96

TABLE 2

	1950s	1960s	1970s
Algeria	35	38	42
India	28	36	41
Egypt	32	41	45
Brazil	53	—	66
Sri Lanka	62	77	83
Poland	72	86	93
France	87	94	97

TABLE 3

	1900	1939	1950	1973
All U.S. Population	63	85	91	96
White Population	65	87	92	97
Other Races	30	71	81	89
Selected States				
Mississippi		81	87	92
New Mexico		69	85	94
Texas		81	87	95
Wisconsin		89	93	97
Minnesota		91	95	98

NOTE: The PQLI ratings (as well as life expectancy, infant mortality, and literacy figures) of all countries are provided in a somewhat revised version in *The United States and World Development: Agenda 1980*, by John W. Sewell and the staff of the Overseas Development Council, 1717 Massachusetts Ave., N.W., Washington, D.C.

intra-country comparisons. Table 3 reveals interesting regional and ethnic contrasts within the United States. This suggests that policy makers can set appropriate targets for improvement in PQLI ratings not only for developing countries but also for developed countries and for regions and specific groups within countries.

Some Concluding Observations

The PQLI is a way of measuring not only the starting level of a country's achievement, but also the rate at which it is able to move toward some attainable level that is more or less fixed. Therefore, it would be a mistake to view the ranking of countries at any point as the result of a competition.

In this sense, PQLI trends suggest something different from the discouraging evidence provided by GNP comparisons over time, which indicate that rich countries are steadily widening the gap between themselves and poor countries. When we measure physical quality of life attainments, the gap between the industrialized countries and most developing countries appears likely to be *narrowed* over time, with a principal issue being: "How long a time?" For example, India's PQLI rose from 28 to 41 between the 1950s and the early 1970s; during the same period, the PQLI in the United States rose from 91 to 96. Nor is the PQLI a falsely optimistic instrument designed to mislead. Rather, it reminds us of some important matters—that rapidly rising GNP may go for fancy gee-gaws, nuclear explosives, or great armies; but the PQLI measures success in attaining certain basic conditions that contribute to a satisfactory quality of human existence.

The evidence that some low-income countries have been able to attain fairly high PQLI rankings suggests that there is hope that substantial improvements in at least these *minimum* human requirements can be provided much more quickly than we realistically can expect per capita GNP to rise under the best of circumstances. We are seeing that low national income need not be synonymous with abject poverty and its consequences. It is now our task to learn what the policies are by which such important achievements can be speedily realized in all poor countries.

Fig. 18

Historically, human development can be viewed as many local experiments at creating social orders of many varieties but usually based on partial concepts; i.e., the social orders worked for *some* people, at the expense of *other* people, and were based on the exploitation of nature. Furthermore, they worked in the *short* term but appear to have failed in the *long* term. Today, all experiments of local and partial human development based on these short-term exploitations, have been failures in one way or another when seen in a planetary perspective. Today, we know that such societies are impossible to maintain and that the destabilization on which they have built themselves are now affecting their internal governmental stability and the global stability of the planet. Interestingly, those instabilities can all be stated in scientific terms:

1. In classical equilibrium thermodynamics, in terms of the first and second laws—the law of conservation and the entropy law—that all human societies (and all living systems) take negentropy (available forms of energy and concentrated materials) and transform it into entropic waste at various rates, and that we can measure these ordering activities and the disorder they create elsewhere. An understanding of this process leads to the realizations that, properly speaking, the U. S. (or any other) Department of Energy should be termed the Department of Entropy. Another example of the workings of the laws of thermodynamics in human societies is the ratio of order/disorder we see within and between societies, e.g., the structuring of European countries in their colonial periods at the price of the concomitant disordering of their colonies, culturally and in terms of indigenous resources.

2. In terms of biology and the evolutionary principle, summed up as: "nothing fails like success," i.e., the trade-offs between short-term and long-term stability and structure; between adaptation and adaptability.

3. In terms of general systems theory, as the phenomenon of suboptimization, i.e., optimizing some systems at the expense of their enfolding systems.

4. In terms of ecology, as violations of the general principle of interconnectedness of ecosystems and the total biosphere; i.e., the continual cycling of all resources, elements, materials, energy, and structures. This interconnectedness of all subsystems on planet earth is much more fundamental than the interdependence of people, nations, cultures, technologies, etc.

Thus, the aspirations for a new world order are not only based on ethical and moral principles, important as these emerging planetary values will be for our species' survival. The need for a new world order can now be *scientifically* demonstrated. We see the principle of interconnectedness emerging out of reductionist science itself, as a basis for it, and the concomitant ecological reality that *redistribution is also a basic principle of nature.* All ecosystems periodically redistribute energy, materials, and structures through biochemical and geophysical processes and cycles; therefore all human species' social systems must also conform to principles of redistribution of these same resources that they use and transform, whether primary energy and materials or derived "wealth" (capital, money, structures, means of production, and "power").

The new scientific understanding of *interconnectedness* and the fundamental processes of *redistribution* are accompanied by the emerging paradigms of *indeterminacy, complementarity,* and *change* as basic descriptions of nature. The five principles operate not only at the phenomenological level of our everyday surface realities and in our observance of nature (in the "middle-range" realm of classical physics) but also at the subatomic level of phenomena of quantum mechanics. The frontier of quantum mechanics is building on the last question raised by Einstein, set forth in his paper written in 1935 with Podolsky and Rosen, "Can the Quantum-Mechanical Description of Physical Reality Be Considered Complete?" The issue concerned the fact that quantum mechanics built on the assumption that causality was *local* (the idea that events happen in certain space/time locations, discretely) and that events and phenomena could not affect each other at a distance, with no intervening medium or means of connectedness. Most physics continues in this assumption, in search of behavior and interaction of ever smaller, more numerous particles, waves, quarks, and improbable phenomena, properties, tendencies, and the like. However, in 1964 J. S. Bell, a physicist at the CERN laboratories in Switzerland, devised a theorem demonstrating the limits of the mathematics used in quantum mechanics and presented physicists with a very fruitful paradox: calling into question this local causality and discreteness on which all physics is founded and leading to the new hypothesis that subatomic events and phenomena are also fundamentally interconnected.[33] Similarly, we have seen how the principles of indeterminacy, complementarity, and change apply not only to quantum levels but to biological, ecological, and social processes and phenomena as well. Thus these five principles emerging in westernized science imply behavioral human adaptation *and* learning *and* social principles:

- *interconnectedness* (planetary cooperation of human societies)
- *redistribution* (justice, equality, balance, reciprocity)
- *change* (redesign of institutions, perfecting means of production, changing paradigms and values)

- *complementarity*   (unity *and* diversity, from either/or to both/and logics)
- *indeterminacy*   (many models and viewpoints, compromise, humility, openness, evolution, "learning societies")

The new world order can be founded both on scientific *and* ethical principles. We are *discovering* the new world order in science and *remembering* that we know it already, since these same five principles are found in all religious and spiritual traditions. Ethical principles have become the frontiers of scientific inquiry. Morality, at last, has become pragmatic, while so-called idealism has become realistic.

But it is equally clear that the necessary global transformation will either occur amid increasing human resistance and social rigidity or it will be accommodated and encouraged by more enlightened and flexible social policies and shifts in human values and behavior. Thus the global politics of reconceptualization involves the emergence of pragmatic strategies, new coalitions, and what might be termed "a new proletariat": not only workers, as Marx preached, but all people who have been tyrannized by arbitrary symbol systems and social designations of their roles; for example, all the world's people whose work has not been monetized, and therefore not valued—rural villagers, subsistence farmers; India's Harijans and all so-called lower-caste workers; ethnic peoples in all nation-states who have been ghettoized in some way, such as the Native American nations, the Aborigines of Australia, and the Ainu in Japan; and all people undervalued by discrimination because of color, sex, race, or religion. In the same way, countries and regions have been subordinated to the tyranny of the global, monetized economy, and their contribution to the world's development and human culture has thereby been devalued. We now see in the emerging paradigms in science that *all these issues flow from erroneous, abstract drawing of boundaries where none exist in nature*. Yet the raising of the conscience of human societies as to these errors is a political task, requiring that these issues and existing power centers be confronted. This is the creative role of the Third World and all subordinated groups. This planetary consciousness-raising activity must be militant, reasoned, and nonviolent, as for example the important new struggle over the finite electromagnetic spectrum.[34] It is now clear that there exists a blatantly unfair monopoly by industrialized nations of this vital global resource as the medium of communications from radio, TV, and telephone, to air and marine navigation, microwave relay, radar, satellites, and other strategic systems. Here the division between nations of this planetary resource is such that, for example, 90 percent of the radio spectrum is monopolized by 10 percent of the world's population. The battleground for the electromagnetic spectrum was joined at the World Administrative Radio Conference, and its implications for the development of more balanced, planetary information sharing are crucial. The lengthy set of issues involved in a new, equitable distribution of the planet's information and communication systems will be another vital arena for the politics of reconceptualization, since only when all cultures have the means to enter the world's dialogue on an equal footing can we hope for a large enough forum for conflict resolution and the creation of new cultural alternatives. Such information-sharing can also rebalance the current tyrannies of *some* cultures' paradigms, symbol systems, and values over those of *other* cultures. Appropriately, UNESCO has become an ongoing forum for dialogue on these issues.

Another key issue that is helping clarify our vision of new planetary realities is arising out of the distillation of the major ideological views of "development" that have operated since the industrial revolution, particularly in communism, socialism, and capitalism and their various expressions: the growing concern for the needs of *real* human beings, and the general issues of *injustice* and *inequality*. These concepts of alternative forms of development, mentioned earlier, are illustrated in three recent, major views of social progress:

1. The Cocoyoc Declaration's definitions of "development," i.e., "the development of human beings, not the development of countries, the production of things, their distribution within social systems, or the transformation of social structures, thus entailing redefinition of the whole purpose of 'development,' which has confused means with ends" (United Nations, General Assembly Document A/C. 2/292).

2. The socialist view, based on Karl Marx's legacy, concerning the issues of oppression of groups in societies by other groups, and Marx's historical documentation of the oppression of working classes by capitalists and property owners as a prime example of this arbitrary oppression.

3. The liberal legacy of the Enlightenment; of political democracy
   (often stated in still-arbitrary languaging as "one man, one
   vote"), which, however, did lead inevitably in the U.S.A. to the
   Carter administration's focus on "human rights" as a major
   policy.[35]

Thus lip service, at least, is now generally paid to this focus on the
rights of the human being, and in our communications-rich world,
few nations can ignore the public-opinion sanctions against blatant
violations of these human rights. Today's leaders have much experience with this double-edged sword of human rights, which exposes
paradoxes of injustice and oppression of minorities within their own
borders and leads to such examples as the pressing of such domestic
grievances of Native Americans and blacks in the U.S.A. and of Soviet feminists and other dissidents in the U.S.S.R. in the world forum
of the United Nations and in an increasingly potent world public
opinion. Thus the general issue of all forms of arbitrary oppression is
coming to the fore, as we have seen vividly in the resurgence of ethnic separatism, and demands for cultural and religious autonomy,
whether in the massive upheaval in Iran, or the guerrilla actions that
brought autonomy to Spain's Basques and in the new protests from
many nonaligned nations at the Soviet troop movement into Afghanistan. A new type of planetary coalition must emerge to undergird politically the pervasive aspirations in industrial and Third World
countries for the new world order: composed of all people tyrannized
by monetization and other arbitrary definitions and political boundaries. Already, such new alignments are visible in the now-frequent
meetings of the planet's culturally disenfranchised ethnic and tribal
peoples, such as those mentioned earlier in Chapter 5. This winning
coalition not only will include workers, as in the earlier concepts
of the Industrial Workers of the World, since workers in industrial
countries are now in the anomalous position in many cases of enjoying better conditions at the expense of more oppressed groups, such
as blacks in the U.S.A. or Pakistanis in Britain, and, unwittingly,
through their corporate employers' exploitation of cheaper labor in
many countries of the southern hemisphere. Similarly, the latest "developing nation," of the world's women, has been overlooked and
their basic roles in production, maintenance, and agriculture uncounted in the economic definitions of capitalism and socialism and

virtually all economic data of their monetized sectors. Revealingly, even the inspired definition of development in the Cocoyoc Declaration grew out of concepts such as those expressed by the Tanzanian Minister of Development, Wilbert K. Chagula: "Our first concern is to redefine the whole purpose of development. This should not be to develop things but to develop *man*" (italics added). Similar statements that *man* has not been able to control *his* technology, and concerning *man's* alienation from nature are precisely correct. The winning planetary coalition must now include woman if it is to be large enough to form a politically viable majority. It is also clear that our chaotic societies now need "mothering" as well as "fathering," in a more balanced sharing of leadership responsibilities. Thus, the meta-issues of human needs, human rights, and ending all arbitrary oppression such as the tyranny of monetization, resting on secure scientific knowledge now emerging, provide the action formula to make operational the ethics of the solar age and actualize its social expressions in a balanced, harmonious, ecologically aligned, new world order. If we in the industrial nations cannot face up to the changes we must make, there are other resource crises waiting just down the road to zap us even harder than the energy crunch: for example, our dependence on foreign materials and metals,[36] and another U.S. resource crisis quietly brewing over our mismanagement of our nation's water supplies. How many more such signals will we need, both from nature and from other countries increasingly impatient with our profligacy, and now rightly demanding a New International Economic Order? Furthermore, these issues of redistribution are *inter*generational: future generations also deserve their share. In the media-rich global village, created, ironically, by their own technology, the industrial nations will be under pressure to conform to these principles of *inter*generational justice, as well as for greater equity in access to resources for all people *alive today,* as outlined by John Rawls's *A Theory of Justice.*[37]

This three-dimensional view of justice and equity is not contradictory, but complementary. For if access to resources, power, and wealth are broadly shared within and between nations, this in itself reduces the dangers of concentrated power and wealth, which leads to overexploitation of resources, human oppression, and the depletion and destruction of future options. Only social systems that learn to use today's resources frugally and fairly can create perpetually renewable, resource-based systems of production managed for long-term sustainability, from the understanding that living in harmony with each other and nature is not merely a moral imperative—it is now the only pragmatic course of action.

Can we unravel the dangerously unstable centralization of political and economic power in nation-states, stem the arms race and nuclear power, bring resource-exploiting multinational corporations under the rule of world law, redirect the macho technologists before it is too late? We must proceed as though it is possible to achieve those tasks—nothing else makes human sense. For example, the peace initiatives of Egypt's Anwar el-Sadat and Israel's Menachem Begin, for which they received the Nobel Peace Prize, were primarily pragmatic, since only peace between their two countries could release sufficient resources to achieve improvement in their domestic, civilian conditions (*Fortune,* January 28, 1980, pp. 68–75). During the coming decades of shifting to renewable, resource-based economies, nations will begin to see more clearly than ever the stark choices between guns and butter, and that militarism on today's scale leads only to national bankruptcy. The same reevaluations are occurring domestically in such enlightened industrial societies as Sweden, with its citizens' changing view of the notion of "standard of living." In a 1979 study by Nordal Ackerman, "Can Sweden Be Shrunk?" opinion polls show these changing values clearly. After the top priority of a secure job, nonmaterial needs are now *preferred* over raises in the "standard of living": less dangerous working environments, more varied housing areas, more rewarding leisure and cultural activities. Of course, these less basic needs could only have emerged after the basics had been assured, but nevertheless they are significant as a changed focus from economic growth as *equated* with "productivist" goals. The study noted that, "In 1974, 81% were already of the opinion that the standard of living and energy consumption should not be raised further. Three years later, only 1% of the public held that the standard was too low, while 60% thought it was too high. Specifically asked about their own standard, a mere 8% thought it should be higher—70% were content with it as it was."[38]

As stated by Mustafa Tolba, Executive Director of the United Nations Environment Program, in the Foreword of John and Magda McHale's *Basic Human Needs; Framework for Action* (Transaction Books, Rutgers University, 1979): "The world has the capacity to achieve sustainable satisfaction of basic human needs for mankind [presumably, *human*kind] as a whole." Thus it is a challenge to many conservationist views that the planet's human population has already exceeded its "carrying capacity" (even though this concept has yet to be formalized). However, an increasing number of world organizations and world citizens are coming to the conclusion that, at least, we must fully explore the possibilities in the famous statement of Mohandas Gandhi that the world has enough for our needs but not for our greeds.

For example, at least one respected political party, New Zealand's Values Party, has adopted the Gandhian approach as a foreign-policy credo and managed to capture 5 percent of the country's vote in their 1975 election. The same idea is stated by Buckminster Fuller as the basic premise behind this World Game, namely, "making the world work for 100% of humanity." This formula lacks a future time dimension, whereas population biologists have argued sincerely and convincingly for the importance of that dimension, doubting even the feasibility of decently providing for the planet's current 4 billion inhabitants, let alone the next doubling of human numbers. But pragmatic political analysts see no sidestepping of the effort to address this immense task of meeting human needs, even though this will require a basic rethinking of all of our traditional ideas about development economics. The key variable in the population-resource equation so often overlooked by those theorists living comfortably in the industrialized northern hemisphere countries is that of per-capita consumption.

Basic Human Needs is one of the best recent efforts to redefine "economic development," rejecting the traditional ideas of Kuznets, and more recently, Walt Rostow in his *Stages of Economic Development* (1960), that "development" is a trickle-down process.

In sum, such a world trading system is not only failing on its own terms, but as it homes in on its own simple price- and GNP-defined goal, it succeeds only in disordering every local social system and every local ecosystem on the planet, reaching an absurd global equilibrium by turning the entire planet into a global "economic behavioral sink"!

Many, including myself, have sought to unravel the cant and mystification surrounding traditional development economics. The task, however, cannot be accomplished only with logical arguments.

In a quantification-fixated policy milieu, the case must be buttressed by new statistics, formulated with diametrically opposed assumptions. Further evidence was unwittingly provided by David Smith in the conservative British journal *Now,* May 30, 1980. In his article "The Invisible Threat to Britain's Fortunes," he noted that Britain's balance of payments could no longer expect its usual bolstering by her "invisible exports" (i.e., insurance, profits from overseas, etc.). After continual surpluses in these "invisibles" for one hundred years, they would fall into deficit in 1980, largely because Britain *herself* would have to start repatriating profits from her North Sea oil to U.S. oil-company investors. In other words, it is better to be an investor in some *other* economy than have others invest in one's *own!* The British should know, since their prosperity during their colonial period rested largely on their foreign investments—a point used by Third World economists whose studies show that foreign investments harm, rather than help, their countries. Such pioneering work as Irma Adelman and Cynthia Taft Morris' *Economic Growth and Social Equity in Developing Countries* (1973), Jan Tinbergen's *Reshaping the International Order* (1976), Frances Moore Lappé's *Food First* (1977), and those new indexes mentioned earlier are examples of these efforts.

The McHales have taken this crucial work much farther, developing a quantitative framework for basic human requirements and operational definitions that, one hopes, will provide new bases for judging the performance of governments all over the world in meeting these basic needs *first.* At last, the old, trickle-down approach is reversed. Such needs for food, clothing, shelter, health, education, equality of minorities and women, human rights, employment, and participation in governance should and can be the benchmarks for a new definition of development. Only when we are able to debate world development in terms of performance of new indicators such as BHN (basic human needs), PQLI, and NNW, or eventually a subjectively polled GSI (general satisfaction index), which I proposed in *Creating Alternative Futures,* can we begin to discredit "development" measured only by GNP.

The new views of the Third World were well summed up in the Founex Report, prepared for the Stockholm conference in 1972 that founded the UN's Environment Program.

In the past, there has been a tendency to equate the development goal with the more narrowly conceived objective of economic growth as measured by the rise in gross national product. It is usually recognized today that high rates of economic growth do not guarantee the easing of urgent social problems. Indeed, in many countries high growth rates have been accompanied by increasing unemployment, rising disparities in income —both between groups and between regions—and the deterioration of social and cultural conditions. A new emphasis is thus being placed on the attainment of social and cultural goals as part of the development process.

In an ironic twist, the nations of the southern hemisphere are now providing moral leadership to the industrial countries caught in the "consumerist trap" and the pathologies of affluence. The Cocoyoc Declaration summarized this new kind of "assistance" for those in "spiritual poverty":

> The world today is not only faced with the anomaly of underdevelopment. We may also talk about overconsumptive types of development that violate the inner limits of men and the outer limits of Nature. . . . Even though the first priority goes to assuring the minima, we shall be looking for those development strategies that also may help the affluent countries, in their enlightened self-interest, in finding more human patterns of life, less exploitive of Nature, of others, of oneself.[38]

NOTES – CHAPTER 13

[1] It is increasingly evident that new value systems are emerging around the new criteria of sustainability and renewable resource-based societies—both in the counter-culture and citizen movements of the industrial nations, and in those of the Third World. It is not surprising that these new value systems are remarkably similar, since they grow out of new shared knowledge of the deeper processes of bio-productivity and appreciation for indigenous cultures and more qualitative forms of development. For example, the manifestoes emerging from Africa (the Declaration of Dakar, January 1980 and the Nairobi Declaration of July 1979) contain the same recommendations for the integration of socio-cultural values in a new, person-centered development and effective participation by all citizens in every phase as are found in the manifestoes of political groups from many European countries, the U.S.A., Canada, Australia, and New Zea-

land. Such documents as the Ecological Manifesto for a Different Europe, drafted by ECOROPA, headquartered in Geneva, The Values Party (New Zealand) manifesto *Beyond Tomorrow,* and the platforms of the U.S.-based Citizens Party and the New World Alliance call for many of the same policies: maximum citizen participation in all levels of government; worker ownership and self-management; new, person-centered development focused on the quality of life and preservation of cultural integrity; control of multinational corporate and financial power; and implementation of technology redirected to serve human purposes. Virtually all these statements refer to ending the arms race and the conversion of military budgets to human needs, and emphasize an end to wasteful consumption and a shift toward spiritual and personal growth. The same startling congruence emerges most clearly in the latest form of "technology transfer" in reverse: the wholesale transmitting of Eastern religious traditions (software!) to the confused cultures of the Occident. Indeed, only if the industrial countries show that they, too, are willing to give up their old preoccupations with materialism, keeping-up-with-the-Joneses, and fascination with "big-bang" technologies, can they hope to share world leadership and moral authority in the Solar Age.

[2] Monograph available from Global Education Associates, 525 Park Avenue, East Orange, NJ 07017, in their series on world order *The Whole Earth Papers,* published quarterly, subscription $10 per year.

[3] It is shameful to find that the United States has been one of the most recalcitrant in several of these new planetary policy debates, for example, foot dragging on the Law of the Seas and stymieing the Moon Treaty, as business lobbies have asserted their private-property prerogatives to the absurd lengths of extending them to exploiting at will the minerals on the deep ocean floors and on the moon (*The Christian Science Monitor,* February 20, 1980).

[4] Monographs and studies available from the Swedish Government, from The Secretariat for Futures Studies, Box 7502, S-103 92 Stockholm, Sweden.

[5] Examples of the imperative for global cooperation for many of the world's problems include a New Economic World Order based on the understanding of the *de facto* collapse of the existing international monetary system, as described, its appalling effects outlined in the Terra Nova Statement on the International Monetary System and the Third World (Kingston, Jamaica, October 1979; *Development Dialogue,* 1980: 1, p. 29), and, most crucial, the widening imbalance between the world's deprived citizens and those who overconsume, and the ever-increasing militarization that diverts funds from human needs. These issues were addressed by the Independent Commission on International Development Issues, chaired by former German Chancellor Willy Brandt, and its report, *North-South: a Program for Survival,* released in February 1980.

The Commission, stressing the mutuality of all of these problems, called for a summit meeting of some twenty-five world leaders "to address the mortal dangers threatening our children and grandchildren." The report called for: 1) an international tax on trade with the heaviest tax on weapons sales; 2) an agreement between oil producers and oil suppliers that would provide regular supplies of oil, ensure predictable prices, encourage conservation, and stimulate oil exploration in the Third World; 3) an international currency to replace the dollar; 4) the creation of a global fund for development; and 5) an $8-billion-per-year fund for food aid to the Third World.

[6] The recent buildup in U.S. military budgets, now rising more steeply than in any comparable period since World War II, will total $1 trillion between

1980 and 1985, while Ronald Reagan increased these funds even further. The new defense posture is a reliance on "smart missiles." Carter's Defense Secretary Harold Brown asserted in 1980 that "Our technology is what will save us," while Pentagon research and development chief William J. Perry claimed "the new missiles will revolutionize war." Incorporating all the latest microprocessor automation, they will take over most human spotting, targeting, and firing, and are nicknamed "the fire-and-forget" weapons. Former Defense Secretary James Schlesinger similarly enthused about these deadly new missiles, stated baldly, "Our forces must be perceived by the Soviets to be growing more fearsome" (*Business Week*, "The New Defense Posture," August 11, 1980, p. 77). Nothing could provide a worse scenario for disaster, since the Soviets, with memories of invasion and staggering losses in World War II, will never allow themselves to fall behind the United States again, a fact Ronald Reagan ignores.

[7] *Teilhard Review*, October 1974, Volume IX, #3, pp. 72–80, London.

[8] The Center for Defense Information, Washington, D.C., in its monthly *Defense Monitor*, August 1980, noted, ". . . an ingrained tendency in our government to overstate Soviet military power and understate U.S. and allied strengths." Rather, the *Monitor* noted, "the total military spending of NATO and the Warsaw Pact should be compared. While NATO spent $215 billion in 1979, its rival spent $175 billion."

[9] *New Directions*, Volume II, No. 2, March–April 1980.

[10] *History of Women's Suffrage*, Volume I, Arno Press, 1969 (originally published in Rochester, N.Y., in 1881 in six volumes, authored by Elizabeth Cady Stanton, Susan B. Anthony, and Matilda Joslyn Gage).

[11] Elise Boulding, *Women in the Twentieth Century World*, Chapter 7 (pp. 167–83), Women and Peace Work, Chapter 8, NBO's and World Problem-Solving (pp. 185–209), Halsted Press, Sage Publications, Inc. 1977.

[12] *Time*, August 4, 1980.

[13] Documentation of this oppression was presented at the UN Conference on the Status of Women in Copenhagen, July 1980, by the International Labor Organization (ILO): the world's women work two thirds of all the hours worked (paid and unpaid), they produce 44 percent of the world's food; however, women are paid only 10 percent of the world's wages and own only 1 percent of the world's land (*The Christian Science Monitor*, July 30 and August 1, 1980). In addition, Western bureaucrats, in attempting to "aid" Third World "development," make matters worse. With their patriarchal assumptions, they skew aid to benefit men exclusively and perpetuate the domination of women while beggaring and depriving them even further, as documented in *The Domestication of Women*, by Barbara Rogers, Kogan Page, London, 1980, and in *The Sisterhood of Man*, by Kathleen Newland, a Worldwatch Institute Book, W. W. Norton, 1979, and earlier studies, such as *Women and World Development*, eds. Irene Tinker and Michele Bo Bramsen, American Association for the Advancement of Science and the Overseas Development Council, 1976, Washington, D.C. Thus, women have the most to gain from a New International Economic Order, and resolutions supporting it were carried at both of the UN conferences on women (Mexico City, 1975, and the 1980 Copenhagen meeting).

[14] "Social Ecology and the Dual Economy," an English excerpt from Anders Arbeiten-Anders Wirtschaften, ed. by Joseph Huber, Fischer-Verlag, Frankfurt, 1979).

[15] Washington *Post*, "The Non-Profit Economy," March 11, 1979.

16 *The Green Revolution,* published monthly by The School of Living, P. O. Box 3233, York, Pa. 17402. Subscription $8 annually.

17 See, inter alia, Elaine Hoffman Baruch, "Women in Utopia," *Alternative Futures,* Volume 2, No. 1, Winter 1979.

18 See, for example, June Singer, *Androgyny: Toward a New Theory of Sexuality,* Anchor Press/Doubleday, 1976, as well as earlier work such as Charlotte Perkins Gilman, *Herland* (1915), reissued by Pantheon in 1979.

19 See, for example, E. C. Riegel's *Private Enterprise Money* (Harbinger, 1944) and Ralph Borsodi's experiment with the introduction of a commodity-backed currency, the *constant,* which I discussed in *Creating Alternative Futures* (p. 166). Such ideas are again sprouting all over the United States. See, for example, Ellery Foster's *The Coming Age of Conscience,* published by Friends of the Peaceful Alternatives, Winona, MN, 1977, which also runs the Free Trade Exchange (a bartering system and credit union). Another bartering cooperative is the Community Service, Inc., P. O. Box 243, Yellow Springs, Ohio, to name but two of the many thousands of these new bartering efforts. Kirkpatrick Sale's *Human Scale* (Coward, McCann & Geoghegan, 1980) is a good source of further information, as is *The Barter Book,* by Dyanne Asimow Simon (E. P. Dutton, 1979).

20 *Women, Minorities and Employment Discrimination,* ed. Phyllis A. Wallace and Annette M. LaMond, Lexington Books, 1977.

21 Philip Slater, *Earthwalk,* Anchor Press/Doubleday, 1974, Chapter 1, "The Extensions of Man," pp. 9–24.

22 The demands of Third World countries for a New International Economic Order specifically challenge the classical economic notions, including that of comparative advantage, i.e., that they should continue exporting their raw materials, while the much more profitable step of manufacturing should remain in the industrial countries. The debate over the NIEO is summed up in *RIO: Reshaping the International Order,* A Report to the Club of Rome, coordinated by Jan Tinbergen (E. P. Dutton, 1976). A Third World view of the unfolding debate is *Towards a New International Economic Order,* by Mohammed Bedjaoui, Algerian ambassador to the UN, published under UNESCO auspices by Holmes & Meier (1980), which focuses on the reshaping of international law as a prerequisite and the democratization of international relations by, among other things, eliminating the veto in the UN exercised by the big powers.

23 See, for example, Bernard-Henri Lévy, *Barbarism with a Human Face,* Harper & Row, 1979 (translated from the French original of 1977), who critiques all the symptoms of the Marxist brand of industrialism: technocracy, authoritarianism, disregard for ecology and human values, etc., without ever being able to name the underlying disease of patriarchy.

24 These alternative paths to endogenous, person-centered development models have been discussed at the UN in a UNITAR-sponsored conference on "Regionalism and the NIEO" in May 1980; at the Third Development Decade debate at the UN in August and September 1980, and at earlier meetings including the UN-sponsored meeting organized by the International Foundation for Development Alternatives (IFDA) of Switzerland, in Scheveningen, Holland, July 1979, and that sponsored by the ILO in January 1976 in Geneva on the Social Implications of a New International Economic Order, at which I presented a paper on "Citizen Movements for Greater Global Equity" (*International Social Science Journal,* Volume XXVIII, No. 4, 1976). The two journals following this debate on alternative development styles are the *IFDA DOSSIER,* 2, Place du

Marché, CH-1260, Nyon, Switzerland, and *Development Dialogue*, the Dag Hammarskjöld Foundation, Ovre Slottesgatan 2, 752-20 Uppsala, Sweden. Both journals are available at no charge.

25 As *Scientific American*, Volume 242, No. 1 (January 1980), documented in its "Science and the Citizen editorial," the Pentagon's "creative accounting" includes pushing many military expenditure items off the DOD budget, such as veterans benefits, employee benefits, and social insurance programs, and onto the Health and Human Services side of the budget, as well as much nuclear weapons research, which it charges to the Department of Energy (where it consumes 48 percent of DOE's budget, as verified by a new study by the Institute for Ecological Policies, of Fairfax, VA, entitled *The Wolf Guarding the Door*, by Vicki F. Tynan (1980). The other large "fudge factor" analyzed by *Scientific American* is the Pentagon's habit of asserting that although military expenditures are rising each year absolutely, they represent a smaller fraction of total GNP, which is also rising. The trick here is that these same defense expenditures are *added* to the GNP, one of the largest factors in its "increase"! Needless to add, this all adds to the general inflation bubble. In fact, in a tragic twist, maybe inflation is our "savior," since all such big-bang technologies, weapons, and other massive projects and thermodynamic follies will produce explosive rates of inflation that will eventually bring them all to a halt—even though many innocent victims will be steamrollered in the process.

26 *Weapons for the World/Update III, the U. S. Corporate Role in International Arms Transfers*. Council on Economic Priorities Publication N O- 3, April 1980. 84 Fifth Avenue, New York, N.Y.

27 As former ambassador to the UN Charles W. Yost put it in an editorial in *The Christian Science Monitor* entitled "Arms: the Easy Answer" (April 11, 1980), "Whenever the United States feels particularly insecure, its instinctive response is to reach for its guns. There is no question that with the world as it is, guns are necessary. However, this is one of the many areas of modern life in which more is not necessarily better—or even safer." He added that, dollar for dollar, foreign-aid programs probably contributed more to our national security.

28 Edmund S. Muskie, in remarks to the Foreign Policy Association, noted, "We have a deep and growing stake in developing countries. We cannot get along without them. We want them to progress, because we care about people. We also want them to succeed, because our economic health is bound up with theirs. Our support for liberty in the world cannot be mounted with weapons alone. I believe that the American people, if they have the facts, will understand that a generous investment in security assistance and economic development abroad is necessary to a strong America" (*The Christian Science Monitor*, August 6, 1980). The message of the Third World's call for the New International Economic Order seemed to be getting through, since the world economy is now such a fragile web that it is no longer clear who is in the driver's seat: creditors or debtors and only mutual re-writing of the rules can save it from breakdown.

29 Kenneth E. Boulding, *Ecodynamics*, Sage Publications, 1979, pp. 214–15. I elaborate on this point in my review of Boulding in the "Symposium on *Ecodynamics*" in the *Journal of Social and Biological Structures*, ed. Harvey Wheeler, Santa Barbara, CA, 1981.

30 The most eloquent statement in 1980 of this new world view came from Boulding himself in his presidential address to the American Association for the Advancement of Science, "Science: Our Common Heritage" (*Science*, Vol-

ume 207, No. 4433 [February 22, 1980], pp. 831–36). It contained such gems as his contention that the distinction between the "hard" and "soft" sciences was illusory, and that it was the social sciences which were the "harder" and more secure, since we humans simply have more knowledge about ourselves than we do about the universe and nature, of which we have only infinitesimal samples and many fewer data. He adds that "many of the great laws of science are empirical truisms, and that it is an illusion to believe that there is a single 'scientific method,' a touchstone that can distinguish what is scientific from what is not." Boulding also argued that science was the result of a change in human values that put a high value on curiosity coupled with a belief in testing, and thus was an uneasy combination of imagination, fantasy, and logic—adding, impishly, that logic, too, was essentially the perception of truisms! He ended by noting that the greatest social cost of science has been its use in the development of weapons, and that nuclear weapons have now destroyed the viability of the nation-state, and that civilian populations are no longer defended by them but are hostages to them.

[31] Attesting to the vigor of these citizens' movements (which have grown as the sagging economy and inflation have increased people's lack of confidence in authority) is a useful new catalog of their publications: *Periodicals of Public Interest Organizations: a Citizen's Guide*, published by the Commission for the Advancement of Public Interest Organizations, the Monsour Medical Foundation, 1875 Connecticut Avenue, Washington, D.C., 1979, $5.

[32] Hugh Davis Graham and Ted Robert Gurr, *The History of Violence in America, a Report to the National Commission on the Causes and Prevention of Violence,* Bantam, 1969, p. 664.

[33] Gary Zukav, *The Dancing Wu Li Masters,* Wm. Morrow, 1979, pp. 301–31.

[34] *Newsweek,* October 8, 1979, p. 52.

[35] In December 1979, a panel headed by former Secretary of Transportation William Coleman, of the United Nations Association, recommended that the human-rights emphasis should be extended with new legislation to cover such violations as systematic discrimination and oppression on the basis of race, gender, or creed, and freedom from prosecution when exercising the right of free speech or association. Significantly, the panel added that the United States should also accept the jurisdiction of the Inter-American Court on Human Rights (New York *Times,* December 11, 1979).

[36] The business press, for obvious reasons, has followed the resource shortages closely. In their July 2, 1979, issue, *Business Week* covered the problems of mineral and metal supplies in their cover story "Now the Squeeze on Metals" (p. 46). In their July 30, 1979, issue, the cover story was "The Oil Crisis Is Real This Time" (p. 44). *Fortune* covered the increasing world competition for raw materials and the problems of U.S. dependency on many key minerals that lie 100 percent in the U.S.S.R. or in Africa, in "Russia's Sudden Reach for Minerals" (July 20, 1980).

[37] Even in the individualistic, private-property-oriented value system of the United States, the new situation has produced a major new legal interpretation of social justice: John Rawls, *A Theory of Justice* (Harvard University Press, 1971). It is a rebuttal of the economists' utilitarianism, Pareto Optimality, and such other operational formulae as cost/benefit analyses, which excessively discount the future, as well as general reliance on the price system to create socially just outcomes. Economists were also outraged by Rawls's principle of justice, relating to the question of equitable distribution of wealth, position, income,

opportunity, which holds that social and economic inequalities are just *only* if they result in compensating benefits for everyone, and in particular for the least-advantaged members of society.

38 *Development Dialogue,* 1979, p. 73, Uppsala, Sweden.

EPILOGUE

The Happier Side of Uncertainty: Everything Can Change in the Twinkling of an Eye!

As 1980, the first year of a crucial decade for humanity, drew to a close, the global crises escalated and were summed up in the now-famous *Global 2000 Report,* issued by the Carter administration in July 1980. The *Report* reviewed all the issues of declining, wasting resources, growing human population, and ecological disruption covered in this book—and found them getting worse. The population, although slowing in growth (from 1.8 percent to 1.7 percent per year) would still reach 6.35 billion by 2000. The gap between the over-consuming industrial nations and the under-consuming Third World was still widening. World food supplies would increase only 15 percent per capita by 2000 and, tragically, would become increasingly maldistributed, leaving millions more than today in Africa and Asia facing famine. Arable land would increase only by 4 percent, and regional water shortages would become severe due to deforestation and pollution. The world's forests are disappearing at the rate of 18 to 20 million hectares a year (an area half the size of California), and soils are being eroded and destroyed, turning formerly fertile areas the size of Maine into deserts each year, while 20 percent of all the species of plants and animals on the planet could be extinct in the next nineteen years. World oil production will peak by 1990, while higher prices would only channel more of it to over-consuming industrial countries—leaving the Third World not only

bereft of oil but facing a shortfall of 25 percent of its needs for traditional wood fuels. A quarter of the world's population living in the
industrial countries could continue consuming three quarters of the
world's mineral production. Meanwhile, industrial processes and fossil-fuel combustion are making rainfall in many areas too acid for
crops and killing fish in lakes while inexorably adding carbon dioxide
to the earth's atmosphere at ever-increasing rates. Many climate experts now are linking the severe weather variability of freak storms,
tornadoes, floods, droughts, freezes, and heat waves during the past
few years to these rising carbon-dioxide levels. Scientists argue
whether the overall trend is to a global warming or another ice age,
and some believe that we have not fifty, but less than twenty, years to
make the transition to the renewable resources of the solar age before these climate changes begin to cut the world's food production
drastically. The still-unsolved problem of storing radioactive wastes
and other hazardous chemicals now present health problems in many
countries.

A crucial finding of the *Global 2000 Report* confirms another central thesis of this book and that of *Creating Alternative Futures:*
that industrial countries have lost control of their affairs. They can
no longer manage these complex sociotechnical societies; indeed,
they cannot even model them and their maze of multidimensional interactions. Their almost total reliance on now-disordering, money-
based statistics and economic forecasts are exacerbating these governments' dilemmas as well as destabilizing the world's monetary and
trade systems. At the same time, industrial countries were falling
behind on their pledges of foreign aid and development assistance to
the Third World, as they became more snarled in their domestic
management problems. Thus we see, in all these crises of industrialism, its failure: the breakdown of its social order, family, and
community cohesion, and its empty promises to the world's aspiring
peoples now revealed in the export only of its crises, wastes, and armaments, culminating in the madness of the nuclear arms race and
growing confrontations over dwindling resources. Thus we see the
end of a two-hundred-year epoch.

How are we to grasp this bewildering world situation—by what
conceptual tools we map these events in order to keep our balance? We can infer that we face a generalized, systemic crisis, since
one of the most important indications of such a crisis is that applying

the conventional remedies will only make matters worse. All the redoubling of old efforts, the technological fixes, the economists' tinkering and adjusting discussed in this book have simply brought industrial countries to the point, as Fritz Schumacher used to say, when "they need a breakthrough a day to keep the crises at bay." The enterprise of industrialism—the obsession with controlling and manipulating nature, and exploiting resources in search of elusive "security" and "equilibrium" through trying to manage ever more of the variables of human existence, together with the compulsive efforts to predict, and reduce the fear of uncertainty—have ended in a paradox: certainty and equilibrium are precise descriptions of *death!*

Thus the pursuit of certainty, order, and equilibrium itself has led industrial societies into their evolutionary cul-de-sac and their current "trial by entropy." But if they can learn to see the multiple crises they face as signals and feedback for learning and restructuring, their transformation to sustainability can be achieved and balance gradually restored in a reshaping of the world order. If we use a model of morphogenetic change, as discussed earlier, we can see that *it is to be expected* that all the planet's subsystems would reach crisis stages *simultaneously,* an indication that such a systemic metamorphosis is occurring. In such morphogenetic change processes, as the ligaments of the old order dissolve, we expect that something is also emerging: as, for example, when the chrysalis turns into a butterfly. Birth is a painful process, as women know, and the new world order *is* being born. For example, in a real sense, there already *is* a New International Economic Order, not only evident in the power Third World debtor countries possess to simply declare the present monetary system illegitimate (as it is) and call the bluff of northern-hemisphere bankers by declaring bankruptcy en masse, but by their control of key resources and growing ability to form regional-trading and mutual-aid blocs. Control of oil has already shifted to the Third World, and many of the key minerals industrial countries import, such as those on which the United States is dependent for 80 percent to 100 percent of its supplies (columbium, sheet mica, strontium, manganese, cobalt, chromium, aluminum, and platinum), come from such Third World countries as Brazil, Thailand, Nigeria, Turkey, Zimbabwe, Jamaica, and Surinam, while others on this list come from the U.S.S.R. and South Africa. Similarly, supplies of phosphates for fertilizers, which come at present from Florida, will be exhausted by 1995, and the United States (with amazing lack of foresight) some years ago contracted to sell more than half of this remaining phosphate to the U.S.S.R. While the price has jumped from eight dollars per ton in 1972 to two hundred dollars a ton today and will rise to as much as four hundred dollars a ton in 1982, the phosphate itself can hardly be valued in mere money. In this case, too, the world's only other major phosphate deposits are in North Africa, where these nations will not only control much petroleum production but also *all* of the world's phosphates. Thus these Third World countries have considerable leverage to save the industrial nations from their own crises and follies. For example, they could help defuse the nuclear arms race and the world trade in armaments by mobilizing support for many of the new proposals for restructuring the international order: international taxation of armaments sales, of the world's depletable resources, of the common heritage of all humanity. For example, a 1 percent tax on the some $450 billion spent annually in military budgets would dwarf all other foreign-aid and development finance. Such proposals, including that of the French-inspired International Disarmament Fund for Development, now are receiving the serious discussion and analysis they deserve (see Horst Paul Wiesebach, "Mobilization of Development Finance: Promises and Problems of Automaticity," *Development Dialogue,* #1, 1980), as well as the many other proposals to restructure a safer, more sustainable world order mentioned in Chapter 13.

Today, we also see that the happier side of accepting the principle of uncertainty and the inevitability of change is that all the dire *trends* discussed in the *Global 2000 Report* are *not* predictions; *not* foreclosed as our destiny, but only important warnings about permitting the historical trends they document from continuing. The crescendo of signals from nature, now being daily amplified by such reports and by mass-media coverage, not to mention citizen movements alerting the body politic to attend to these dangers, is now *also* producing a crescendo of new concern, political action, and reconceptualization. The many examples of this rising human concern and action discussed in Chapter 13 are being augmented daily in ways cynics have held were impossible, such as the successful efforts of the United Nations Environment Program to bring the eighteen nations that share the coastline of the dying Mediterranean into signing a comprehensive treaty to clean it up. The bill will be $10 billion, mostly paid by the countries most responsible for polluting it: France, Spain, and Italy (*The Christian Science Monitor,* July 18, 1980). The role of citizen movements as the key levers for bringing about such beneficial changes is increasingly acknowledged by governments otherwise constrained by powerful interest groups in their own policies. As the December 1979 *Center Report* of the United Nations Environment Program noted, the efforts of these nongovernmental organizations are key to the successful transformation to sustainable forms of development. They perform certain functions that cannot be filled by governments, corporations, or UN agencies.

Citizen movements and nongovernmental organizations are now officially credited with predicting the energy crisis as early as 1969; promoting alternative, solar energy, and sustainable-development strategies; and forcing issues of disarmament and peace, human rights and social justice onto national and international agendas. The citizens' forums organized around the United Nations conferences on environment in 1972 (and subsequently on population, food, habitat, employment), and those organized around the August 1980 debate on the Third Development Decade, all provided a rich new yeast of alternative policies and concepts vigorously asserted to challenge conventional, bureaucratic wisdom. The dramatic success of the consumer boycott of companies selling infant formulas in Third World countries led to a strict set of recommendations at the joint UNICEF/World Health Organization meeting in Geneva in 1979, including a provision to ban advertising of such infant formulas and curb their sales and promotion. UN analyst Thierry Lemaresquir assessed the growing clout and enormous untapped potential of citizen movements and all nongovernmental organizations in providing a new force for levering national governments and international deliberations in "The Case for Another Relationship between NGOs and the United Nations System" (*Development Dialogue,* #1, 1980).

Indeed, how could we have expected the needed alternatives to come from the dominant cultures and existing control centers, since individuals learn faster than institutions and it is always the dinosaur's brain that is the last to get the new messages! If there is to be a third way, the Third World will devise it, together with the subordinated populations in industrial countries: blacks, native peoples, ethnic minorities, and all those tyrannized by monetization, as the world's women have been for so many generations. It is only the oldest law of evolution coming into play once again: nothing fails like success, or the continuing trade-off between adaptation and adaptability. Anthropologists call it the law of the retarding lead, where cultures best adapted to *past* conditions fall behind those more flexible and less specialized, which then can adapt best to *new* conditions. Theologians sum things up as "the last shall be first."

If we can recognize that change and uncertainty are basic principles, we can greet the future and the transformation we are undergoing with the understanding that *we do not know enough to be pessimistic.* The life force within each of us can then focus on the *possible* and the *potentialities.* One can call it faith in the future, or the acknowledgment that *we are not in charge,* and that the planet is not a spaceship that we humans are "steering" or "managing." This old-fashioned image has served its purpose, but it encouraged our childish fascination with vehicles, transportation, speed, and power. The maturing understanding, growing out of both scientific research and folk knowledge, confirmed by age-old religious and mythic traditions, is that we are a conscious part of the earth—no mechanical spaceship, but a living planet, a total, teeming, pulsating, living organism: Gaia, the mysterious, self-organizing Earth Mother, nurturer of us—and all life.

Index